Effects of Environmental Exposure on Host and Microbial Metabolism

Effects of Environmental Exposure on Host and Microbial Metabolism

Guest Editors

Pengcheng Tu
Bei Gao

MDPI

Basel • Beijing • Wuhan • Barcelona • Belgrade • Novi Sad • Cluj • Manchester

Guest Editors

Pengcheng Tu
Department of
Environmental Health
Zhejiang Provincial Center
for Disease Control and
Prevention
Hangzhou
China

Bei Gao
School of Environmental
Science and Engineering
Nanjing University of
Information Science and
Technology
Nanjing
China

Editorial Office
MDPI AG
Grosspeteranlage 5
4052 Basel, Switzerland

This is a reprint of the Special Issue, published open access by the journal *Metabolites* (ISSN 2218-1989), freely accessible at: https://www.mdpi.com/journal/metabolites/special_issues/VKC5O283QF.

For citation purposes, cite each article independently as indicated on the article page online and as indicated below:

Lastname, A.A.; Lastname, B.B. Article Title. *Journal Name* **Year**, *Volume Number*, Page Range.

ISBN 978-3-7258-6173-6 (Hbk)
ISBN 978-3-7258-6174-3 (PDF)
https://doi.org/10.3390/books978-3-7258-6174-3

Contents

About the Editors

Pengcheng Tu

Pengcheng Tu is an accomplished research scientist specializing in cutting-edge environmental health issues. He obtained his Ph.D. from the University of North Carolina at Chapel Hill (UNC) and currently works at the Zhejiang Provincial Center for Disease Control and Prevention. His research primarily focuses on elucidating the toxicity mechanisms and health impacts of emerging pollutants, with a specialized emphasis on microplastics. Dr. Tu spearheaded the establishment of a pioneering cohort study for monitoring internal human exposure to microplastics. This initiative provides critical biomonitoring data and a foundational resource for investigating the health implications of microplastics.

Dr. Tu has built an impressive publication portfolio, having authored over 60 papers. His work has gained substantial recognition in the scientific community, as reflected by a total citation count of over 3000, underscoring the broad influence and relevance of his research contributions.

Dr. Tu has presided over grants from prestigious funding bodies including the National Natural Science Foundation of China and the Natural Science Foundation of Zhejiang Province. In acknowledgment of his significant contributions to medical research, he has been honored with the "Zhejiang Medical Rising Star" award.

Dr. Tu also actively engages in scholarly communication and peer review, serving as an Associate Editor for *Frontiers in Nutrition* and as an Editorial Board Member for *Metabolites*.

Bei Gao

Bei Gao is an Associate Professor at Nanjing University of Information Science and Technology, specializing in environmental toxicology and human health. She laid a robust foundation for her research career by earning a Ph.D. in Environmental Health Science from the University of Georgia, where she was mentored by the distinguished environmental toxicologist, Professor Kun Lu. Dr. Gao further honed her expertise through prestigious postdoctoral fellowships at the West Coast Metabolomics Center at the University of California, Davis, under the guidance of Professor Oliver Fiehn, a global leader in metabolomics, and subsequently at the University of California, San Diego, with Professor Bernd Schnabl, an established leader in gut microbiome–liver research.

Her research program is highly interdisciplinary, effectively bridging the fields of environmental exposure, metabolomics, and microbiome science to elucidate the health impacts of environmental contaminants. Dr. Gao has authored over 80 peer-reviewed publications in international journals, including *Nature Communications*, *Environmental Health Perspectives*, and *Environmental Science & Technology*, with more than 5400 citations and an H-index of 37. Her consistent scholarly influence has been recognized through her inclusion in the "World's Top 2% Scientists" list from 2022 to 2025. She currently serves as an Associate Editor for *Frontiers in Nutrition* and holds Editorial Board Member positions for *Environmental Health Perspectives* and *Metabolites*.

metabolites

MDPI

Editorial

Effects of Environmental Exposure on Host and Microbial Metabolism

Bei Gao [1] and Pengcheng Tu [2,*]

[1] School of Environmental Science and Engineering, Nanjing University of Information Science and Technology, Nanjing 210044, China; bgao@nuist.edu.cn
[2] Zhejiang Provincial Center for Disease Control and Prevention, 3399 Binsheng Road, Hangzhou 310051, China
* Correspondence: tupengcheng1@163.com

Trillions of microorganisms are living in our gastrointestinal tract, known as the gut microbiota, which plays an essential role in human health and disease. The gut microbiota could be affected by many factors, including environmental exposure. Disruptions of the gut microbiota by environmental exposure may exert adverse effects on human health by affecting host metabolism, intrinsic metabolism in the gut microbiota, and/or gut microbiota–host co-metabolism. This Special Issue, "Effects of Environmental Exposure on Host and Microbial Metabolism", contains six original research articles and five review articles, which covers the effects of both emerging environmental pollutants, including microplastics, per- and polyfluoroalkyl substances (PFASs), bisphenol S (BPS), and traditional fluoride and fungicide epoxiconazole (EPX) contaminants, as well as dietary factors such as non/low-caloric artificial sweeteners (NAS), probiotics, prebiotics, and natural compounds. The contributions are listed below.

1. Tu, P.; Xue, J.; Niu, H.; Tang, Q.; Mo, Z.; Zheng, X.; Wu, L.; Chen, Z.; Cai, Y.; Wang, X. Deciphering Gut Microbiome Responses upon Microplastic Exposure via Integrating Metagenomics and Activity-Based Metabolomics. *Metabolites* **2023**, *13*, 530. https://doi.org/10.3390/metabo13040530.
2. Niu, H.; Yang, Y.; Zhou, Y.; Ma, X.; Ding, Z.; Xu, M.; Wu, L.; Li, X.; Xing, M.; Zhang, Q.; et al. Differential Impacts of Environmentally Relevant Microplastics on Gut Barrier Integrity in Mice Fed High-Fat Diet Versus Normal Chow Diet. *Metabolites* **2025**, *15*, 557. https://doi.org/10.3390/metabo15080557.
3. Xu, M.; Niu, H.; Wu, L.; Xing, M.; Mo, Z.; Chen, Z.; Li, X.; Lou, X. Impact of Microplastic Exposure on Blood Glucose Levels and Gut Microbiota: Differential Effects under Normal or High-Fat Diet Conditions. *Metabolites* **2024**, *14*, 504. https://doi.org/10.3390/metabo14090504.
4. Niu, H.; Liu, S.; Jiang, Y.; Hu, Y.; Li, Y.; He, L.; Xing, M.; Li, X.; Wu, L.; Chen, Z.; et al. Are Microplastics Toxic? A Review from Eco-Toxicity to Effects on the Gut Microbiota. *Metabolites* **2023**, *13*, 739. https://doi.org/10.3390/metabo13060739.
5. Cox, A.; Nowshad, F.; Callaway, E.; Jayaraman, A. Integrated Metagenomic and Metabolomic Analysis of In Vitro Murine Gut Microbial Cultures upon Bisphenol S Exposure. *Metabolites* **2024**, *14*, 713. https://doi.org/10.3390/metabo14120713.
6. Ma, X.; Cai, D.; Chen, Q.; Zhu, Z.; Zhang, S.; Wang, Z.; Hu, Z.; Shen, H.; Meng, Z. Hunting Metabolic Biomarkers for Exposure to Per- and Polyfluoroalkyl Substances: A Review. *Metabolites* **2024**, *14*, 392. https://doi.org/10.3390/metabo14070392.
7. Li, X.; Niu, H.; Huang, Z.; Zhang, M.; Xing, M.; Chen, Z.; Wu, L.; Xu, P. Deciphering the Role of the Gut Microbiota in Exposure to Emerging Contaminants and Diabetes: A Review. *Metabolites* **2024**, *14*, 108. https://doi.org/10.3390/metabo14020108.

8. Weng, Y.; Xu, T.; Wang, C.; Jin, Y. Oral Exposure to Epoxiconazole Disturbed the Gut Micro-Environment and Metabolic Profiling in Male Mice. *Metabolites* **2023**, *13*, 522. https://doi.org/10.3390/metabo13040522.
9. Mo, Z.; Wang, J.; Meng, X.; Li, A.; Li, Z.; Que, W.; Wang, T.; Tarnue, K.F.; Ma, X.; Liu, Y.; et al. The Dose-Response Effect of Fluoride Exposure on the Gut Microbiome and Its Functional Pathways in Rats. *Metabolites* **2023**, *13*, 1159. https://doi.org/10.3390/metabo13111159.
10. Feng, J.; Peng, J.; Hsiao, Y.-C.; Liu, C.-W.; Yang, Y.; Zhao, H.; Teitelbaum, T.; Wang, X.; Lu, K. Non/Low-Caloric Artificial Sweeteners and Gut Microbiome: From Perturbed Species to Mechanisms. *Metabolites* **2024**, *14*, 544. https://doi.org/10.3390/metabo14100544.
11. Chu, X.; Xing, H.; Chao, M.; Xie, P.; Jiang, L. Gut Microbiota Modulation in Osteoporosis: Probiotics, Prebiotics, and Natural Compounds. *Metabolites* **2025**, *15*, 301. https://doi.org/10.3390/metabo15050301.

Regarding the effects of microplastics, Contribution 1 found that polystyrene (PS) microplastic exposure (0.1 mg/day) perturbed the gut microbiota composition, diversity, and functional pathways in C57BL/6 mice. Metabolomic analysis revealed that cholesterol metabolism, primary and secondary bile acid biosynthesis, taurine and hypotaurine metabolism, as well as short-chain fatty acid metabolism, were perturbed upon microplastic exposure. Contribution 2 found that the impact of PS microplastics on the intestinal barrier function was different between mice fed a high-fat diet and a normal diet. Contribution 3 found that PS microplastics exacerbate high-fat diet-induced glucose metabolism disorder in male C57BL/6 mice. Moreover, Contribution 4 reviewed studies on the adverse effects of micro- and nanoplastics on invertebrates and vertebrates, as well as on the gut microbiota and its metabolites. Contribution 5 investigated the impact of BPS on gut microbiota using in vitro cultures and found that compared to controls, BPS did not overtly distort profiles of the metagenome and metabolome of the microbial cultures even at a supraphysiologic dose. Contribution 6 summarized recent metabolomics-based studies and elucidated the potential biomarkers of PFAS exposure. Contribution 7 discussed the link between emerging pollutants and glucose metabolism and the potential role played by the gut microbiota. These studies provide insights into the toxicity of emerging contaminants and the underlying mechanisms associated with microbial changes.

In addition to emerging contaminants, the effects of traditional contaminants on gut microbiota were also investigated in our Special Issue. EPX is a triazole fungicide, which is widely used to control pests in agriculture. Contribution 8 found that EPX exposure disrupted intestinal barrier function and altered the composition and abundance of gut microbiota and the glycolipid metabolism in male C57BL/6 mice. Contribution 9 found fluoride exposure induced a profound shift in the gut microbial composition, with dose-dependent responses observed in key genera. Microbial functional pathways were also disturbed by fluoride exposure, including D-lyxose ketol-isomerase and DNA polymerase III subunit gamma/tau.

Dietary factors are key factors that influence the gut microbiota. NAS are chemical additives that substitute sugars to avoid sugar-derived diseases. Contribution 10 implemented a comprehensive two-stage literature analysis and found that specific NAS exhibited discrepant impacts on the gut microbiota, and the overlapping microbial genera and species were identified. Impairment of glucose tolerance in the host was induced by some NAS, but the key metabolites and underlying mechanisms varied. Contribution 11 reviewed the recent findings on the role of prebiotics, probiotics, and natural bioactive substances in the modulation of gut microbiota to improve the health of the bone.

In conclusion, this Special Issue of *Metabolites* highlights the significant progress made in understanding the complex interactions between environmental exposures, host metabolism, and microbial communities. These studies are pivotal in unraveling the intricate metabolic processes that shape both human health and the dynamics of our microbial partners. However, further research is essential to elucidate the causal roles that gut microbiota may play in disease development under varying environmental conditions. The findings presented here underscore the need for continued investigation into these interdependencies, ultimately guiding the development of more targeted and effective strategies for disease prevention and treatment.

Author Contributions: Writing—original draft preparation, B.G.; writing—review and editing, P.T.; All authors have read and agreed to the published version of the manuscript.

Funding: B.G. is supported by the National Natural Science Foundation of China (grant No. 42107459), Key Technology R&D Program of Jiangsu Province (BE2022788), Science and Technology Innovation Project for Returned Overseas Individuals of Nanjing City (R2022LZ06), and startup funding of Nanjing University of Information Science and Technology. P.T. is funded by the Natural Science Foundation of Zhejiang Province (LQ24H260004), the National Natural Science Foundation of China (grant No. 42407553), and the Zhejiang Provincial Project for Medical Research and Health Sciences (2024KY898).

Conflicts of Interest: The authors declare no conflicts of interest.

metabolites

MDPI

Article

Differential Impacts of Environmentally Relevant Microplastics on Gut Barrier Integrity in Mice Fed High-Fat Diet Versus Normal Chow Diet

Huixia Niu [1,2,†], Ying Yang [1,†], Yuting Zhou [2], Xue Ma [2], Zhehao Ding [2,3], Manjin Xu [2], Lizhi Wu [2], Xueqing Li [2], Mingluan Xing [2], Qin Zhang [1], Hao Chen [1], Xiongwei Tao [1], Zhe Mo [2], Zhijian Chen [2], Pengcheng Tu [2,*] and Xiaoming Lou [2,*]

1 Center for Disease Control and Prevention of Jinyun County, 89 Cuizhu Road, Jinyun 321400, China; niuhuixia1232025@163.com (H.N.); yangying1232025@163.com (Y.Y.); zhangqin123202506@163.com (Q.Z.); chenhao123202506@163.com (H.C.); t173328120@163.com (X.T.)

2 Zhejiang Provincial Center for Disease Control and Prevention, 3399 Binsheng Road, Hangzhou 310051, China; zzzyt@stu.xmu.edu.cn (Y.Z.); xma@cdc.zj.cn (X.M.); 3190104398@zju.edu.cn (Z.D.); margexmj@outlook.com (M.X.); lzhwu@cdc.zj.cn (L.W.); xqli@cdc.zj.cn (X.L.); mlxing@cdc.zj.cn (M.X.); zhmo@cdc.zj.cn (Z.M.); zhjchen@cdc.zj.cn (Z.C.)

3 School of Medicine, Zhejiang University, 866 Yuhangtang Road, Hangzhou 310012, China

* Correspondence: pchtu@cdc.zj.cn (P.T.); xmlou@cdc.zj.cn (X.L.)

† These authors contributed equally to this work.

Abstract

Background: Despite escalating global pollution from microplastics (MPs) and the concurrent surge in high-fat food consumption, the health impacts of MP exposure on individuals under different dietary patterns remain poorly understood. Methods: This study investigated the differential effects of environmentally relevant concentrations of polystyrene microplastics (5 μm, 8 mg/kg) on gut barrier function in mice fed either a normal chow diet (CD) or a high-fat diet (HFD). Results: Key findings revealed that, in HFD-fed mice, MP exposure significantly reduced ($p < 0.05$) the transcriptional levels of genes encoding the tight junction proteins (ZO-1, Occludin, and Claudin-1), as well as the mucin protein Muc-2, accompanied by decreased protein expression levels of these markers in both colonic and ileal tissues. In contrast, no significant differences were observed in CD-fed mice exposed to MPs. Analysis of the gut microbiota and measurement of short-chain fatty acid (SCFA) metabolites showed that MPs induced significant alterations in the composition and diversity indices of the gut microbiota, along with a marked decrease ($p < 0.05$) in the levels of the characteristic metabolite butyrate in HFD-fed mice. Conversely, butyrate levels remained unchanged in CD-fed mice following MP exposure. Quantitative PCR (qPCR) and immunofluorescence staining of colonic tissues demonstrated that MP exposure significantly downregulated ($p < 0.05$) both the transcription and protein expression of peroxisome proliferator-activated receptor γ (PPARγ) in HFD-fed mice. Again, no significant changes were detected in CD-fed mice. Conclusions: These results collectively indicate that the impact of microplastics on the intestinal barrier differs significantly between mice fed normal and high-fat diets. The gut microbiota and its metabolites, particularly butyrate, may play a critical role, possibly through modulating PPARγ signaling. This study contributes valuable insights into understanding the toxicity profiles of microplastics and establishing crucial links between dietary patterns and the health effects of emerging pollutants.

Keywords: emerging pollutants; microplastics; gut barrier; gut microbiota

1. Introduction

In 2004, Thompson et al. introduced the concept of microplastics, defining them as plastic particles with a diameter of less than 5 mm [1]. Microplastics (MPs) are ubiquitous in the environment and have been widely detected in food, water, and everyday consumer products [2–6]. These microplastics can enter the human body through ingestion, inhalation, and dermal contact. Studies indicate that ingestion is the primary route of exposure, with numerous investigations confirming the presence of microplastics in daily diets [7]. For instance, in 2015, Yang et al. detected microplastics in sea salt, lake salt, and rock/well salt [8]. Mattsson et al. found that microplastics in marine environments can transfer through the algae–daphnia–fish food chain [9]. Furthermore, microplastics can enter the human body through trophic transfer and nutritional uptake. Cauwenberghe et al. studied the potential risks of microplastics in seafood to humans, using mussels as a model to observe the direct impacts of microplastics via the food chain [10]. Their research revealed that consuming an average portion (250 g wet weight) of mussels could result in the ingestion of approximately 90 microplastic particles. Processed foods appear to contain higher levels of microplastics compared to unprocessed foods. Additionally, the presence of microplastics has been detected in human placenta and whole blood, suggesting that some inhaled microplastics are bioavailable: Microplastic particles are capable of being absorbed into the circulatory system, subsequently translocating to and accumulating within various organs [11,12]. Consequently, the potential health risks posed by microplastics have garnered increasing attention. Research has extensively examined the histopathological changes, oxidative stress, inflammatory responses, neurotoxicity, and reproductive toxicity induced by microplastics in biological systems. These include liver sinusoidal dilation, disordered hepatocyte arrangement, intestinal villi rupture, enterocyte lysis, elevated reactive oxygen species (ROS) levels, and upregulated expression of tumor necrosis factor alpha (TNF-α) and interleukin 6 (IL-6) [13–17].

The intestinal barrier function, comprising physical, chemical, immune, and microbial barriers, is critical in preventing the invasion of harmful substances from the gut and maintaining homeostasis between the internal and external environments [18]. The microbial barrier, in particular, consists of the microbiota that maintain cellular junctions and promote epithelial repair [19]. The human gut microbiota predominantly resides in the colon, with approximately 10^{11}–10^{12} bacteria per gram of colonic tissue [20]. Changes in the composition and relative abundance of the gut microbiota can alter the metabolic products that enter the human body. Microbial-derived lipopolysaccharides (LPSs) and other endotoxins can leak into the bloodstream through a compromised gut barrier, leading to metabolic endotoxemia. LPS can reduce the expression of tight junction proteins between intestinal epithelial cells, increasing gut permeability. Conversely, short-chain fatty acids and indole metabolites produced by the gut microbiota can enhance the physical barrier of the gut by increasing the expression of tight junction proteins and cytoskeleton-associated proteins [21]. Studies have shown that sodium butyrate, within the concentration range of 1–10 mM, can increase Muc-2 protein levels, significantly improving epithelial function in human colonic epithelial cells [22]. Tang et al. also concluded that butyrate promotes gut barrier function [23]. Moreover, butyrate can bind to peroxisome proliferator-activated receptor γ (PPARγ), activating pathways that strengthen the gut barrier [24]. Simeoli et al. found that after administering Dextran Sulfate Sodium Salt (DSS) to mice for 7–12 days, PPARγ gene transcription levels significantly decreased, leading to impaired gut barrier function. Treatment with *Lactobacillus paracasei* improved gut barrier function and increased PPARγ gene transcription levels [25]. However, the effects of microplastic exposure on butyrate and PPARγ expression in colonic tissue and their roles in modulating gut barrier function remain to be further explored.

In recent years, the intake of high-fat foods worldwide has been steadily increasing. The potential health hazards of microplastics have attracted widespread attention both internationally and domestically. However, there is a lack of in-depth understanding of the differential health impacts of microplastic exposure under different dietary structures. Therefore, this study investigates the effects of environmentally relevant concentrations of microplastics on gut barrier function in mice fed either a normal or high-fat diet and examines whether these effects differ between the dietary groups. This research aims to enhance our understanding of the toxicity of environmental microplastics and their potential health impacts on humans.

2. Materials and Methods

2.1. Chemicals

Monodisperse polystyrene microspheres with a particle size of 5 μm were obtained from Tianjin BaseLine ChromTech Research Center (Tianjin, China). The stock solution (2.5% *w*/*v*) appeared as a milky-white suspension, with an initial concentration of 250 mg per 10 mL aqueous suspension. The characterized using a Coulter particle size analyzer [26]. The characterization of the polystyrene microplastics is presented in Supplementary Figure S1. Prior to use, the microplastic solution was diluted to a working concentration of 1.6 mg/mL using deionized water and stored at 4 °C for further experiments.

2.2. Animal and Experimental Scheme

Male C57BL/6 mice, aged 4–6 weeks, were procured from SLAC Laboratory Animal Co., Ltd. (Shanghai, China). All experimental procedures were conducted in accordance with protocols approved by the Animal Management and Ethics Committee of Zhejiang Chinese Medical University (No. 20230213-07). The mice were housed in an animal facility with controlled temperature (22 °C) and humidity (40–70%) levels, under a 12 h light/dark cycle. After a one-week acclimatization period, the mice were randomly divided into four groups (n = 10 per group; total N = 40) as follows [27,28]: (1) normal chow diet (CD) + water group (10 kcal%: fat-derived energy constitutes 10% of the total dietary energy in the feed), (2) CD + 8 mg/kg MP group (10 kcal%), (3) high-fat diet (HFD) + water group (60 kcal%: fat-derived energy constitutes 60% of the total dietary energy in the feed), and (4) HFD + 8 mg/kg MP group (60 kcal%).

Polystyrene microplastics with a particle size of 5 μm were selected for exposure at a dose of 8 mg/kg body weight per day. The particle size selection was based on the study by Jin et al. [27]. The exposure dose was chosen to simulate environmental microplastic concentrations. According to Deng et al., the estimated daily intake of microplastics for adults ranges from 13 to 39.3 mg/d [29]. Assuming an average adult body weight of 60 kg, the daily intake of microplastics per kg body weight is approximately 0.22–0.66 mg/kg. Using classical pharmacological and toxicological formulas, the estimated microplastic exposure dose for mice ranges from 2.7 to 8.1 mg/kg body weight [30]. Considering the daily food intake and average body weight of mice, the gavage dose in this study was set at 8 mg/kg body weight per day.

2.3. Tissue Collection and Histological Observation

Mice were fasted overnight and then euthanized by cervical dislocation [31]. Blood samples were collected prior to sacrifice. The serum was separated by centrifugation and stored at −80 °C for the subsequent analysis of inflammatory cytokine concentrations. Colonic tissues were collected, fixed in 4% paraformaldehyde, and embedded in paraffin for histopathological examination. Tissue sections (4 μm thick) were observed under an optical microscope. Another portion of the colonic tissue was immediately frozen in liquid nitrogen and stored at −80 °C until analysis.

Colonic tissue sections were subjected to hematoxylin and eosin (HE) staining, immunohistochemical staining, and immunofluorescence staining.

2.4. Biochemical Analysis

The levels of LPS, IL-1β, IL-10, TNF-α, and IL-6 in the serum were determined using enzyme-linked immunosorbent assay (ELISA) kits from Mei Mian Biotech Co., Ltd. (Yancheng, China).

2.5. qPCR Analysis

The remaining colonic tissue was used for qPCR analysis. Total RNA was extracted from the colonic tissue using a tissue homogenizer with TRIzol. The complementary DNA (cDNA) was synthesized using 4× genomic DNA (gDNA) wiper rMix and 5× HiScriptII qRT SuperMix II. Real-time PCR was performed with Forward Primer (10 μM), Reverse Primer (10 μM), 2× Taq Pro Universal SYBR qPCR Master Mix, and Template DNA/cDNA. Primer sequences are listed in Supplementary Table S1.

2.6. Gut Microbiota Analysis

Genomic DNA from fecal samples was extracted using the Mag-bind Soil DNA Kit. The V3-V4 region of the 16S ribosomal RNA (rRNA) of the gut microbiota was amplified using universal 16S rRNA primers. PCR products were recovered from 2% agarose gels, purified using the AxyPrep DNA Kit, and detected by 2% agarose electrophoresis. Library construction was performed using the TruSeq DNA PCR-Free Sample Preparation Kit. Sequencing was conducted on the Illumina NovaSeq 6000 platform (PE250). Operational taxonomic units (OTUs) with a similarity of over 97% were clustered, and the abundance of each OTU was statistically analyzed. The species composition and relative abundance of the gut microbiota at the phylum and genus levels were determined. Principal Coordinate Analysis (PCoA) was used to analyze community composition similarities and differences. Venn diagrams were used to count the shared and unique species among different experimental groups.

2.7. SCFAs Analysis

Metabolites were extracted using 20% phosphoric acid and 4-methylvaleric acid. Separation was carried out on an Agilent DB-FFAP capillary column (30 m × 250 μm × 0.25 μm) using a gas chromatography system. Fecal samples were detected in SCAN/SIM mode. The chromatographic peak areas were extracted using Mass Hunter software (MSD Chemstation Data Analysis 1701FA F.01.00) to construct standard curves, and the SCFA content in the samples was calculated based on these curves. The analysis of short-chain fatty acids (SCFAs) was performed by Shanghai Applied Protein Technology Co., Ltd. (Shanghai, China).

2.8. Statistical Analysis

Statistical analyses were performed using SPSS 22.0 (IBM, Inc., New York, NY, USA) and GraphPad Prism 8.4.2 software from GraphPad Software, version 8.4.2 for Windows, GraphPad Software, San Diego, CA, USA. Continuous data are presented as mean ± standard deviation ($\bar{x} \pm$ SD). Between-group comparisons were analyzed by one-way ANOVA, with data transformation applied or non-parametric tests when normality assumptions were violated. All figures are two-tailed cutoffs, and significance was set at $p < 0.05$, unless otherwise stated.

3. Results

3.1. Levels of Inflammatory Factors and Barrier Protein Gene Transcription

The intestinal barrier functions to segregate the gut microbiota and their metabolites from the host body. The physical barrier, composed of intestinal epithelial cells and the tight junction proteins that connect these cells, plays a critical role in preventing harmful substances from entering the body. To investigate the impact of polystyrene microplastic exposure on the transcription levels of tight junction proteins (ZO-1, Occludin, and Claudin-1) in the colonic tissue of mice, qPCR was performed (Figure 1A–C). Compared with the CD + water group, the HFD + water group showed significantly decreased transcription levels of ZO-1 and Occludin ($p < 0.05$), with no significant change in Claudin-1. The CD + 8 mg/kg MP group exhibited reduced transcription levels of ZO-1 and Occludin compared to the CD + water group ($p < 0.05$), with no significant change in Claudin-1. Additionally, the HFD + 8 mg/kg MP group demonstrated significantly reduced transcription levels of ZO-1, Occludin, and Claudin-1 compared to the HFD + water group ($p < 0.05$). These results indicate that microplastic exposure alone reduces the transcription levels of tight junction proteins in colonic tissue, and co-exposure with a high-fat diet exacerbates these effects.

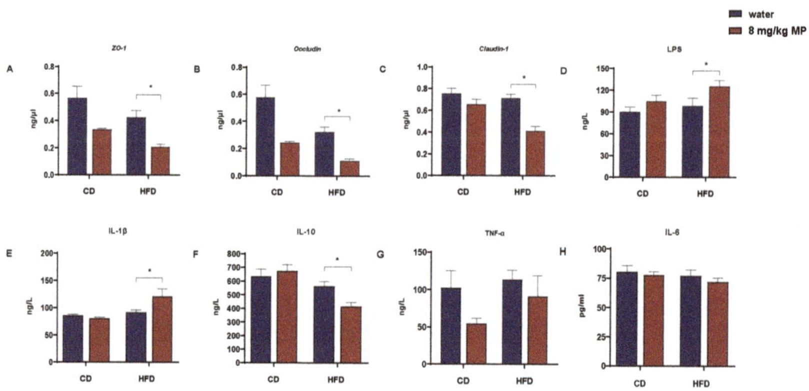

Figure 1. Polystyrene microplastic-induced changes in serum inflammatory factor and intestinal barrier protein gene expression of mice fed normal or high-fat diets: (**A**) transcription of ZO-1 proteins located in the colon, (**B**) transcription of Occludin proteins located in the colon, (**C**) transcription of Claudin-1 proteins located in the colon, (**D**) LPS content in mice serum, (**E**) IL-1β content in mice serum, (**F**) IL-10 content in mice serum, (**G**) TNF-α content in mice serum, and (**H**) IL-6 content in mice serum. * $p < 0.05$.

Increased intestinal barrier permeability allows LPS and inflammatory factors from the gut to enter the circulatory system. ELISA kits were used to measure the levels of LPS and inflammatory factors in the serum of mice (Figure 1D–H). After 14 weeks of polystyrene microplastic exposure, serum IL-6 and TNF-α levels showed no significant changes. While IL-1β levels in the HFD + water group showed an increasing trend compared to the CD + water group, there was no significant statistical difference. There was no significant change in IL-1β levels between the CD + water and CD + 8 mg/kg MP groups. However, the HFD + 8 mg/kg MP group exhibited significantly elevated IL-1β levels compared to the HFD + water group ($p < 0.05$). For IL-10, the HFD + water group showed a decreasing trend compared to the CD + water group, with no significant difference between the CD + water and CD + 8 mg/kg MP groups. The HFD + 8 mg/kg MP group displayed significantly reduced IL-10 levels compared to the HFD + water group ($p < 0.05$). LPS levels showed an increasing trend in the HFD + water group compared to the CD + water group. Similar trends were observed in the CD + 8 mg/kg MP group compared to the CD + water group, with significantly elevated LPS levels in the HFD + 8 mg/kg MP group compared to the HFD + water group ($p < 0.05$).

3.2. Morphology in the Colon

HE staining was performed to observe the morphological changes in the colonic tissue of mice following polystyrene microplastic exposure (Figure 2A). The results indicated that the colonic tissue of the CD + water group exhibited intact glands, clear crypt structures, regular arrangement, and continuous mucosal epithelium. There were no significant morphological changes in the colonic tissue of the CD + 8 mg/kg MP group compared to the control group, with normal tissue structure. However, the HFD + water group showed disrupted crypt structures, irregular crypt surfaces, and uneven mucosal surfaces. These effects were more pronounced in the HFD + 8 mg/kg MP group, indicating that microplastic exposure alone had no significant impact on colonic morphology, but co-exposure with a high-fat diet exacerbated the morphological changes induced by the high-fat diet. Immunohistochemical staining was performed to observe the expression of tight junction proteins in the colonic tissue following polystyrene microplastic exposure (Figure 2B–D). The CD + water group showed uniform and continuous positive brown areas on the surface of colonic epithelial cells, with no significant differences in the CD + 8 mg/kg MP group. The HFD + water group exhibited discontinuous and diffuse distribution, with significantly reduced expression areas of ZO-1, Occludin, and Claudin-1 proteins in the HFD + 8 mg/kg MP group. These results suggest that, while microplastic exposure alone does not significantly affect the expression of tight junction proteins, co-exposure with a high-fat diet exacerbates the impact of the high-fat diet on tight junction protein expression.

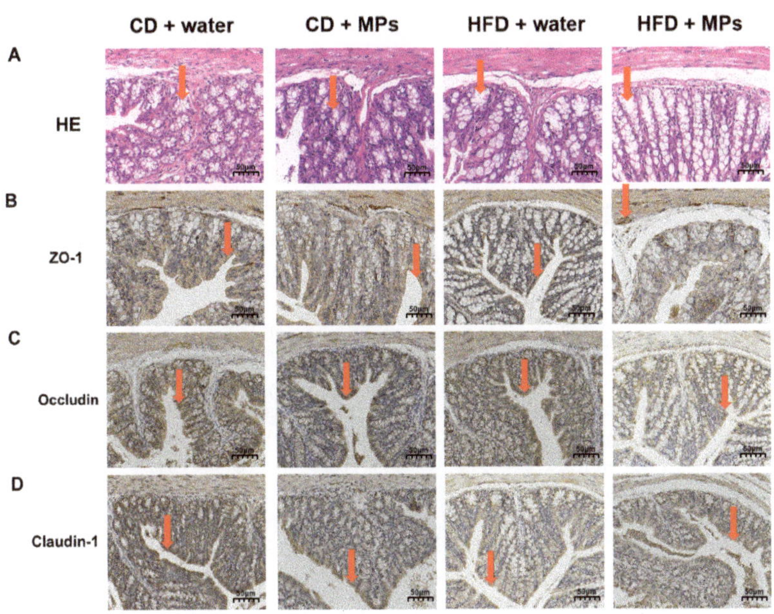

Figure 2. Polystyrene microplastic-induced changes in the colon morphology of mice fed normal or high-fat diets: (**A**) HE staining of colon tissue, (**B**) expression of ZO-1 proteins located in the colon, (**C**) expression of Occludin proteins located in the colon, and (**D**) expression of Claudin-1 proteins located in the colon. Red arrows: denote critical pathological features ((**A**) crypt structures; (**B**) ZO-1 protein expression; (**C**) Occludin protein expression; (**D**) Claudin-1 protein expression). Scale bar: 50 μm.

3.3. Composition and Diversity of the Gut Microbiota

The gut microbiota plays a vital role in maintaining host metabolism and health and regulating intestinal barrier function. To investigate changes in the gut microbiota composition following polystyrene microplastic exposure, the composition at the phylum level was analyzed based on the 16S rRNA sequencing results (Figure 3A). The results showed that *Firmicutes, Bacteroidota, Actinobacteriota, Desulfobacterota, Campilobacterota,* and *Deferribacterota* were the most abundant phyla in the mouse gut microbiota. The HFD + water group exhibited significantly increased levels of *Firmicutes, Desulfobacterota,* and *Campilobacterota* and decreased levels of *Bacteroidota* and *Actinobacteriota* compared to the CD + water group. The CD + 8 mg/kg MP group also showed increased relative abundance of *Firmicutes, Desulfobacterota,* and *Campilobacterota,* with a decreasing trend in the abundance of *Bacteroidota* and *Actinobacteriota* compared to the CD + water group. Similar trends were observed in the HFD + 8 mg/kg MP group compared to the HFD + water group.

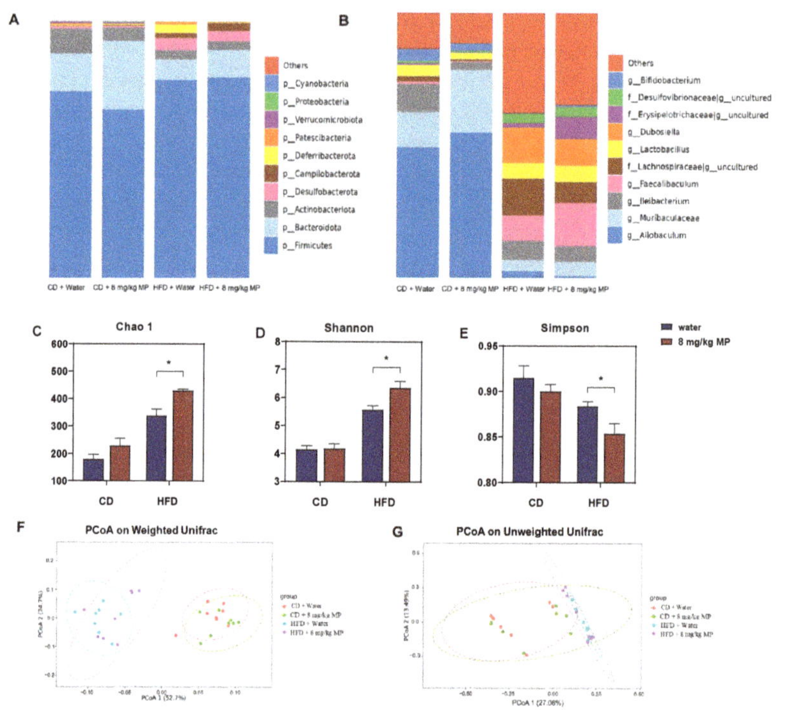

Figure 3. Polystyrene microplastic-induced changes in the composition and diversity of gut microbiota of mice fed normal or high-fat diets (**A**) Changes in gut microbiota composition at the phylum level. (**B**) Changes in gut microbiota composition at the genus level. (**C**) The Chao 1 index analyzes changes in gut microbiota diversity in mice. (**D**) The Shannon index analyzes changes in gut microbiota diversity in mice. (**E**) The Simpson index analyzes changes in gut microbiota diversity in mice. (**F**) Changes in gut microflora diversity in mice analyzed by Principal Coordinate Analysis (PCoA) based on weighted UniFrac distance. (**G**) Changes in gut microflora diversity in mice analyzed by PCoA based on unweighted UniFrac distance. * $p < 0.05$.

To reveal more detailed changes in the gut microbiota composition, genus-level analysis was performed (Figure 3B). The most abundant genera included *Allobaculum, Muribaculaceae, Ileibacterium, Faecalibaculum, Lachnospiraceae, Lactobacillus,* and *Dubosiella*. Compared to the CD + water group, the HFD + water group showed significantly increased levels of *Faecalibaculum, Lachnospiraceae, Lactobacillus,* and *Dubosiella,* with decreased relative abundance of *Allobaculum* and *Muribaculaceae*. The HFD + 8 mg/kg MP group exhibited similar trends compared to the HFD + water group.

Previous studies have shown that the diversity of the gut microbiota can be influenced by environmental chemicals. Analysis of the 16S rRNA sequencing results revealed the impact of microplastic exposure on the diversity of the gut microbiota (Figure 3C–G). The Chao 1 index, which estimates species richness, was significantly higher in the HFD + water group compared to the CD + water group ($p < 0.05$). While there was no significant difference between the CD + 8 mg/kg MP group and the CD + water group, an increasing trend in the index values was observed. The HFD + 8 mg/kg MP group showed significantly higher Chao 1 indices compared to the HFD + water group ($p < 0.05$). The Shannon and Simpson indices, which estimate microbial diversity, indicated increased diversity when the Shannon index was high and the Simpson index was low. The HFD + water group showed significantly increased Shannon indices and decreased Simpson indices compared to the CD + water group ($p < 0.05$). There were no significant differences between the CD + 8 mg/kg MP group and the CD + water group, but the HFD + 8 mg/kg MP group exhibited increased Shannon indices and decreased Simpson indices compared to the HFD + water group ($p < 0.05$). These results suggest that polystyrene microplastic exposure significantly increases the α-diversity of the gut microbiota.

Principal Coordinate Analysis (PCoA) of the 16S rRNA sequencing results was performed. Based on the weighted UniFrac distance algorithm, there were significant differences in the OTU levels between normal-diet-fed mice and high-fat-diet-fed mice, with no overlap. Microplastic-exposed mice and control mice showed significant differences in OTU levels regardless of diet, with more pronounced differences in the high-fat-diet-fed mice. The unweighted UniFrac distance algorithm also revealed significant differences in OTU levels between normal-diet-fed and high-fat-diet-fed mice, with minimal overlap. Significant differences in OTU levels were observed between microplastic-exposed and control mice regardless of diet, with more pronounced differences in high-fat-diet-fed mice. These results indicate that polystyrene microplastic exposure significantly increases the β-diversity of the gut microbiota.

3.4. Changes in the Gut Microbiota

To further assess the impact of polystyrene microplastic exposure on the composition of the gut microbiota in mice, LEfSe analysis was used to identify specific changes. Compared to the CD + water group, the relative abundances of *g__Allobaculum*, *o__Oscillospirales*, *f__Oscillospiraceae*, *g__Butyricimonas*, and *g__Adlercreutzia* significantly increased in the CD + 8 mg/kg MP group after polystyrene microplastic exposure. In comparison to the HFD + water group, the relative abundance of *g__uncultured* significantly increased in the HFD + 8 mg/kg MP group. These results indicate that exposure to microplastics significantly alters the composition of the gut microbiota in mice (Figure 4).

3.5. Short-Chain Fatty Acids

Due to the close relationship between short-chain fatty acids (SCFAs) and the intestinal barrier, changes in the content of SCFAs in fecal samples were analyzed after microplastic exposure, as shown in Figure 5A–H. Analysis revealed no significant difference in total SCFA content between the CD + water group and the HFD + water group, nor between the CD + water group and the CD + 8 mg/kg MP group. However, compared to the HFD + water group, the total SCFA content significantly decreased in the HFD + 8 mg/kg MP group ($p < 0.05$). Although differences in specific SCFA contents between the CD + water group and the HFD + water group were not statistically significant, propionate, valerate, and caproate showed a decreasing trend, while acetate and isovalerate exhibited an increasing trend. Compared to the CD + water group, the CD + 8 mg/kg MP group showed a significant decrease in valerate, but other SCFAs showed only changing trends without statistical significance. Compared to the HFD + water group, the HFD + 8 mg/kg MP group showed no significant differences in isobutyrate and isovalerate but had significantly increased acetate and significantly decreased propionate, butyrate, valerate, and caproate ($p < 0.05$). These results suggest that polystyrene microplastic exposure alone does not significantly affect SCFA metabolism, but combined exposure with a high-fat diet exacerbates the impact on SCFA metabolism.

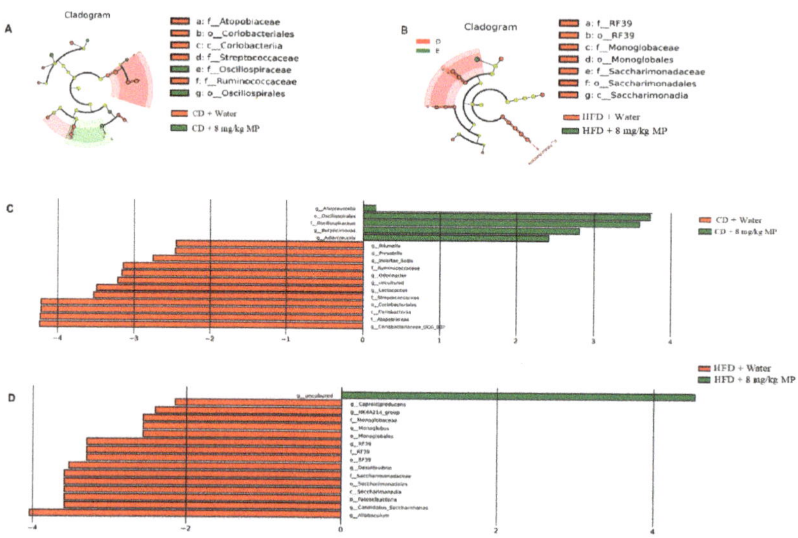

Figure 4. Polystyrene microplastic-induced changes in gut microbiota composition of mice fed normal or high-fat diets: the differences in microbiota between groups after exposure to polystyrene microplastics are shown by LEfSe analysis. (**A**) Cladogram analysis of the gut microbiota of mice in CD + water and CD + 8 mg/kg MP groups. (**B**) Histogram of the distribution of linear discriminant analysis (LDA) values of the gut microbiota of mice in CD + water and CD + 8 mg/kg MP groups. (**C**) Cladogram analysis of the gut microbiota of mice in HFD + water and HFD + 8 mg/kg MP groups. (**D**) Histogram of the distribution of LDA values of the gut microbiota of mice in HFD + water and HFD + 8 mg/kg MP groups.

Butyrate may affect intestinal barrier function by binding to PPARγ, so this study aims to explore whether the impact of polystyrene microplastic exposure on glucose metabolism in mice is related to intestinal barrier dysfunction caused by reduced butyrate levels. According to the above results, the butyrate content in fecal samples significantly decreased in the HFD + 8 mg/kg MP group compared to the HFD + water group. Therefore, quantitative analysis of butyrate-producing bacteria in fecal samples was conducted, as shown in Figure 5I,J. The results showed that, compared to the CD + water group, the relative abundances of *Clostridium* and *Eubacterium* significantly decreased in the fecal samples of the CD + 8 mg/kg MP group ($p < 0.05$). Moreover, compared to the HFD + water group, the relative abundances of *Clostridium* and *Eubacterium* further decreased in the HFD + 8 mg/kg MP group. These results indicate that polystyrene microplastics affect the metabolic pathway of butyrate by reducing the relative abundance of butyrate-producing bacteria in the gut microbiota of mice, ultimately leading to a decrease in butyrate content.

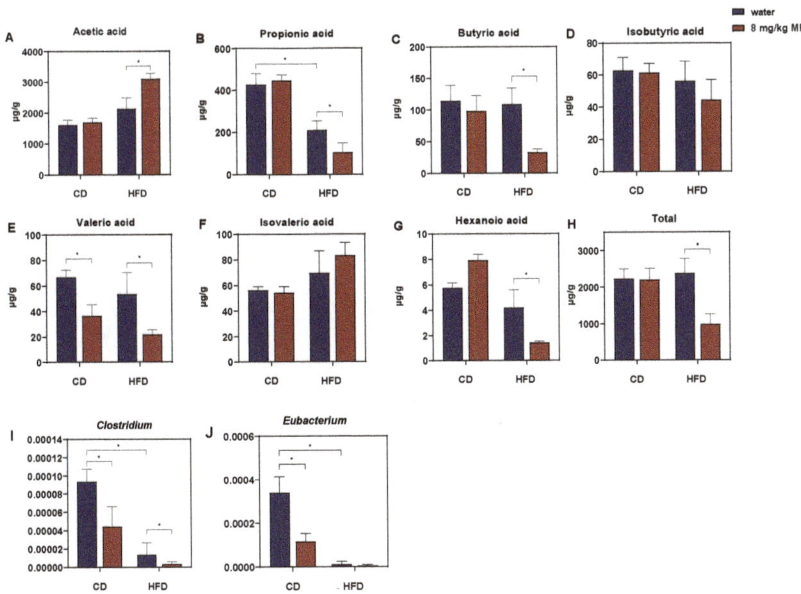

Figure 5. Polystyrene microplastic-induced changes in fecal short-chain fatty acid and butyrate-producing gut microbiota of mice fed normal chow diets (CDs) or high-fat diets (HFD). (**A**) Acetic acid in mouse feces. (**B**) Propionic acid in mouse feces. (**C**) Butyric acid in mouse feces. (**D**) Isobutyric acid in mouse feces. (**E**) Valeric acid in mouse feces. (**F**) Isovaleric acid in mouse feces. (**G**) Hexanoic acid in mouse feces. (**H**) Total short-chain fatty acid in mouse feces. (**I**) Changes in relative abundance of *Clostridium*. (**J**) Changes in relative abundance of *Eubacterium*. * $p < 0.05$.

3.6. Expression of PPARγ

The nuclear transcription factor PPARγ binds to butyrate and participates in the regulation of intestinal barrier function. Therefore, qPCR was used to observe the transcription of the PPARγ gene in the colon tissues of mice after polystyrene microplastic exposure, as shown in Figure 6A. Compared to the CD + water group, the transcription level of the PPARγ gene significantly decreased in the colon tissues of the CD + 8 mg/kg MP group ($p < 0.05$). Similarly, compared to the HFD + water group, the transcription level of the PPARγ gene significantly decreased in the HFD + 8 mg/kg MP group ($p < 0.05$). These results indicate that microplastic exposure significantly reduces the transcription level of the PPARγ gene in colon tissues, and when combined with a high-fat diet, microplastic exposure exacerbates the impact on the transcription level of the PPARγ gene.

Immunofluorescence staining of PPARγ protein in colon tissues was used to observe the effect of polystyrene microplastic exposure on the expression of the PPARγ gene in colon tissues. The results showed that compared to the CD + water group, the expression of PPARγ significantly decreased in the HFD + water group. There was no significant difference in the expression of PPARγ in colon tissues between the CD + water group and the CD + 8 mg/kg MP group. However, compared to the HFD + water group, the expression of PPARγ decreased even more significantly in the HFD + 8 mg/kg MP group. These results indicate that microplastic exposure alone does not significantly affect the expression of PPARγ in colon tissues, but when combined with a high-fat diet, microplastic exposure exacerbates the impact on the expression of PPARγ in the colon tissues of mice.

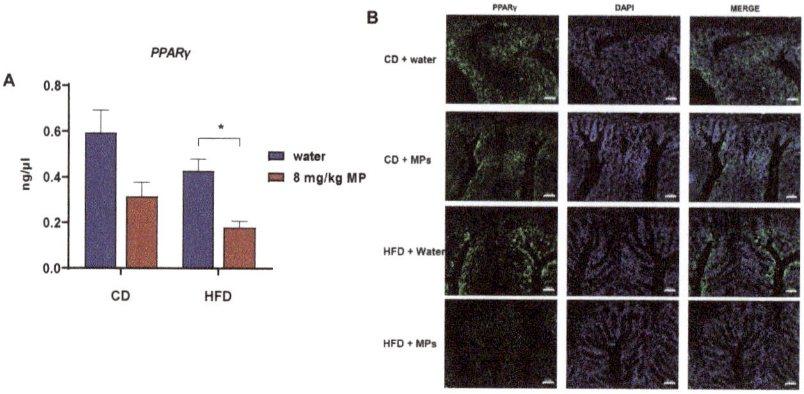

Figure 6. Polystyrene microplastic-induced changes in PPARγ expression levels in the colon tissue of mice fed normal or high-fat diets (**A**) qPCR analysis of PPARγ gene transcription. (**B**) Immunofluorescence staining of PPARγ protein expression (CD + water: normal chow diet + water group; CD + MPs: normal chow diet + 8 mg/kg MP group; HFD + water: high-fat diet + water group; HFD + MPs: high-fat diet + 8 mg/kg MP group). Scale bar: 50 μm, * $p < 0.05$.

4. Discussion

Human activities have led to the pervasive presence of microplastics in the environment, which can enter the human body through ingestion, inhalation, and dermal contact. Consequently, increasing research attention has been directed towards the potential hazards posed to human health by microplastic exposure. Given the ubiquitous presence of microplastics and high-fat diets in human life, this study utilizes mice fed normal and high-fat diets to investigate the effects of microplastic exposure on gut barrier function, as well as the differential impacts of these dietary patterns. Our findings indicate that environmental concentrations of microplastic exposure exacerbate high-fat-diet-induced gut barrier dysfunction and increased intestinal permeability. The underlying mechanism may involve microplastic-induced gut dysbiosis, significant reductions in the relative abundance of the butyrate-producing bacteria *Clostridium* and *Eubacterium*, decreased butyrate levels, and reduced PPARγ receptor expression, ultimately leading to impaired gut barrier function.

Organisms are continually exposed to microplastics, with humans facing complex and varied exposure pathways that make intake assessments highly variable [32]. In this study, we selected a microplastic exposure dose of 8 mg/kg body weight/day, derived from environmental microplastic concentrations converted using classical pharmacological and toxicological formulas. Our findings indicate that although the high-fat diet mice did not reach statistical significance, they still exhibited a discernible trend of change. Similarly, numerous studies have demonstrated that dietary fat can directly modulate the structural integrity of the intestinal barrier, thereby influencing intestinal permeability. For instance, long-term high-fat diet feeding has been shown to downregulate the expression of genes encoding tight junction proteins [33]. Kirpich et al. reported that continuous feeding of a diet rich in unsaturated fatty acids significantly reduced the expression of tight junction proteins and increased the flux of fluorescein isothiocyanate (FITC)-conjugated 4 kDa dextran [34]. Combined exposure to a high-fat diet and microplastics results in the significantly reduced expression of tight junction proteins (ZO-1, Occludin, and Claudin-1) in the colonic tissues of mice, leading to disrupted and diffuse distribution and impaired physical gut barriers. The results indicate that microplastics may exacerbate intestinal barrier dysfunction induced by a high-fat diet. Yan et al. similarly observed that polystyrene microplastic exposure in rats resulted in shallower crypt structures and damaged mucosal barriers. Additionally, gut permeability analysis revealed significantly elevated plasma D-lactate and diamine oxidase (DAO) levels in microplastic-exposed mice, biomarkers indicative of leaky gut syndrome [35]. These findings were corroborated by Okamura's experiments, where fluorescein isothiocyanate (FITC)-dextran assay results demonstrated higher green fluorescent protein (GFP) signals in the high-fat diet and microplastic-exposed rats compared to the high-fat diet control group [36]. In our study, microplastics may affect people on a high-fat diet more than those on a normal diet, which may be attributed to the synergistic effects between microplastics and a high-fat diet upon co-exposure: specifically, the combined impact of HFD and microplastics exceeds the sum of their individual effects, thereby exacerbating the influence of HFD on intestinal barrier function and gut microbiota in mice. The lack of significant effects of microplastic exposure on mice under a normal diet in this study may be attributed to the selected particle size, dosage, or duration of exposure. However, upon retrospective analysis of the results, we observed that although the differences were not statistically significant, normal-diet mice exposed to microplastics exhibited trends consistent with those observed in high-fat-diet mice.

The disruption of tight junctions increases intestinal permeability, allowing harmful substances such as bacteria and endotoxins to enter systemic circulation, adversely affecting organismal health [37,38]. Serum analysis revealed elevated levels of LPS and the pro-inflammatory cytokine IL-1β, along with decreased levels of the anti-inflammatory cytokine IL-10 in the HFD + 8 mg/kg MP group compared to the HFD + water group. LPS, a major component of the outer membrane of Gram-negative bacteria, showed increased relative abundance in response to microplastic exposure, significantly elevating endotoxin levels in the serum and indicating compromised gut barrier function. In the bloodstream, LPS forms trimers with lipopolysaccharide-binding protein, signaling through Toll-like receptors and pattern recognition receptors to activate multiple inflammatory pathways and promote chronic systemic inflammation [39,40].

The gut microbiota plays a crucial role in maintaining gut barrier function, making it essential to examine the impact of microplastic exposure on the gut microbiota. While no significant differences were observed in alpha diversity between the normal diet microplastic group and the normal diet control group, combined exposure to a high-fat diet and microplastics resulted in increased alpha diversity and significant differences in OTU levels between exposed and control mice. These findings are consistent with numerous studies indicating that microplastic exposure disrupts gut microbiota diversity and composition. For instance, Marine medaka (*Oryzias melastigma*) exposed to 50 nm and 45 μm polystyrene microplastics exhibited altered gut microbiota diversity and composition, with increased alpha diversity and changes in principal components [41]. Jin et al. also reported significant alterations in gut microbiota composition at the genus level, with 15 bacterial taxa showing notable changes following microplastic exposure [27]. Li's mouse model study demonstrated that environmental microplastic exposure (6, 15, 60, and 600 μg/d) increases total gut microbiota, bacterial abundance, and diversity at high concentrations [14]. In contrast to control mice, exposed mice showed a significant increase in *Staphylococcus* and a decrease in *Parabacteroides* relative abundance. Another study found that exposure to 0.5 μm and 50 μm polystyrene microplastics significantly reduced gut microbiota diversity, with 310 OTUs affected at the phylum level, and notable reductions in fecal Firmicutes and α-Proteobacteria relative abundance in the 0.5 μm exposure group [42]. However, the authors did not provide a detailed explanation of the specific mechanisms underlying the reduction in gut microbiota diversity following microplastic exposure in the referenced literature. We speculate that this phenomenon may be associated with microplastic-induced damage to the intestinal physical barrier and mucus layer, oxidative stress and inflammatory responses, direct toxicity to gut microbiota, as well as alterations in intestinal physiology and function. Regarding the observed increase in gut microbiota diversity in the high-fat diet + microplastic group in our study, a plausible explanation is that the co-exposure to microplastics and a high-fat diet may lead to an elevated diversity of certain harmful bacterial species in the gut microbiota of mice, ultimately contributing to an overall increase in microbial diversity. Li et al. also reported microplastic-induced gut dysbiosis associated with intestinal inflammation in mice: The treatment of microplastics can significantly increase the relative abundance of *Staphylococcus* genera in the gut microbiota of mice, and the upregulation of *Staphylococcus* abundance may potentially induce an elevation in IL-1α levels [14]. Furthermore, another study has demonstrated that *Staphylococcus aureus* infection can induce the expression of the pro-inflammatory cytokine IL-1α [43].

Butyrate, a critical short-chain fatty acid, binds to PPARγ receptors in the gut to activate pathways that regulate gut barrier function [25,44]. In this study, combined exposure to a high-fat diet and microplastics resulted in significantly decreased butyrate levels and a reduced relative abundance of butyrate-producing bacteria. The expression of PPARγ in colonic tissues mirrored the trend in butyrate levels, with significantly decreased transcription and expression in the HFD + 8 mg/kg MP group compared to the HFD + water group. Okamura similarly concluded that short-chain fatty acid concentrations in fecal samples were significantly lower in the high-fat diet and microplastic-exposed mice compared to high-fat diet controls [36]. Kundu's study demonstrated that gavage with *Salmonella typhimurium* suppressed PPARγ expression in gut epithelial cells, inducing acute infectious colitis [45]. Exposure to DSS also significantly reduced PPARγ transcription levels and impaired gut barrier function [25]. These results suggest that butyrate production enhances gut barrier function by binding to PPARγ receptors, mitigating microplastic-induced intestinal inflammation.

5. Conclusions

This study demonstrates that environmental concentrations of microplastic exposure can induce gut dysbiosis, alter butyrate levels, and reduce PPARγ receptor expression, leading to gut barrier dysfunction in mice. Furthermore, the impact of microplastic exposure on gut barrier function varies with different dietary structures. In high-fat diet-fed mice, the co-exposure to microplastics was particularly notable, exacerbating the adverse effects on the gut microbiota, butyrate levels, and intestinal barrier function. Considering the unavoidable human exposure to microplastics and the critical relationship between gut barrier function and human health, these findings contribute to a better understanding of the health effects of microplastics and their toxicity in relation to dietary structures.

Supplementary Materials: The following supporting information can be downloaded at https://www.mdpi.com/article/10.3390/metabo15080557/s1, Figure S1: Coulter particle size analysis of polystyrene microspheres (offered by Tianjin BaseLine ChromTech Research Center); Table S1: Primer sequence list.

Author Contributions: Conceptualization: H.N., Y.Z., P.T. and X.L. (Xiaoming Lou); methodology: H.N., X.M., Z.D., H.C., Q.Z., M.X. (Manjin Xu) and L.W.; writing—original draft: H.N., M.X. (Manjin Xu), M.X. (Mingluan Xing), Z.M., Z.C., X.L. (Xueqing Li), Y.Y., X.T. and P.T.; writing—review and editing: all authors. All authors have read and agreed to the published version of the manuscript.

Funding: This research was supported by the Natural Science Foundation of Zhejiang Province [Grant number LQ24H260004], the National Natural Science Foundation of China [Grant number 42407553], the Zhejiang Provincial Project for Medical Research and Health Sciences [Grant number 2024KY898], and the Central Guiding Local Science and Technology Development Fund Projects

[Grant number 2023ZY1024], the Zhejiang CDC Science and Technology Talent Incubation Project [Grant number 2023-B-04].

Institutional Review Board Statement: The animal study protocol was approved by the Ethics Committee of NAME OF Animal Management and the Ethics Committee of Zhejiang Chinese Medical University (protocol code No.: 20230213-07 and 13 March 2023 of approval).

Informed Consent Statement: Not applicable.

Data Availability Statement: 3 Data available on request due to restrictions eg privacy or ethical. The data presented in this study are available on request from the corresponding author. The data are not publicly available due to institutional animal ethics committee restrictions requiring controlled access to primary experimental datasets involving vertebrate animals (Approval No. 20230213-07).

Conflicts of Interest: All authors declare that they have no conflict of interest.

Abbreviations

The following abbreviations are used in this manuscript:

MPs	Microplastics
ROS	Reactive oxygen species
CD	Chow diet
HFD	High-fat diet
PPARγ	Peroxisome proliferator-activated receptor γ
LPS	Lipopolysaccharides
ELISA	Enzyme-linked immunosorbent assay
PCoA	Principal Coordinate Analysis
SCFAs	Short-chain fatty acids

References

1. Thompson, R.C.; Olsen, Y.; Mitchell, R.P.; Davis, A.; Rowland, S.J.; John, A.W.G.; McGonigle, D.; Russell, A.E. Lost at sea: Where is all the plastic? *Science* **2004**, *304*, 838. [CrossRef]
2. Koelmans, A.A.; Mohamed Nor, N.H.; Hermsen, E.; Kooi, M.; Mintenig, S.M.; De France, J. Microplastics in freshwaters and drinking water: Critical review and assessment of data quality. *Water Res.* **2019**, *155*, 410–422. [CrossRef]
3. Bäuerlein, P.S.; Hofman-Caris, R.C.H.M.; Pieke, E.N.; Ter Laak, T.L. Fate of microplastics in the drinking water production. *Water Res.* **2022**, *221*, 118790. [CrossRef]
4. Makhdoumi, P.; Hossini, H.; Pirsaheb, M. A review of microplastic pollution in commercial fish for human consumption. *Rev. Environ. Health* **2023**, *38*, 97–109. [CrossRef]
5. Acarer, S. Abundance and characteristics of microplastics in drinking water treatment plants, distribution systems, water from refill kiosks, tap waters and bottled waters. *Sci. Total Environ.* **2023**, *884*, 163866. [CrossRef]
6. Sun, Q.; Ren, S.-Y.; Ni, H.-G. Incidence of microplastics in personal care products: An appreciable part of plastic pollution. *Sci. Total Environ.* **2020**, *742*, 140218. [CrossRef]
7. Jones, L.R.; Wright, S.J.; Gant, T.W. A critical review of microplastics toxicity and potential adverse outcome pathway in human gastrointestinal tract following oral exposure. *Toxicol. Lett.* **2023**, *385*, 51–60. [CrossRef]
8. Yang, D.; Shi, H.; Li, L.; Li, J.; Jabeen, K.; Kolandhasamy, P. Microplastic Pollution in Table Salts from China. *Environ. Sci. Technol.* **2015**, *49*, 13622–13627. [CrossRef]
9. Mattsson, K.; Ekvall, M.T.; Hansson, L.-A.; Linse, S.; Malmendal, A.; Cedervall, T. Altered behavior, physiology, and metabolism in fish exposed to polystyrene nanoparticles. *Environ. Sci. Technol.* **2015**, *49*, 553–561. [CrossRef] [PubMed]
10. Van Cauwenberghe, L.; Janssen, C.R. Microplastics in bivalves cultured for human consumption. *Environ. Pollut.* **2014**, *193*, 65–70. [CrossRef] [PubMed]
11. Ragusa, A.; Svelato, A.; Santacroce, C.; Catalano, P.; Notarstefano, V.; Carnevali, O.; Papa, F.; Rongioletti, M.C.A.; Baiocco, F.; Draghi, S.; et al. Plasticenta: First evidence of microplastics in human placenta. *Environ. Int.* **2021**, *146*, 106274. [CrossRef]
12. Leslie, H.A.; van Velzen, M.J.M.; Brandsma, S.H.; Vethaak, A.D.; Garcia-Vallejo, J.J.; Lamoree, M.H. Discovery and quantification of plastic particle pollution in human blood. *Environ. Int.* **2022**, *163*, 107199. [CrossRef]

13. Zhao, L.; Shi, W.; Hu, F.; Song, X.; Cheng, Z.; Zhou, J. Prolonged oral ingestion of microplastics induced inflammation in the liver tissues of C57BL/6J mice through polarization of macrophages and increased infiltration of natural killer cells. *Ecotoxicol. Environ. Saf.* **2021**, *227*, 112882. [CrossRef] [PubMed]
14. Li, B.; Ding, Y.; Cheng, X.; Sheng, D.; Xu, Z.; Rong, Q.; Wu, Y.; Zhao, H.; Ji, X.; Zhang, Y. Polyethylene microplastics affect the distribution of gut microbiota and inflammation development in mice. *Chemosphere* **2020**, *244*, 125492. [CrossRef]
15. Guo, X.; Lv, M.; Li, J.; Ding, J.; Wang, Y.; Fu, L.; Sun, X.; Han, X.; Chen, L. The distinct toxicity effects between commercial and realistic polystyrene microplastics on microbiome and histopathology of gut in zebrafish. *J. Hazard. Mater.* **2022**, *434*, 128874. [CrossRef]
16. Wang, J.; Li, Y.; Lu, L.; Zheng, M.; Zhang, X.; Tian, H.; Wang, W.; Ru, S. Polystyrene microplastics cause tissue damages, sex-specific reproductive disruption and transgenerational effects in marine medaka (*Oryzias melastigma*). *Environ. Pollut.* **2019**, *254*, 113024. [CrossRef] [PubMed]
17. Xia, X.; Sun, M.; Zhou, M.; Chang, Z.; Li, L. Polyvinyl chloride microplastics induce growth inhibition and oxidative stress in Cyprinus carpio var. larvae. *Sci. Total Environ.* **2020**, *716*, 136479. [CrossRef] [PubMed]
18. Cui, Y.; Wang, Q.; Chang, R.; Zhou, X.; Xu, C. Intestinal Barrier Function-Non-alcoholic Fatty Liver Disease Interactions and Possible Role of Gut Microbiota. *J. Agric. Food Chem.* **2019**, *67*, 2754–2762. [CrossRef]
19. Ghosh, S.; Whitley, C.S.; Haribabu, B.; Jala, V.R. Regulation of Intestinal Barrier Function by Microbial Metabolites. *Cell Mol. Gastroenterol. Hepatol.* **2021**, *11*, 1463–1482. [CrossRef]
20. Niu, H.; Liu, S.; Jiang, Y.; Hu, Y.; Li, Y.; He, L.; Xing, M.; Li, X.; Wu, L.; Chen, Z.; et al. Are Microplastics Toxic? A Review from Eco-Toxicity to Effects on the Gut Microbiota. *Metabolites* **2023**, *13*, 739. [CrossRef]
21. Deleu, S.; Machiels, K.; Raes, J.; Verbeke, K.; Vermeire, S. Short chain fatty acids and its producing organisms: An overlooked therapy for IBD? *EBioMedicine* **2021**, *66*, 103293. [CrossRef]
22. Nielsen, D.S.G.; Jensen, B.B.; Theil, P.K.; Nielsen, T.S.; Knudsen, K.E.B.; Purup, S. Effect of butyrate and fermentation products on epithelial integrity in a mucus-secreting human colon cell line. *J. Funct. Foods* **2018**, *40*, 9–17. [CrossRef]
23. Tang, G.; Du, Y.; Guan, H.; Jia, J.; Zhu, N.; Shi, Y.; Rong, S.; Yuan, W. Butyrate ameliorates skeletal muscle atrophy in diabetic nephropathy by enhancing gut barrier function and FFA2-mediated PI3K/Akt/mTOR signals. *Br. J. Pharmacol.* **2022**, *179*, 159–178. [CrossRef]
24. Zhao, M.; Jiang, Z.; Cai, H.; Li, Y.; Mo, Q.; Deng, L.; Zhong, H.; Liu, T.; Zhang, H.; Kang, J.X.; et al. Modulation of the Gut Microbiota during High-Dose Glycerol Monolaurate-Mediated Amelioration of Obesity in Mice Fed a High-Fat Diet. *MBio* **2020**, *11*, e00190-20. [CrossRef]
25. Simeoli, R.; Mattace Raso, G.; Lama, A.; Pirozzi, C.; Santoro, A.; Di Guida, F.; Sanges, M.; Aksoy, E.; Calignano, A.; D'Arienzo, A.; et al. Preventive and therapeutic effects of Lactobacillus paracasei B21060-based synbiotic treatment on gut inflammation and barrier integrity in colitic mice. *J. Nutr.* **2015**, *145*, 1202–1210. [CrossRef] [PubMed]
26. Kou, G.; Yao, S.; Ullah, A.; Fang, S.; Guo, E.; Bo, Y. Polystyrene microplastics impair brown and beige adipocyte function via the gut microbiota-adipose tissue crosstalk in high-fat diet mice. *J. Hazard. Mater.* **2025**, *492*, 138225. [CrossRef]
27. Jin, Y.; Lu, L.; Tu, W.; Luo, T.; Fu, Z. Impacts of polystyrene microplastic on the gut barrier, microbiota and metabolism of mice. *Sci. Total Environ.* **2019**, *649*, 308–317. [CrossRef] [PubMed]
28. Huang, D.; Zhang, Y.; Long, J.; Yang, X.; Bao, L.; Yang, Z.; Wu, B.; Si, R.; Zhao, W.; Peng, C.; et al. Polystyrene microplastic exposure induces insulin resistance in mice via dysbacteriosis and pro-inflammation. *Sci. Total Environ.* **2022**, *838*, 155937. [CrossRef]
29. Deng, Y.; Chen, H.; Huang, Y.; Zhang, Y.; Ren, H.; Fang, M.; Wang, Q.; Chen, W.; Hale, R.C.; Galloway, T.S.; et al. Long-Term Exposure to Environmentally Relevant Doses of Large Polystyrene Microplastics Disturbs Lipid Homeostasis via Bowel Function Interference. *Environ. Sci. Technol.* **2022**, *56*, 15805–15817. [CrossRef] [PubMed]
30. Nair, A.B.; Jacob, S. A simple practice guide for dose conversion between animals and human. *J. Basic Clin. Pharm.* **2016**, *7*, 27–31. [CrossRef]
31. Ahmadi, H.; Bognar, Z.; Csabai-Tanics, T.; Obodo, B.N.; Szekeres-Bartho, J. Allergic Disposition of IVF-Conceived Mice. *Int. J. Mol. Sci.* **2024**, *25*, 12993. [CrossRef]
32. Prata, J.C.; da Costa, J.P.; Lopes, I.; Duarte, A.C.; Rocha-Santos, T. Environmental exposure to microplastics: An overview on possible human health effects. *Sci. Total Environ.* **2020**, *702*, 134455. [CrossRef] [PubMed]
33. Cani, P.D.; Bibiloni, R.; Knauf, C.; Waget, A.; Neyrinck, A.M.; Delzenne, N.M.; Burcelin, R. Changes in gut microbiota control metabolic endotoxemia-induced inflammation in high-fat diet-induced obesity and diabetes in mice. *Diabetes* **2008**, *57*, 1470–1481. [CrossRef]
34. Kirpich, I.A.; Feng, W.; Wang, Y.; Liu, Y.; Barker, D.F.; Barve, S.S.; McClain, C.J. The type of dietary fat modulates intestinal tight junction integrity, gut permeability, and hepatic toll-like receptor expression in a mouse model of alcoholic liver disease. *Alcohol. Clin. Exp. Res.* **2012**, *36*, 835–846. [CrossRef] [PubMed]

35. Yan, J.; Pan, Y.; He, J.; Pang, X.; Shao, W.; Wang, C.; Wang, R.; He, Y.; Zhang, M.; Ye, J.; et al. Toxic vascular effects of polystyrene microplastic exposure. *Sci. Total Environ.* **2023**, *905*, 167215. [CrossRef] [PubMed]
36. Okamura, T.; Hamaguchi, M.; Hasegawa, Y.; Hashimoto, Y.; Majima, S.; Senmaru, T.; Ushigome, E.; Nakanishi, N.; Asano, M.; Yamazaki, M.; et al. Oral Exposure to Polystyrene Microplastics of Mice on a Normal or High-Fat Diet and Intestinal and Metabolic Outcomes. *Environ. Health Perspect.* **2023**, *131*, 27006. [CrossRef]
37. Aleman, R.S.; Moncada, M.; Aryana, K.J. Leaky Gut and the Ingredients That Help Treat It: A Review. *Molecules* **2023**, *28*, 619. [CrossRef]
38. Camilleri, M. Leaky gut: Mechanisms, measurement and clinical implications in humans. *Gut* **2019**, *68*, 1516–1526. [CrossRef]
39. Ciesielska, A.; Matyjek, M.; Kwiatkowska, K. TLR4 and CD14 trafficking and its influence on LPS-induced pro-inflammatory signaling. *Cell Mol. Life Sci.* **2021**, *78*, 1233–1261. [CrossRef]
40. Li, P.; Ye, J.; Zeng, S.; Yang, C. Florfenicol alleviated lipopolysaccharide (LPS)-induced inflammatory responses in Ctenopharyngodon idella through inhibiting toll/NF-κB signaling pathways. *Fish Shellfish Immunol.* **2019**, *94*, 479–484. [CrossRef]
41. Kang, H.-M.; Byeon, E.; Jeong, H.; Kim, M.-S.; Chen, Q.; Lee, J.-S. Different effects of nano- and microplastics on oxidative status and gut microbiota in the marine medaka Oryzias melastigma. *J. Hazard. Mater.* **2021**, *405*, 124207. [CrossRef] [PubMed]
42. Lu, L.; Wan, Z.; Luo, T.; Fu, Z.; Jin, Y. Polystyrene microplastics induce gut microbiota dysbiosis and hepatic lipid metabolism disorder in mice. *Sci. Total Environ.* **2018**, *631–632*, 449–458. [CrossRef]
43. Kielian, T.; Bearden, E.D.; Baldwin, A.C.; Esen, N. IL-1 and TNF-alpha play a pivotal role in the host immune response in a mouse model of Staphylococcus aureus-induced experimental brain abscess. *J. Neuropathol. Exp. Neurol.* **2004**, *63*, 381–396. [CrossRef]
44. Louis, P.; Flint, H.J. Diversity, metabolism and microbial ecology of butyrate-producing bacteria from the human large intestine. *FEMS Microbiol. Lett.* **2009**, *294*, 1–8. [CrossRef]
45. Kundu, P.; Ling, T.W.; Korecka, A.; Li, Y.; D'Arienzo, R.; Bunte, R.M.; Berger, T.; Arulampalam, V.; Chambon, P.; Mak, T.W.; et al. Absence of intestinal PPARγ aggravates acute infectious colitis in mice through a lipocalin-2-dependent pathway. *PLoS Pathog.* **2014**, *10*, e1003887. [CrossRef] [PubMed]

metabolites

MDPI

Article

Integrated Metagenomic and Metabolomic Analysis of In Vitro Murine Gut Microbial Cultures upon Bisphenol S Exposure

Amon Cox, Farrhin Nowshad, Evelyn Callaway and Arul Jayaraman *

Artie McFerrin Department of Chemical Engineering, Texas A&M University, College Station, TX 77843, USA; ancox@tamu.edu (A.C.)
* Correspondence: arulj@tamu.edu

Abstract: Background: The gut microbiota are an important interface between the host and the environment, mediating the host's interactions with nutritive and non-nutritive substances. Dietary contaminants like Bisphenol A (BPA) may disrupt the microbial community, leaving the host susceptible to additional exposures and pathogens. BPA has long been a controversial and well-studied contaminant, so its structural analogues like Bisphenol S (BPS) are replacing it in consumer products, but have not been well studied. Methods: This study aimed to determine the impact of BPS on C57BL/6 murine gut microbiota using shotgun metagenomic sequencing and the metabolomic profiling of in vitro anaerobic cultures. Results: The results demonstrated that a supraphysiologic BPS dose did not overtly distort the metagenomic or metabolomic profiles of exposed cultures compared to controls. A distinct BPS-associated metabolite profile was not observed, but several metabolites, including saturated fatty acids, were enriched in the BPS-exposed cultures. In the absence of a BPS-associated enterotype, *Lactobacillus* species specifically were associated with BPS exposure in a discriminant model. Conclusions: Our study provides evidence contrasting the effects of BPS in the gut microbiome to its predecessor, BPA, but also emphasizes the role of inter-animal variation in microbiome composition, indicating that further study is needed to characterize BPS in this context.

Keywords: bisphenols; gut microbiome; BPA; BPS; metabolomics; metagenomics

1. Introduction

The microbiota in the mammalian digestive track are exposed to a broad range of dietary and non-nutritive compounds which can influence their composition and function [1]. Dietary contaminants can infiltrate food during production, processing and packaging, or preparation, and are introduced to the microbiota once ingested by the host. Within the gut, these contaminants may prompt the dysbiosis of the microbial community, disrupt metabolic function, or be transformed by the gut flora into products with toxicological relevance distinct from the parent compound [2]. Dysbiosis has also been linked to inflammatory bowel disease, metabolic disorders, and obesity as both a cause and consequence of these diseases [3,4].

A prevalent class of dietary contaminants are bisphenols—plasticizers used to produce polycarbonate plastics and epoxy resins which contact food by way of plastic packaging and canned goods. Bisphenols are diphenylmethane derivatives characterized by their two hydroxyphenyl functionalities, with the most well-known member being Bisphenol A (BPA). BPA and its analogues have been a part of food packaging materials since the 1960s and can leach into food, particularly after inappropriate heat exposure [5]. Biomonitoring studies such as the National Health and Nutrition Examination Survey (NHANES) typically use urinary concentrations to determine bisphenol exposure, and commonly report concentrations of bisphenol analogues in the 1–10 nM range present within ~90% of the population [6–8]. BPA is well-established as an endocrine-disrupting chemical (EDC) and now regarded as an archetypal dietary EDC. Increased regulations on BPA use have led to the development of many structural analogues, of which Bisphenol S (BPS) is the most prominent. BPS differs from BPA in that its hydroxyphenyl groups are linked by a sulfonyl bridge. While considered to be a "weak" estrogenic compound when comparing nuclear receptor binding against estradiol (E_2), BPA and its analogues BPS and Bisphenol F (BPF) exhibit a similar magnitude of several hormonal effects, with BPA and BPS being similar to E_2 in nongenomic estrogenic activity [9,10]. Thoene et al. (2020) proposed that BPS be regulated to the same degree as BPA, citing several instances where BPS appears to operate by pathways distinct from BPA to produce similar obesogenic effects, and where BPS correlates with metabolic disorders that BPA does not [11]. Furthermore, the European Food Safety Authority recently announced a revised opinion on BPA and lowered its tolerable daily intake (TDI) from 4 μg/kg body weight (bw) per day to 0.2 ng/kg bw per day, placing the mean and 95th percentile exposure estimates for all age groups from its 2015 opinion at two to three orders of magnitude above the current TDI [12]. As a dietary contaminant, and one whose metabolic conjugates likely undergo enterohepatic circulation, it is important to evaluate if and to what degree the gut microbiome mediates bisphenols' health effects [13,14].

The majority of animal studies on bisphenols that focus on the microbiome have used in vivo oral exposure to BPA and characteristically feature subtle perturbations to the microbial community, as opposed to broad dysbiotic effects [2]. In Dutch-belted rabbits exposed to 200 μg/kg body weight/day (oral exposure for dams and perinatal exposure for offspring), beta-diversity analyses clustered the 16S rRNA gene sequencing profiles by BPA exposure for fecal, colon, and cecal communities, but statistical analyses did not detect significant differences between the exposed and control offspring profiles [15]. A similar perinatal exposure study using California mice (*Peromyscus californicus*) found no patterns among the 16S rRNA gene sequencing profiles in regard to exposure, but differential analyses of specific sex, generation, and exposure intersections identified microbes enriched in the exposed groups, including members from *Sutterella* spp., Clostridiales, *Mogibacteriacae*, Mollicutes, *Prevotellaceae*, and *Bifidobacterium* spp. [16]. BPA exposure studies across several animal models commonly did not observe overt microbial changes due to BPA, but instead identified some differentially abundant features of the community [2]. However, a more recent study using male CD-1 mice exposed to 50 μg/kg body weight/day BPS through the diet for 24 weeks found that the exposed mice had increased intestinal permeability, reduced gut microbial diversity, and disrupted microbial composition [17]. While several studies identified a reduced *Akkermansia* abundance in response to BPA, a microbial enterotype indicative of bisphenol exposure has yet to be identified [2,17–19]. Increased Bacillota abundances have also been associated with BPA exposure in several studies, including a finding that suggests Bacillota members are among the most bisphenol-tolerant microbes [2,20]. BPA exposure produced a unique metabolomic profile in ovariectomized female C57BL/6 mice fed 50 μg/kg body weight/day and reduced the concentrations of tryptophan and aromatic amino acids [21].

In vitro approaches to investigate the effect of bisphenols on the human gut microbiota include inoculating bioreactors with whole communities from a fecal sample or combining representative strains to form mock communities. One such study used the simulator of the human intestinal microbial ecosystem (SHIME), which consists of five bioreactors simulating the human GI tract, and tested the response of the microbiota to BPA at ranges from 25 to 2500 μg/L for 10 days each [22]. Metagenomic profiling revealed that BPA generally reduced alpha diversity measures in the BPA-exposed cultures compared to controls, and implicated *Lactobacillus*, *Acidovorax*, *Stenotrophomonas*, and *Mycobacterium* members as potential BPA degraders by way of their enriched abundance [22]. Another study used a 45 μM BPS exposure in the extended simplified human intestinal microbiota (SIHUMIx) system and observed no changes in short-chain fatty acid production, microbial growth and total biomass, or long-term compositional differences [23].

To date, the effect of BPS on the gut microbiome has not been investigated to the same extent as that of BPA. A study in zebrafish compared the dysbiotic potential of bisphenol analogues and found that, while BPS exposure was the lowest-ranking bisphenol analogue in regard to developmental toxicity (no observed effect at 45 μM), it ranked with BPA and BPF in producing significant concentration-dependent perturbations in the microbial community composition [24]. Perinatal exposure to BPS reduced the beta diversity of the fecal microbiota in CD-1 mice, with discriminant analyses marking *Lactobacillus* genus as a biomarker for the BPS-exposed group [25]. Additionally, BPS upregulated the fecal bile acid levels and decreased the acetic acid levels in the study [25].

The present study aims to characterize the effect of BPS on the gut microbiome. To reduce the influence of the host metabolism on BPS–microbiota interactions so as to better identify the microbial and metabolic features which may be especially responsive to BPS, an in vitro anaerobic batch culture model [26] of murine fecal slurries was used. To accelerate the microbial perturbations which may occur over long-term exposure to low levels of BPS, a supraphysiologic dose of 10 μM BPS was chosen to observe changes in the abundance of bisphenol-sensitive members of the murine fecal community within 48 h of exposure. Shotgun metagenomics and untargeted LC-MS/MS metabolomics were used for identifying the microbes and metabolites whose abundances were altered upon BPS exposure.

2. Materials and Methods

2.1. Chemicals

Bisphenol S (4,4′-Sulfonfyldiphenol; CAS #80-09-1) was purchased from Sigma-Aldrich (St. Louis, MO, USA; Cat #103039) and stored at room temperature.

2.2. Animals and Experimental Design

Female C57BL/6 mice (n = 5) were housed in the same cage, with free water and standard chow access, at the Texas A&M University Laboratory Animal Resources and Research (LARR) facility (College Station, TX, USA), with breeding and animal care services provided by the facility. At six weeks post-weaning, all five mice were briefly placed in individual sterilized, BPA-free (polypropylene) jars and allowed to roam the jars while depositing fecal pellets. Fresh fecal pellets were collected using sterilized metal tweezers and transported in Anaerobic Tissue Transport Medium Surgery Packs (ATTMSP) from Anaerobe Systems (Morgan Hill, CA, USA; fulfilled through Fisher Scientific, Cat # NC9647836). The fecal material was transferred into a vinyl anaerobic chamber (Coy Lab, Grass Lake, MI, USA) maintained at 0 ppm oxygen and 2.4% hydrogen, then suspended in pre-reduced phosphate-buffered saline (PBS) with cysteine. A 10 mM stock solution of BPS dissolved in DMSO was prepared and stored in the chamber 24 h prior to the experiment. Fecal slurries for each mouse were generated by vortexing the suspensions for 2 min and aliquoting the slurries into polypropylene culture tubes, dispensing 40–45 mg feces per tube. Each culture then received 2 mL of PBS spiked with 10 μM BPS or 0.1% DMSO as a vehicle control. Cultures were incubated at 37 °C for 24 h or 48 h. For the 0 h baseline time point, cultures were set up as described above and immediately removed from the anaerobic chamber for processing. At each time point, the cultures were removed from the chamber and centrifuged at 20,000× *g* for 15 min at 4 °C, then the supernatants and culture pellets were stored separately at −80 °C until further processing.

2.3. Shotgun Metagenomic Sequencing

DNA extraction and shotgun metagenomic sequencing were carried out at the Texas A&M Institute for Genome Sciences & Society (TIGSS; College Station, TX, USA). DNA was isolated from frozen bacterial pellets using a FastDNA Spin Kit (MP Biomedicals, Santa Ana, CA, USA). As the samples were cell pellets prepared from bacterial cultures, including the day 0 inoculum samples, no host DNA removal step was performed. The Swift 2s Turbo Whole Genome Library Prep (Swift Biosciences, Ann Arbor, MI, USA) kit was used to prepare whole-genome libraries from the isolated DNA, and the quality and size distribution of the prepped DNA libraries were analyzed on an Agilent 2200 TapeStation system

(Agilent, Santa Clara, CA, USA). DNA libraries were sequenced using a NovaSeq 6000 System with a NovaSeq S4 flow cell and XP workflow for 2 × 150 paired-end sequencing (Illumina, San Diego, CA, USA). The data were received as fastq files, which were then submitted to the Metagenomic Phylogenetic Analysis (MetaPhlAn2) webtool hosted on Galaxy Europe, for compositional profiling of microbial communities [27]. The webtool Format MetaPhlAn2 on Galaxy Europe was used to export the results for all taxonomic levels as relative abundance data tables for further analysis in R. Fastq files were additionally analyzed by the HMP Unified Metabolic Analysis Network (HUMAnN) tool hosted on Galaxy Europe to determine the functional potential of the microbial communities [28]. The pathway abundance files were renormalized and unstratified to output the relative abundances of pathways per culture for analysis in R.

2.4. Untargeted LC-MS/MS Metabolomic Analysis

In preparation for metabolomic analysis, the culture supernatants were thawed and a 100 µL aliquot was diluted 9:1 with ice-cold methanol, before vortexing for 1 min. Samples were then centrifuged for 10 min at 4 °C at 20,000× *g*. Next, 800 µL of the sample was transferred to fresh microcentrifuge tubes, centrifuged again, before 100 µL of each methanol-diluted sample was aliquoted into autosampler vials. Metabolomic analysis was carried out at the Texas A&M University Integrated Metabolomics Analysis Core (IMAC) (College Station, TX, USA). Untargeted liquid chromatography high-resolution accurate mass spectrometry (LC-HRAM) analysis was performed in negative ionization mode on a Q Exactive Plus orbitrap mass spectrometer (Thermo Scientific, Waltham, MA, USA) coupled to a binary pump HPLC (UltiMate 3000, Thermo Scientific). Full MS followed by ddMS2 scans were obtained at resolutions of 35,000 (MS1) and 17,500 (MS2) with a 1.5 m/z isolation window and a stepped NCE (20, 40, 60). All samples were maintained at 4 °C before injection. The injection volume was 10 µL. Chromatographic separation was achieved on a Kinetix 2.6 um 100 × 2.1 mm Polar C18 column (Phenomenex, Torrance, CA, USA) maintained at 30 °C using a reverse-phase solvent gradient method. Solvent A was 5 mM ammonium formate in water. Solvent B was methanol. The gradient method used was 0–0.5 min (20% B to 60% B), 0.5–0.6 min (60% B to 95% B), 0.6–3 min (95% B), 3–4.1 min (95% B to 20% B), and 4.1–5 min (20% B). The flow rate was 0.5 mL min^{-1}. Sample acquisition was performed using Xcalibur (Thermo Scientific). Pooled samples were used as quality controls (QCs) spaced at one QC per fifteen samples. The sample run order was scrambled to reduce bias. Compound Discoverer v3.0 was used to build the analysis workflow for annotating the metabolite features against the mzCloud and ChemSpider databases. Negative ionization mode was chosen in anticipation of gut microbial products belonging to short-chain fatty acids and other organic acids, as well as potential BPS biodegradation products.

2.5. Statistical and Multivariate Analyses

Data analyses, statistical tests, and visualizations were conducted in R using custom scripts developed in-house. Metabolomic data were log base 2 transformed to stabilize variances of metabolites and mitigate skew. Additionally, the entry for BPS itself was omitted from the metabolomics analyses to avoid confounding multivariate and discriminant analyses. Metagenomics data were analyzed as the relative abundance of features per sample. All statistical analyses excluded the inoculum (0 h) samples to focus on the paired BPS-exposed and control culture profiles. Multivariate Analysis of Variance (MANOVA) was conducted on the components from a Principal Components Analysis (PCA) of each dataset. Data were mean-centered and scaled to unit variance prior to PCA, then the number of components required to hit 90% explained variance were assessed for normality via the Shapiro–Wilk test. PCs that passed the assumptions of normality were submitted to MANOVA for an initial analysis of BPS vs. control group differences, omitting inoculum cultures. An empirical Bayes moderated t-test was used to test for metabolites which differed in concentration between groups and was corrected for false discovery rate (alpha 0.05). Partial Least Squares Discriminant Analysis (PLS-DA) was applied to both metabolomics and metagenomics datasets to classify the culture profiles by BPS exposure and assess the weights or "loadings" (contributions) of each variable to the classification model, incorporating a sparsity step for feature selection (sPLS-DA). sPLS-DA was also applied to the metagenomics data to classify profiles and identify discriminating features. Hierarchical clustering from Euclidean distances was used to visualize and compare the culture profiles by metabolite z-scores in a heatmap. The alpha diversity of species-level metagenomic profiles was scored with the Shannon and Inverse Simpson indices to account for species richness and evenness. The beta-diversity of the metagenomic profiles was assessed using the Bray–Curtis index to generate a dissimilarities matrix from the relative abundance data, then visualized as a Principal Coordinates Analysis (PCoA). A Permutational Multivariate Analysis of Variance (PERMANOVA) was applied to the dissimilarities matrix to assess the differences between BPS-exposed and control profiles at 24 h and 48 h. To identify differentially abundant species, a Linear Discriminant Analysis Effect Size (LEfSe) was conducted using the LEfSe webtool hosted on the Huttenhower Lab Galaxy server [29]. A random forests test was conducted using the randomForest package for R [30]. The correlations between differential features from both the metagenomic and metabolomic datasets were identified using a Canonical Correlations Analysis (CCA), also featuring a sparsity step, using the PMA package for R. Differential metabolic pathways from the functional profiling were identified using sPLS-DA and Microbiome Multivariable Association with Linear Models (MaAsLin2) [31]. All plots were produced using the ggplot2 package for R.

3. Results

3.1. Supraphysiologic BPS Exposure Does Not Induce Overt Differences in Microbial Community Composition and Metabolome

To assess the potential of BPS to disrupt the gut microbiome, shotgun metagenomic sequencing was used to determine the compositional changes in anaerobic fecal cultures exposed to BPS. Metagenomic data were used to generate a compositional view of the microbial members, as well as a functional profile of the communities. In parallel, an untargeted LC-MS/MS metabolomic analysis was used to analyze changes in the culture supernatants. After 48 h of anaerobic culture with 10 μM BPS, neither the microbial abundances nor the metabolite profiles clustered by BPS exposure in a reduced dimensional space based on the first two components of a Principal Components Analysis (PCA; Figure 1A,C). For the metabolite profiles, the dimension capturing the most variance in the data (principal component 1) heavily emphasized time-dependent changes in metabolite concentrations, whereas the second component separated the data by mouse replicate (Figure 1A). In Figure 1C, the first two components of the PCA on the metagenomic profiles emphasize diverging enterotypes of the cultures over the incubation time, forming clusters around one set of three and the remaining two mice. As the first two components for the PCAs only captured between 30 and 50% of the total variance in the data, MANOVA was applied to the first 18 components of the metabolomics PCA and to the first 3 components of the metagenomics PCA, which met the condition of a normal distribution out of the total PCs required for a cumulative ~90% variance from each dataset, excluding the PCA scores from the inoculum group. The MANOVA did not indicate differences in response to BPS exposure for either the metabolomic (p = 0.605) or metagenomic (p = 0.989) culture profiles. The culture scores for the most variable components of each dataset were plotted over time to determine if any of the dimensions separated the profiles by BPS exposure, as subtle changes are not always captured in the first two components of PCA and related multivariate methods. None of the most variable components differentiated the BPS-treated cultures from the controls for the metagenomic (Figure 1D), functional (Figure S3A), and metabolomic (Figure 1B) profiles, instead illustrating inter-animal variation.

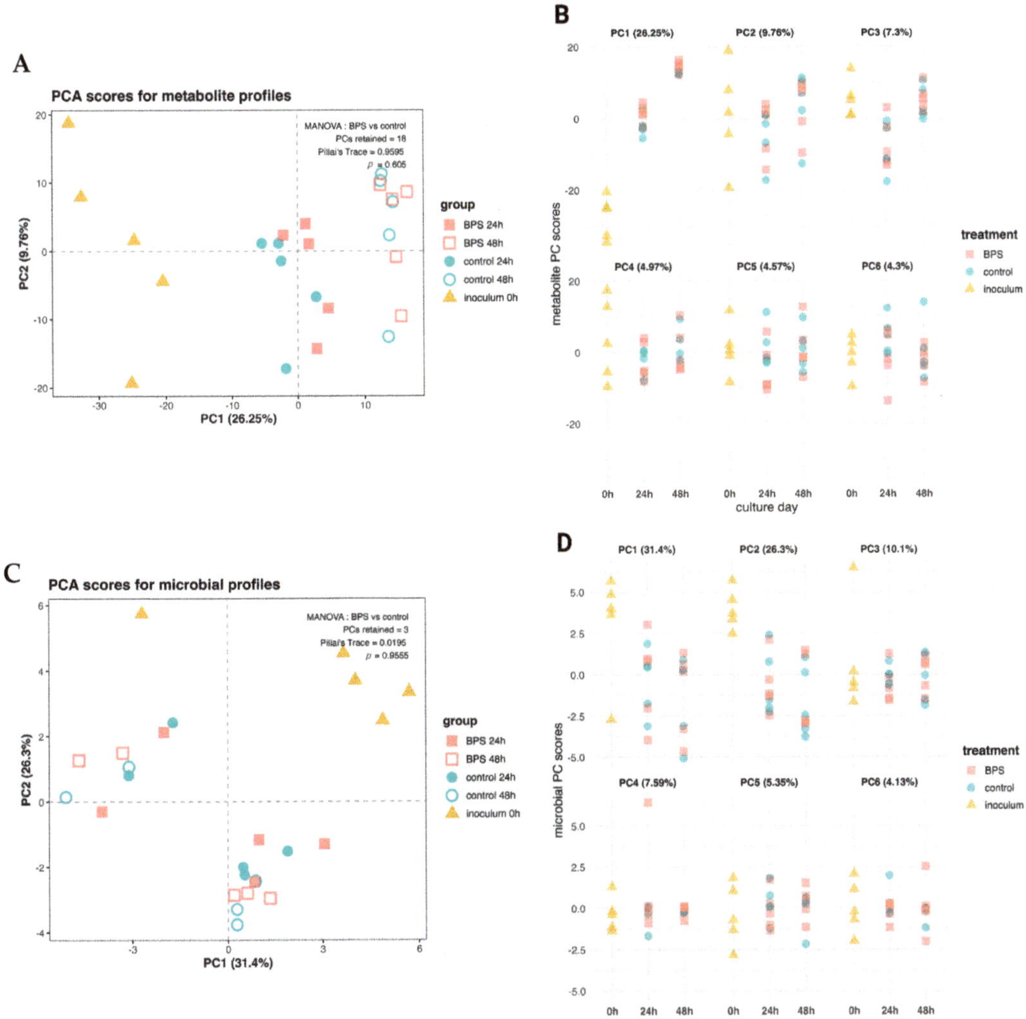

Figure 1. Overview of multi-omics analyses. PCA score plots of metabolomic (**A**) and metagenomic (**C**) profiles with results from MANOVA on BPS-treated cultures vs. controls (n = 5; excluding inoculum). PCA scores by first six components over time for metabolomic (**B**) and metagenomic (**D**) profiles.

3.2. Supraphysiologic BPS Exposure Does Not Perturb the Microbial Community Composition

To assess the dysbiotic potential of BPS, shotgun metagenomic sequencing was used to taxonomically profile the murine fecal cultures down to the species level. Exposure to 10 μM BPS for up to 48 h in anaerobic conditions did not alter the diversity of the murine fecal cultures. The Shannon and inverse Simpson indices consider both species richness and evenness, with higher values indicating a greater diversity. As shown in Figure 2A, the inoculum demonstrated higher diversity scores, which decreased over time for both the BPS and control cultures. Additionally, Kruskal–Wallis tests comparing the alpha scores for the BPS-exposed profiles against controls did not indicate differences by the Shannon (24 h $p = 0.754$; 48 h $p = 0.251$) or inverse Simpson (24 h $p = 0.602$; 48 h

p = 0.465) indices. To determine any effect from BPS on the microbial composition over time, Principal Coordinates Analysis (PCoA) was used to visualize and assess the beta-diversity relationships between the microbial profiles (Figure 2B). Given that relative abundance data are compositional in nature, the Bray–Curtis index was used to calculate the dissimilarity matrix for PCoA. As shown in Figure 2B, the PCoA emphasized differences between two clusters of biological replicates, cultures from mice A, B, and C in one cluster, and those from mice D and E in the other. PERMANOVA tests were applied to the dissimilarity matrices for the 24 h and 48 h profiles and did not indicate differences based on BPS exposure (24 h p = 0.989; 48 h p = 0.815). Consistent between both the BPS-exposed and control cultures, *Anaerotruncus* sp. G3(2012) dominated the cultures at all timepoints, with a mean relative abundance between 35 and 45% (Figure 2C), while *Bacteroides thetaiotaomicron*, *Akkermansia muciniphila*, and *Enterohabdus caecimuris* were enriched over time. Patterns of diminished diversity over time are reflected in other taxonomic levels, with the family-level relative abundances for group means and by biological replicate shown in Figure S1. Altogether, the data indicate a reduction in species evenness over time, a divergence in composition based on two clusters of biological replicates, and no community-wide responses to BPS exposure.

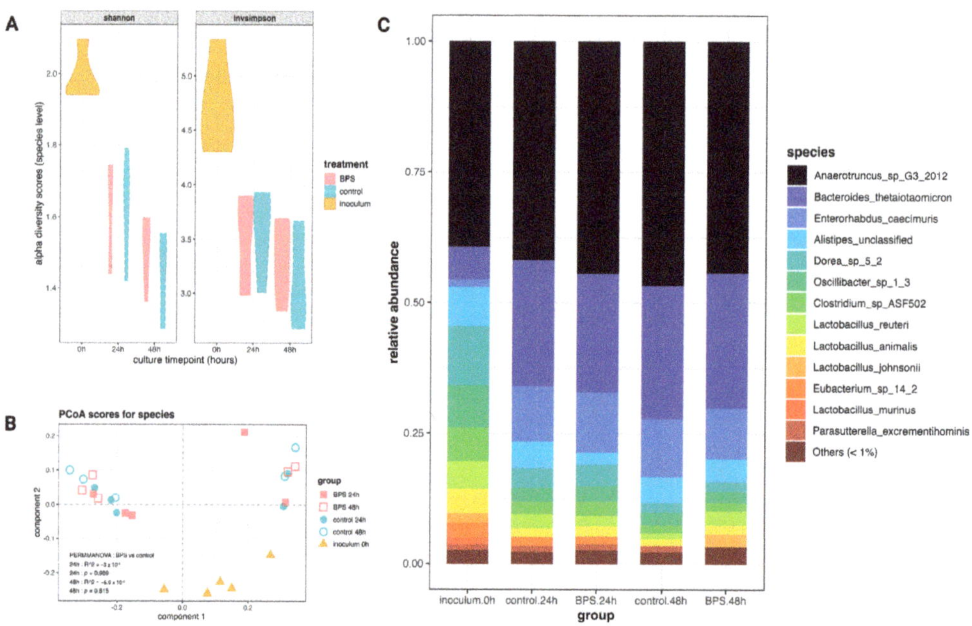

Figure 2. Diversity measures of metagenomic profiles at species level. (**A**) Violin plot of alpha diversity indices (Shannon and Inverse Simpson) for 0 h, 24 h, and 48 h profiles by group. (**B**) Beta diversity of metagenomic profiles as a Principal Coordinates Analysis scores plot, with PERMANOVA on 24 h and 48 h profiles (n = 5). (**C**) Relative abundance of species means per group.

In addition to profile-based analyses, we applied discriminant analyses to identify the microbial features which may respond to BPS and whose fluctuations would not be captured from data overviews. Linear Discriminant Analysis (LDA) Effect Size (LEfSe) [29] did not indicate any differentially abundant species in the BPS-exposed cultures compared to controls at 24 h and 48 h; Kruskal–Wallis tests, as part of the LEfSe methodology, could not identify significantly different features. A random forest [30] model conducted on the data demonstrated a poor performance in classifying the profiles as BPS-exposed or control (out-of-bag error rate = 84.62%; accuracy = 0% on test dataset), and, thus, was discounted for selecting differential species. sPLS-DA retained most species in its components for each culture day and discriminated those species as being BPS-associated or control-associated (Figure 3B), albeit poorly separating the profiles by BPS exposure (Figure 3A). The sPLS-DA variable weights for classification ("loadings") contrasted some species' associations between culture days, including *Anaerotruncus* and *Eubacterium* members, *Lactobacillus reuteri*, and *Enterococcus faecalis*. Relative abundance changes between groups and mouse replicates (Figure 3C) suggest that these conflicting classifications were due to inter-animal variation, which was also demonstrated in the fluctuations of very-low-abundance (<1%) features across biological replicates (Figure S2). From evaluating the behavior of related taxa, members from *Lachnospiraceae* collectively demonstrated an association with BPS exposure at the 24 h time point, and *Lactobacillus* members were collectively enriched in the BPS-exposed cultures at the 48 h time point. The sPLS-DA for 24 h featured a greater number of species associated with BPS exposure, with some of the greatest loadings belonging to *Ruminococcus torques*, *Staphylcoccus xylosus*, *Burkholderiales bacterium_1_1_47*, and *Eubacterium plexicaudatum*. The sPLS-DA for 48 h featured fewer BPS-associated species, but with greater loadings, for taxa such as the *Lactobacillus* members. These data indicate that, while BPS did not induce broad compositional changes in the microbiota, there were subtle time-dependent abundance changes for a select number of species.

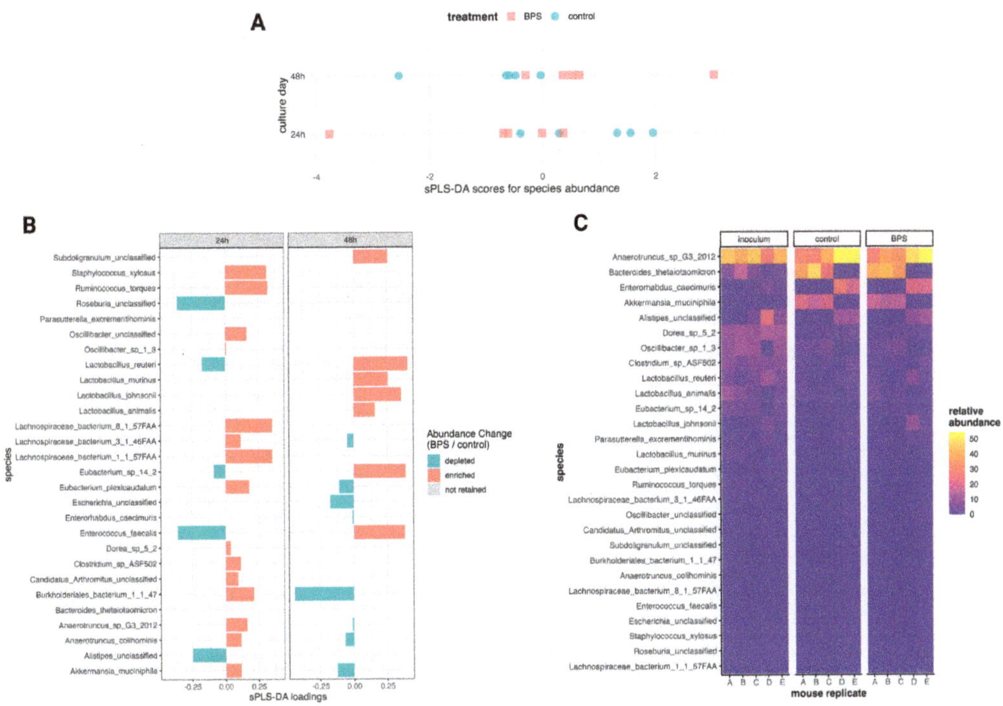

Figure 3. Discriminant analysis of microbial species. (**A**) sPLS-DA scores for species relative abundance by time point. (**B**) sPLS-DA loadings for bacterial species by time point. Loadings signs for 24 h were flipped to align with 48 h. (**C**) Heatmap displaying relative abundances by group per biological replicate at the species level, including species with mean relative abundances < 1%.

3.3. *Supraphysiologic BPS Exposure Perturbed Low-Abundance Microbial Pathways*

Shotgun sequencing data for the 24 h and 48 h time points were analyzed using HMP Unified Metabolic Analysis Network (HUMAnN) to profile the functional metabolic potential of the microbiota and to provide a breakdown of gene families organized into pathways with the associated abundances per metagenome. These data were filtered and reorganized into the relative abundances of metabolic pathways per culture, providing 356 annotated pathways for analysis. The most abundant pathways covered a maximum of a ~1.6% relative abundance, with the majority of pathways exhibiting abundances less than 1%. When comparing the average relative abundances of all pathways, the median was 0.09% and the mean was 0.28%. The distributions of the ten most abundant pathways (Table S1), with their averages ranging from 1.3 to 1.65%, did not differentiate by exposure to BPS (Figure S3B). Sparse PLS-DA differentiated control from BPS-exposed metabolic pathway profiles in both the 24 h and 48 h models (Figure 4A). Of the initial 365 pathways, each model retained 36 pathways in differentiating the profiles (Figure 4B,C). After removing outlier pathways present in only one or two replicates, the resulting differentially abundant metabolic pathways numbered 14 at 24 h and 25 at 48 h, with an overlap of 3 pathways perturbed at both time points (Table S2). The pathways perturbed at both time points were PWY-6138: CMP-N-acetylneuraminate biosynthesis I (eukaryotes), GLCMANNANAUT-PWY: superpathway of N-acetylglucosamine, N-acetylmannosamine and N-acetylneuraminate degradation, and PWY-6531: mannitol cycle, each of which demonstrated enrichment in the BPS-exposed cultures over the controls.

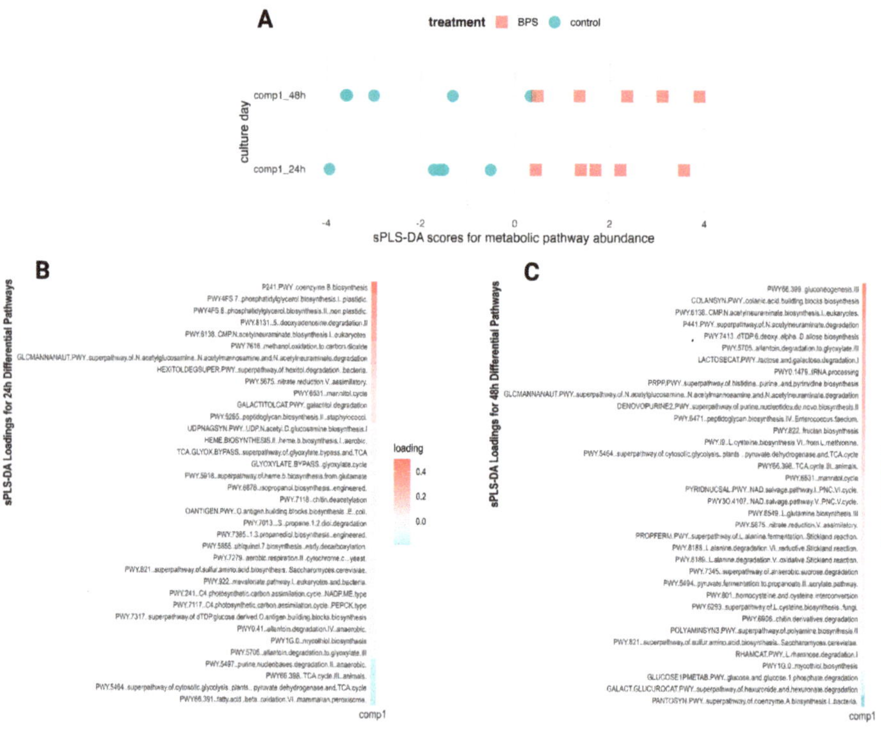

Figure 4. Discriminant analysis of microbial metabolic pathways. (**A**) sPLS-DA scores for metabolic pathway relative abundance by time point. sPLS-DA loadings for differential metabolic pathways retained in the 24 h hour model (**B**) and 48 h model (**C**).

3.4. *Supraphysiologic BPS Exposure Induces Subtle Changes in the Microbial Metabolome*

To investigate subtle or feature-specific metabolomic differences in the BPS-treated cultures, differential metabolomic features were identified by a fold change greater than 2.0 or less than 0.5 and a p-value of < 0.05 for the 24 h and 48 h metabolomic profiles (Figure 5A,B). There were 61 metabolomic features for the 24 h profiles and 37 features for the 48 h ones. As a parallel method to identify the features which are associated with BPS exposure, sPLS-DA was applied to classify the 24 h and 48 h profiles by BPS exposure, incorporating sparsity parameters for feature selection. The sPLS-DA selected 50 metabolomic features per time point to classify the profiles, successfully separating them by the treatment variable (Figure 5C). Of the significant metabolomic features displayed in the Figure 5A,B volcano plots and those retained by the sPLS-DA for discrimination, the metabolite features meeting those criteria are listed in Table 1 alongside their putative annotations. The scarcity of annotated metabolites within the differential features prevents commentary on metabolite classes, but it may be worth noting that several saturated fatty acids were included. Specifically, margaric acid showed depletion in the BPS-exposed cultures, while montanic acid, docosanoic acid, and lignoceric acid increased in the BPS-exposed cultures. However, the identified differential metabolites were too sparse and dissimilar to elucidate BPS' effects on specific biological pathways. The hierarchical clustering of metabolomic profiles demonstrated tight clustering of the biological replicates for the inoculum (culture time of 0 h), followed by clustering by 24 h and 48 h, with some differentiation between replicate mice A, B, and C and replicates D and E (Figure 5D).

Table 1. Differential metabolites in BPS-exposed murine fecal cultures, including significant ($p < 0.05$) features and metabolites retained from sparse PLS-DA.

Putative Annotation	Formula	*m/z*	rt (min) [1]	Fold Change [2]	adj. *p*-Value [3]	Timepoint (Hours)
1,5-Isoquinolinediol	$C_9H_7NO_2$	160.03948	1.371	2.85737425	0.00093819	24
1-[5-(4-Hydroxy-2-butanyl)-2-methyltetrahydro-2-furanyl]-4-methyl-2-pentanone	$C_{15}H_{28}O_3$	255.19743	2.124	0.44354233	0.0491865	24
Emedastine	$C_{17}H_{26}N_4O$	301.20224	1.794	2.09983088	0.00978352	24
(4-Methylumbelliferone)-β-D-glucopyranoside	$C_{16}H_{18}O_8$	337.09131	3.199	0.38710528	0.01007916	24
Sucrose	$C_{12}H_{22}O_{11}$	341.10817	3.303	0.45945739	0.01401479	24
Botrydial	$C_{17}H_{26}O_5$	309.17221	3.028	2.51502097	0.00574063	24
4-nitrocatechol	$C_6H_5NO_4$	200.01961	1.145	0.46534155	0.00059632	24
butyrin	$C_{15}H_{26}O_6$	301.16626	1.92	3.97519788	0.04796165	48
Margaric acid	$C_{17}H_{34}O_2$	269.24915	2.592	0.47287413	0.01975824	48
4-Methylene-2-oxoglutarate	$C_6H_6O_5$	157.01341	1.215	0.3701104	0.00035043	48
4-Hydroxy-2-oxoglutaric acid	$C_5H_6O_6$	161.00828	0.903	0.43793592	5.0864×10^{-6}	48
--	--	--	--	--	Loading [4]	--
Montanic acid	$C_{28}H_{56}O_2$	423.42125	2.688	1.28422989	0.2462051	24
Docosanoic Acid	$C_{22}H_{44}O_2$	339.32695	3.193	1.24776908	0.08635497	24
2-Oxo-3-(phosphonooxy)propyl decanoate	$C_{13}H_{25}O_7P$	323.12509	1.396	1.07451343	0.18333194	48
Lignoceric Acid	$C_{24}H_{48}O_2$	367.35833	3.852	1.19045219	0.13535602	48
1-[5-(4-Hydroxy-2-butanyl)-2-methyltetrahydro-2-furanyl]-4-methyl-2-pentanone	$C_{15}H_{28}O_3$	255.19743	2.124	1.27565896	0.12206422	48
Sparfloxacin	$C_{19}H_{22}F_2N_4O_3$	391.15988	1.371	1.3391899	0.0693887	48
Metirosine	$C_{10}H_{13}NO_3$	194.08168	1.674	0.73220113	−0.0645802	48

[1] rt—retention time (minutes). [2] Fold change with a cutoff at ≥2 indicating enriched features and ≤0.5 indicating depleted features for BPS-exposed cultures. [3] *p*-values from empirical Bayes moderated t-test, adjusted for false discovery rate. [4] loadings >0 indicate features with greater concentrations in BPS-exposed cultures relative to controls, and loadings <0 indicate features with lesser concentrations in BPS-exposed cultures. The signs of loadings for the 24 h sPLS-DA have been corrected to align with those for 48 h, for consistency.

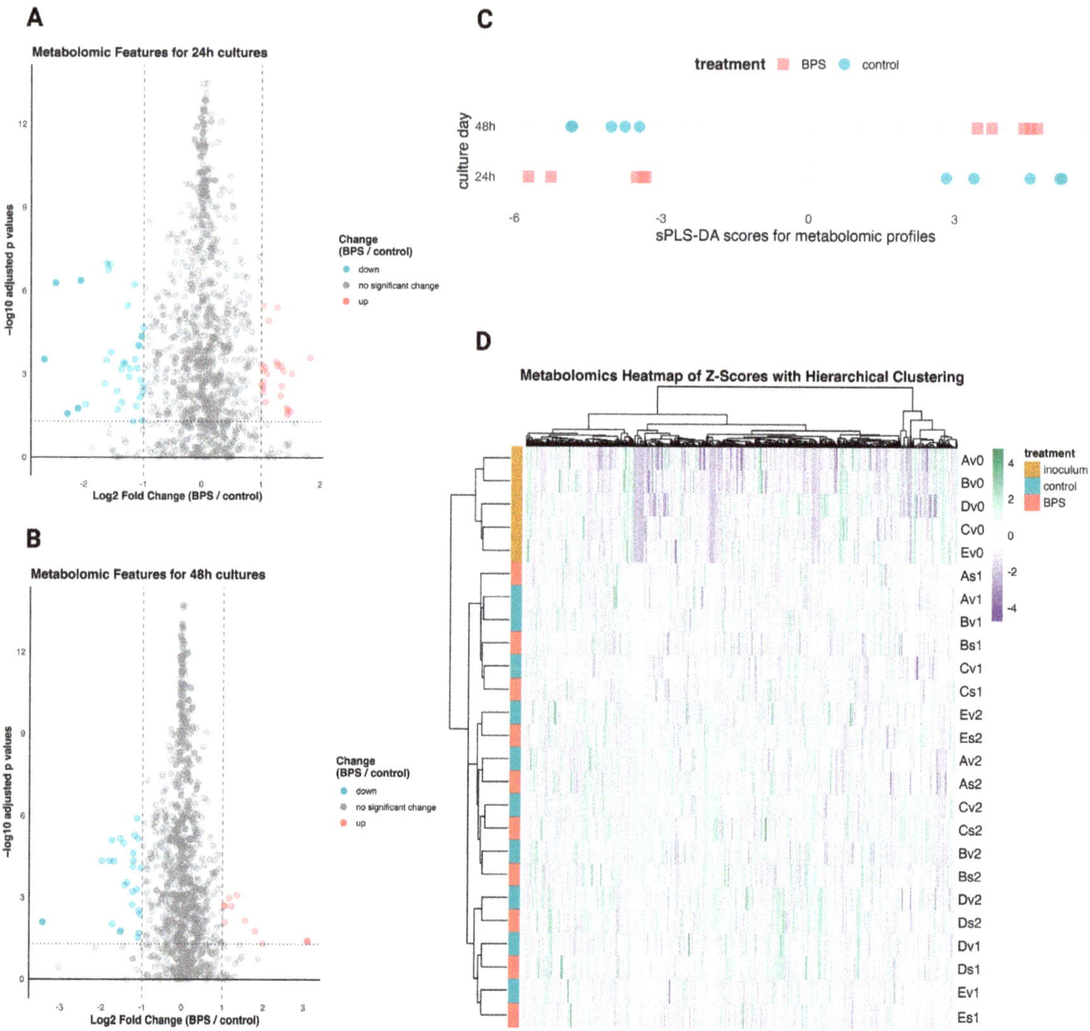

Figure 5. Differentially altered metabolomic features. Log2 fold changes (BPS-treated/control) against negative log 10 transformed adjusted *p*-values from empirical Bayes moderated *t*-test (n = 5), for 24 h (**A**) and 48 h (**B**) profiles. (**C**) Scores plot of sPLS-DA, considering only the 24 h and 48 h profiles and respective top 50 metabolomic features contributing the greatest variance. (**D**) Heatmap of metabolomic feature z-scores with hierarchical clustering. Heatmap row names represent culture replicates; "A–E" denote mouse, "s" or "v" denote BPS-exposed and control, respectively; 0–2 denote culture time points 0 h, 24 h, and 48 h.

3.5. Features Which Differentiate by BPS Exposure Poorly Explain Variance Across Metagenome and Metabolome

To determine the relationship between differential metabolomic features and microbial fluctuations, sparse CCA was applied to the most variable features from each dataset which were differentiated, to some degree, by BPS exposure. CCA is akin to PCA in that it finds linear combinations (canonical variates; CV) of the original variables, but does so to maximize the correlation of the components generated for each dataset. Species retained by the sPLS-DA on metagenomics were used as the metagenomics input to sCCA. Annotated metabolites from the fold change and significance cutoffs were combined with those retained by sPLS-DA for the metabolites input. Across the differential analyses for both time points, 26 species and 17 metabolites were used as the inputs to sCCA. The scaled scores from the sCCA are plotted in Figure 6A,B, where the two CVs of a given dataset make up the x and y axes and the first CV of the complementary dataset colors the points to aid in observing correlations. The sCCA primarily contrasted the inoculum profiles from those at 24 h and 48 h and had correlations of 0.7424 between the first CVs and 0.5784 between the second CVs of each dataset. Additionally, sCCA retained a limited selection of features from each dataset (five metabolites and nine species) to generate the correlated components, suggesting that features which differentiated by BPS exposure had little relationship with each other between the metagenomic and metabolomic datasets (Figure 6C,D). An sCCA on the full 'omics datasets shows a high correlation between the CVs of each dataset (CV1: 0.9296; CV2: 0.9818), suggesting that feature fluctuations from each dataset collaboratively contrast inoculum cultures from later time points, and differentiate cultures by biological replicate (Figure S4).

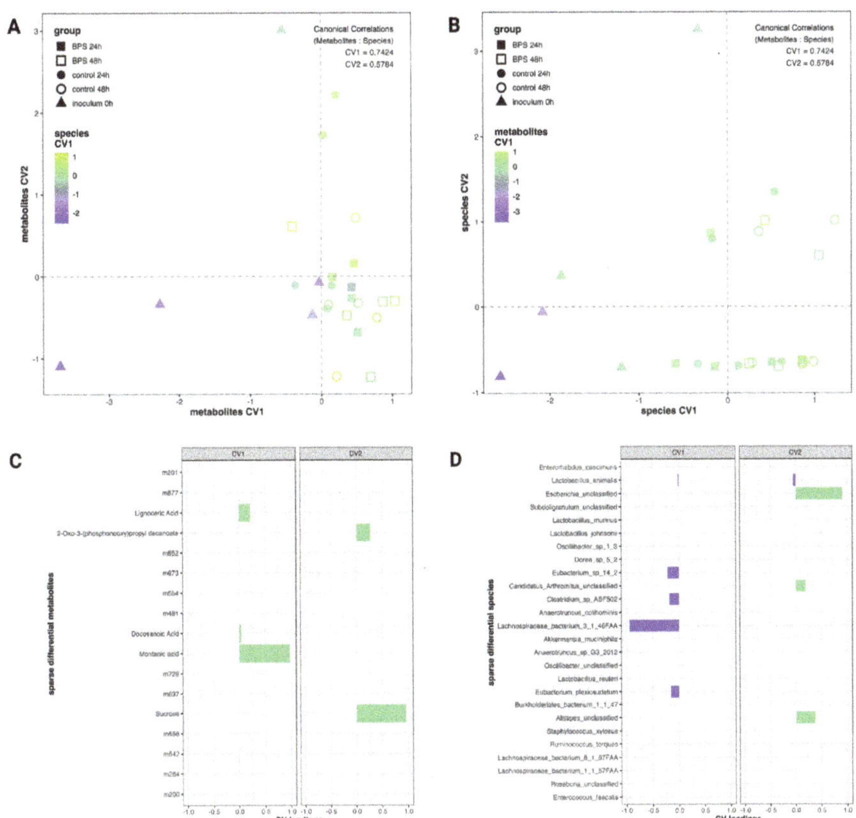

Figure 6. Canonical correlations between differential 'omics features. Sparse Canonical Correlations Analysis (sCCA) scores plots of differential annotated metabolites with color scaled by species component 1 (**A**) and of sPLS-DA-selected species colored by metabolites component 1 (**B**). sCCA loadings for metabolites (**C**) and species (**D**); metabolomic features with non-zero loadings are denoted by their annotation, whereas features with loadings equal to zero are only listed by their analysis code.

4. Discussion

The gut microbiota are widely studied for their role as a mediator between the host and environmental exposures. As a major component in plastics production and food contact materials, bisphenols are facing increased scrutiny from health-regulating organizations [12]. The discussion around BPA and its analogues as endocrine disruptors is expanding to include evidence for cardiovascular effects and metabolic dysregulation, with bisphenols' impact on hosts and the gut microbiome similarly gaining interest over the past decade [32–34]. To fully characterize the hazard posed by dietary contaminants like bisphenols, it is crucial to understand their impact on the gut microbiome, without confounding influences of the host.

To identify the microbial members and their metabolites which may be most sensitive to BPS, we exposed in vitro cultures of murine fecal microbiota to 10 μM BPS over for 2 days. This in vitro batch culture model was used over the oral exposure of mice to focus explicitly on the relationship between the microbiota and the contaminant [26]. Cultures were prepared using whole fecal material suspended in anoxic phosphate-buffered saline (PBS), as opposed to a nutritive media, to preserve a familiar nutrient environment and prevent domination by fast-growing members. Prior in vitro and in vivo studies on bisphenols employed doses in the supraphysiologic range of 10–100 μM to identify sensitive toxic endpoints, which, in the context of the microbiome, are microbiota which dramatically change in abundance or metabolic function following exposure. Our results showed that acute exposure to supraphysiologic levels of BPS did not produce an enterotype that differentiated the exposed cultures from controls. Instead, the data demonstrated differences between cultures by biological replicate, with those differences becoming more prominent over the 48 h culture time (Figure 1C,D). No broad alterations in the functional or metabolite profiles were identified which discriminated the BPS-exposed cultures from controls. As with the compositional changes, the greatest variance originated from changes in the cultures over time, and secondly due to inter-animal biological variation, as evidenced by the PCAs in Figure S3A and Figure 1A,B.

Alpha and beta diversity analyses emphasized the role of inter-animal biological variation in determining the microbial community composition, as shown in Figure 2. While some differential analyses did not identify differentially abundant species between the BPS-exposed and control cultures, the sPLS-DAs by time point discriminated abundance profiles by BPS exposure, and in that process, selected some species as BPS- or control-associated. The sPLS-DA on the 48 h profiles attributed relatively large loadings to four *Lactobacillus* species as classifying cultures in the BPS-exposed group. *Lactobacillus* is a well-known mutualistic, biofilm-forming genus in the human and animal microbiomes [35]. This association corroborates the findings of Gomez et al. (2021) in indicating that an increased *Lactobacillus* abundance may be a biomarker for BPS exposure [25]. The functional profiles of the microbial metabolic pathways similarly draw attention to inter-animal variation and time-dependent changes. By count, ~10% of these pathways were differentially abundant between the BPS-exposed and control cultures; however, each of the 39 perturbed pathways made up less than 1% relative abundance of all identified pathways. The identification of the CMP-N-acetylneuraminate biosynthesis I (eukaryotes) pathway, as opposed to its bacterial variant, likely reflects the residual host DNA captured by the sequencing process. Some pathogenic and symbiotic bacteria possess CMP-N-acetylneuraminate biosynthesis II (bacteria) and use the cell surface sugar to blend in with mammalian cells [36]. CMP-N-acetylneuraminate biosynthesis was identified as being enriched in the BPS-exposed cultures after 24 h and 48 h of exposure, and the identification of the eukaryotic pathway over the bacterial one may indicate it as a marker of interest for host-focused studies on BPS. These perturbations demonstrate that, like with the microbial members, low-abundance features may be sensitive to BPS in the absence of profile-dominating effects.

Although metabolome-wide changes were not observed, we did observe specific changes in metabolite features upon BPS exposure. Assessing fold changes and significant differences revealed ~100 metabolomic features with abundance differences between the

BPS-treated and control cultures. Similarly, an sPLS-DA model selected 50 metabolomic features per time point to discriminate the profiles by BPS exposure; however, the scarcity of annotated metabolites among the differential metabolomic features (Table 1) precluded analyses based on comparisons of chemical classes. These putative annotations also included compounds whose presence was unexpected in these samples, further limiting the utility of feature-specific information in identifying effects due to BPS. Emedastine and Sparfloxacin are orally administered drugs for anti-histamine and antibiotic applications, respectively, and were not administered to the mice prior to fecal collection. Their detection may come from contaminated animals or samples, but the features are likely mis-annotations, instead reflecting other endogenous metabolites with similar m/z to the drugs, but either of a lower priority or undocumented within the databases used. The metabolomic profiles of the BPS-exposed cultures did not identify any candidate bisphenol breakdown products such as those put forth by Li et al. (2020), Kyrila et al. (2021), and Li et al. (2023) regarding BPA microbial biodegradation [37–39]; however, the structural differences between BPA and BPS, specifically BPS' characteristic sulfonyl bridge between phenols, would reasonably lead to divergent degradation pathways. The lack of key gut biomarkers of damage or disease amongst the putative annotations and the absence of potential bisphenol breakdown products within the differential metabolites precluded follow-up targeted LC-MS/MS analysis.

An sCCA was applied to the differential metabolomic features and species retained in the metagenomics sPLS-DA to identify any relationships between the metagenomic and metabolomic datasets. Figure 6C,D show that the majority of the differential features did not exhibit variance, which correlated with variances in the complementary dataset. Interestingly, saturated fatty acids made up three of the five metabolites retained in the sCCA, and all appeared in the first canonical variate. Given that the correlation between the CV1 of each dataset was 0.7424, this would suggest that increased concentrations of lignoceric acid, docosanoic acid, and montanic acid were correlated with the decreased *Lachnospiraceae bacterium_3_1_46FAA*, *Clostridium* sp. ASF502, *E. plexicaudatum*, and *E. sp. 14_2* abundances in our model. *Eubacterium* members are known for producing butyrate and play a role within the gut in bile acid and cholesterol transformation, while *Lachnospiraceae* are a prominent and controversial family within the human microbiome that ferment plant polysaccharides to short-chain fatty acids [40,41]. Similarly, *Clostridium* is a genus of gut colonizers whose species can have beneficial or disruptive effects on human hosts [42].

BPA has been shown to perturb the mammalian gut microbiome in terms of metabolite production and community composition, although consistent effects on specific microbial abundances have yet to be identified [2,17,21]. Studies on BPS as a dysbiotic agent are not as numerous as BPA, with insufficient data to identify cross-species or cross-model patterns in microbial responses to BPS exposure. A study using zebrafish determined that BPS was among the most dysbiotic bisphenols tested, while perinatal BPS exposure prominently disrupted the fecal microbiome of CD-1 mice [24,25]. In contrast, the results from our isolated culture system do not implicate BPS as a dysbiotic agent, instead drawing attention specifically to *Lactobacillus* members as having some time-dependent association with BPS exposure in the murine fecal cultures. The current literature gap regarding BPS in microbial systems provides a challenge for identifying and corroborating metagenomic, functional, and metabolomic perturbations in response to this compound.

It is important to note two factors which may have obfuscated effects from the BPS exposure. First, inter-animal variation among the biological replicate cultures was substantial. As part of the experimental design, feces were collected from five female C57BL/6 mice specifically housed in the same cage, as closely housed mice demonstrate similar microbial profiles and a reduced variance in the data related to individual differences [43,44]. However, the metagenomic data demonstrated a clear split among the replicate cultures, evident in the PCA in Figure 1C, the PCoA in Figure 2B, and especially the heatmap of species relative abundances in Figure 3C. Considering that inter-animal biological variation remained a dominant factor, despite collecting feces explicitly from co-housed mice, an alternative approach to addressing this variation may be to use lower-density cages. Recent work from Russell et al. (2022) suggests that a decreased housing density increases murine gut microbiota sensitivity to selective pressures and chemical challenges [43]. While those findings are primarily oriented toward in vivo studies featuring exposures or treatment through diet, their recommendation may also apply to in vitro microbial culture experiments by avoiding experiment-wide enterotypes which resist perturbations from short-term chemical challenges.

A second factor in the experimental design that may have obfuscated effects from BPS is the omission of an acclimation phase for the microbiota within the culture system. BPS exposure was initiated at the time of inoculation in the in vitro culture system to ensure that BPS exposure was to a microbial community that was representative of the host community; however, it is likely that the change in environment from the animal host to the in vitro culture system resulted in a rebalancing of the microbial community, even when using minimal medium to avoid community domination by fast-growing members and to identify potential BPS degraders in the depleted nutrient environment. This rebalancing is evident in the differences between the inoculum (0 h) cultures and those of later time points in our multivariate analyses and diversity indices. Specifically, this occurrence made interpretation of the metagenomic sPLS-DA difficult, in that some species, such as *E. plexicaudatum,* were considered to be indicative of control cultures at one time point then were used to classify cultures as BPS-exposed in the other. The reason for these contradictions may be that, since the overall diversity of these cultures declined over time, reflecting a reduced abundance of several members, the sPLS-DA instead reflected differing rates of species "die-off" and not enrichment in the presence of BPS. An alternative may be to acclimatize the microbiota in a dilute nutritional media for 12–24 h before BPS exposure, effectively trading host representativeness for a more stable baseline to identify perturbations.

Even with the above caveats, the absence of broad disruptive effects in our model suggests that the effect of BPS, and potentially other bisphenols, may originate primarily from contaminant–host interactions. Our results would position the microbiome as a downstream target of bisphenols' effects instead of as a mediator, and raises the same consideration for similar kinds of dysbiotic agents such as phthalates (e.g., DEHP). Future work can also employ in vitro co-culture methods to partition host and microbiome influences. For example, intestinal epithelial or hepatocyte cell culture models can be exposed to substrates like BPS to first generate host-biotransformed products and endogenous metabolites, which could then be used to test in the fecal microbial culture model for dysbiosis. In this way, the contaminant–microbiome and contaminant–host aspects of exposure can be evaluated side-by-side to determine the relative importance of each compartment in mediating the effects of dysbiotic agents, without the use of live animal models.

5. Conclusions

In summary, a supraphysiologic (10 μM) BPS exposure did not perturb the overarching murine fecal microbial composition, function, nor metabolome. A handful of metabolomic features demonstrated differential abundances dependent on time and BPS exposure, and *Lactobacillus* species showed a relatively greater abundance in BPS-exposed cultures compared to controls after 48 h of BPS exposure. The results from this in vitro culture model do not indicate that BPS is a direct dysbiotic agent, contrasting it against BPA. However, inter-individual and inter-species biological variation play substantial roles in microbiome composition, necessitating further studies to characterize the interactions that bisphenol analogues like BPS may have with the gut microbiota.

Supplementary Materials: The following supporting information can be downloaded at: https://www.mdpi.com/article/10.3390/metabo14120713/s1, Figure S1: Family-Level Relative Abundances; Figure S2: Relative Abundances of Very-Low-Abundance Species; Table S1: Ten most abundant HUMAnN functional pathways of BPS-exposed and unexposed murine fecal cultures; Figure S3: Overview of Metabolic Pathway Abundances; Table S2: Differential metabolic pathways from sparse

PLS-DA on HUMAnN functional profiling of BPS-exposed and un-exposed murine fecal cultures; Figure S4: Canonical Correlations Between Full 'Omic Datasets.

Author Contributions: Conceptualization, A.C., F.N. and A.J.; methodology, A.C., F.N. and E.C.; investigation, A.C.; project administration, E.C.; formal analysis, A.C.; resources, A.J.; writing—original draft preparation, A.C.; writing—review and editing, A.C. and A.J. All authors have read and agreed to the published version of the manuscript.

Funding: This research was funded by the National Institute of Environmental Health Sciences (NIEHS), number 1R56ES033052-01A1, and partially supported by the Ray B. Nesbitt Endowed Chair. Trainee support for A.C. during this work was provided by NIEHS via a T32 Ruth L. Kirschstein Institutional Research Service Award to Texas A&M University's Toxicology program, number 2T32ES026568-06A1.

Institutional Review Board Statement: The animal study protocol was approved by the Institutional Animal Care and Use Committee (IACUC) of Texas A&M University (AUP 2022-111, 11 December 2020).

Informed Consent Statement: Not applicable.

Data Availability Statement: The data supporting these results can be found at: 10.5281/zenodo.13917959. Shotgun sequencing raw sequence reads and LC-MS/MS spectra can be made available by the authors on request.

Acknowledgments: Use of the TAMU Integrated Metabolomics Analysis Core is acknowledged.

Conflicts of Interest: The authors declare no conflicts of interest.

References

1. Singh, R.; Chang, H.; Yan, D.; Lee, K.; Ucmak, D.; Wong, K.; Abrouk, M.; Farahnik, B.; Nakamura, M.; Zhu, T.; et al. Influence of diet on the gut microbiome and implications for human health. *J. Transl. Med.* **2017**, *15*, 73. [CrossRef]
2. Cox, A.; Bomstein, Z.; Jayaraman, A.; Allred, C. The intestinal microbiota as mediators between dietary contaminants and host health. *Exp. Biol. Med.* **2023**, *248*, 2131–2150. [CrossRef] [PubMed]
3. DeGruttola, A.K.; Low, D.; Mizoguchi, A.; Mizoguchi, E. Current Understanding of Dysbiosis in Disease in Human and Animal Models. *Inflamm. Bowel Dis.* **2016**, *22*, 1137–1150. [CrossRef]
4. Weiss, G.A.; Hennet, T. Mechanisms and consequences of intestinal dysbiosis. *Cell. Mol. Life Sci.* **2017**, *74*, 2959–2977. [CrossRef] [PubMed]
5. U.S. Food & Drug Administration. Bisphenol A (BPA): Use in Food Contact Application. Available online: https://www.fda.gov/food/food-additives-petitions/bisphenol-bpa-use-food-contact-application (accessed on 11 December 2023).
6. Lehmler, H.J.; Liu, B.; Gadogbe, M.; Bao, W. Exposure to Bisphenol A, Bisphenol F, and Bisphenol S in U.S. Adults and Children: The National Health and Nutrition Examination Survey 2013–2014. *ACS Omega* **2018**, *3*, 6523–6532. [CrossRef] [PubMed]
7. Geens, T.; Aerts, D.; Berthot, C.; Bourguignon, J.; Goeyens, L.; Lecomte, P.; Maghuin-Rogister, G.; Pironnet, A.; Pussemier, L.; Scippo, M.; et al. A review of dietary and non-dietary exposure to bisphenol-A. *Food Chem. Toxicol.* **2012**, *50*, 3725–3740. [CrossRef]
8. Colorado-Yohar, S.M.; Castillo-González, A.C.; Sánchez-Meca, J.; Rubio-Aparicio, M.; Sánchez-Rodríguez, D.; Salamanca-Fernández, E.; Ardanaz, E.; Amiano, P.; Fernández, M.F.; Mendiola, J.; et al. Concentrations of bisphenol-A in adults from the general population: A systematic review and meta-analysis. *Sci. Total Environ.* **2021**, *775*, 145755. [CrossRef] [PubMed]
9. Rochester, J.; Bolden, A. Bisphenol S and F: A Systematic Review and Comparison of the Hormonal Activity of Bisphenol A Substitutes. *Environ. Health Perspect.* **2015**, *123*, 643–650. [CrossRef]
10. Viñas, R.; Watson, C.S. Bisphenol S Disrupts Estradiol-Induced Nongenomic Signaling in a Rat Pituitary Cell Line: Effects on Cell Functions. *Environ. Health Perspect.* **2013**, *121*, 352–358. [CrossRef]
11. Thoene, M.; Dzika, E.; Gonkowski, S.; Wojtkiewicz, J. Bisphenol S in Food Causes Hormonal and Obesogenic Effects Comparable to or Worse than Bisphenol A: A Literature Review. *Nutrients* **2020**, *12*, 532. [CrossRef]
12. EFSA Panel on Food Contact Materials; Lambré, C.; Barat Baviera, J.M.; Bolognesi, C.; Chesson, A.; Cocconcelli, P.S.; Crebelli, R.; Gott, D.M.; Grob, K.; Lampi, E.; et al. Re-evaluation of the risks to public health related to the presence of bisphenol A (BPA) in foodstuffs. *EFSA J.* **2023**, *21*, e06857. [CrossRef]
13. Sakamoto, H.; Yokota, H.; Kibe, R.; Sayama, Y.; Yuasa, A. Excretion of bisphenol A-glucuronide into the small intestine and deconjugation in the cecum of the rat. *Biochim. Biophys. Acta* **2002**, *1573*, 171–176. [CrossRef] [PubMed]
14. Völkel, W.; Colnot, T.; Csanády, G.; Filser, J.; Dekant, W. Metabolism and kinetics of bisphenol a in humans at low doses following oral administration. *Chem. Res. Toxicol.* **2002**, *15*, 1281–1287. [CrossRef]
15. Reddivari, L.; Veeramachaneni, D.N.R.; Walters, W.A.; Lozupone, C.; Palmer, J.; Hewage, M.K.K.; Bhatnagar, R.; Amir, A.; Kennett, M.J.; Knight, R.; et al. Perinatal Bisphenol A Exposure Induces Chronic Inflammation in Rabbit Offspring via Modulation of Gut Bacteria and Their Metabolites. *mSystems* **2017**, *2*, 10-1128. [CrossRef] [PubMed]

16. Javurek, A.B.; Spollen, W.G.; Johnson, S.A.; Bivens, N.J.; Bromert, K.H.; Givan, S.A.; Rosenfeld, C.S. Effects of exposure to bisphenol A and ethinyl estradiol on the gut microbiota of parents and their offspring in a rodent model. *Gut Microbes* **2016**, *7*, 471–485. [CrossRef]
17. Feng, D.; Zhang, H.; Jiang, X.; Zou, J.; Li, Q.; Mai, H.; Su, D.; Ling, W.; Feng, X. Bisphenol A exposure induces gut microbiota dysbiosis and consequent activation of gut-liver axis leading to hepatic steatosis in CD-1 mice. *Environ. Pollut.* **2020**, *265*, 114880. [CrossRef] [PubMed]
18. Ni, Y.; Hu, L.; Yang, S.; Ni, L.; Ma, L.; Zhao, Y.; Zheng, A.; Jin, Y.; Fu, Z. Bisphenol A impairs cognitive function and 5-HT metabolism in adult male mice by modulating the microbiota-gut-brain axis. *Chemosphere* **2021**, *282*, 130952. [CrossRef] [PubMed]
19. Rodrigues, V.F.; Elias-Oliveira, J.; Pereira, Í.S.; Pereira, J.A.; Barbosa, S.C.; Machado, M.S.G.; Carlos, D. Akkermansia muciniphila and Gut Immune System: A Good Friendship That Attenuates Inflammatory Bowel Disease, Obesity, and Diabetes. *Front. Immunol* **2022**, *13*, 934695. [CrossRef]
20. López-Moreno, A.; Torres-Sánchez, A.; Acuña, I.; Suárez, A.; Aguilera, M. Representative Bacillus sp. AM1 from Gut Microbiota Harbor Versatile Molecular Pathways for Bisphenol A Biodegradation. *Int. J. Mol. Sci.* **2021**, *22*, 4952. [CrossRef] [PubMed]
21. DeLuca, J.; Allred, K.; Menon, R.; Riordan, R.; Weeks, B.; Jayaraman, A.; Allred, C. Bisphenol-A alters microbiota metabolites derived from aromatic amino acids and worsens disease activity during colitis. *Exp. Biol. Med.* **2018**, *243*, 864–875. [CrossRef] [PubMed]
22. Wang, Y.; Rui, M.; Nie, Y.; Lu, G. Influence of gastrointestinal tract on metabolism of bisphenol A as determined by in vitro simulated system. *J. Hazard. Mater.* **2018**, *355*, 111–118. [CrossRef] [PubMed]
23. Schäpe, S.; Krause, J.; Masanetz, R.; Riesbeck, S.; Starke, R.; Rolle-Kampczyk, U.; Eberlein, C.; Heipieper, H.; Herberth, G.; von Bergen, M.; et al. Environmentally Relevant Concentration of Bisphenol S Shows Slight Effects on SIHUMIx. *Microorganisms* **2020**, *8*, 1436. [CrossRef]
24. Catron, T.; Keely, S.; Brinkman, N.; Zurlinden, T.; Wood, C.; Wright, J.; Phelps, D.; Wheaton, E.; Kvasnicka, A.; Gaballah, S.; et al. Host Developmental Toxicity of BPA and BPA Alternatives Is Inversely Related to Microbiota Disruption in Zebrafish. *Toxicol. Sci. Off. J. Soc. Toxicol.* **2019**, *167*, 468–483. [CrossRef]
25. Gomez, M.V.; Dutta, M.; Suvorov, A.; Shi, X.; Gu, H.; Mani, S.; Yue Cui, J. Early Life Exposure to Environmental Contaminants (BDE-47, TBBPA, and BPS) Produced Persistent Alterations in Fecal Microbiome in Adult Male Mice. *Toxicol. Sci.* **2021**, *179*, 14–30. [CrossRef]
26. Lei, M.; Menon, R.; Manteiga, S.; Alden, N.; Hunt, C.; Alaniz, R.C.; Lee, K.; Jayaraman, A. Environmental Chemical Diethylhexyl Phthalate Alters Intestinal Microbiota Community Structure and Metabolite Profile in Mice. *Msystems* **2019**, *4*, 10-1128. [CrossRef]
27. Truong, D.T.; Franzosa, E.A.; Tickle, T.L.; Scholz, M.; Weingart, G.; Pasolli, E.; Tett, A.; Huttenhower, C.; Segata, N. MetaPhlAn2 for enhanced metagenomic taxonomic profiling. *Nat. Methods* **2015**, *12*, 902–903. [CrossRef] [PubMed]
28. Beghini, F.; McIver, L.J.; Blanco-Míguez, A.; Dubois, L.; Asnicar, F.; Maharjan, S.; Mailyan, A.; Manghi, P.; Scholz, M.; Thomas, A.M.; et al. Integrating taxonomic, functional, and strain-level profiling of diverse microbial communities with bioBakery 3. *eLife* **2021**, *10*, e65088. [CrossRef]
29. Segata, N.; Izard, J.; Waldron, L.; Gevers, D.; Miropolsky, L.; Garrett, W.S.; Huttenhower, C. Metagenomic biomarker discovery and explanation. *Genome Biol.* **2011**, *12*, R60. [CrossRef] [PubMed]
30. Liaw, A.; Wiener, M. Classification and Regression by randomForest. *R News* **2002**, *2*, 18–22.
31. Mallick, H.; Rahnavard, A.; McIver, L.J.; Ma, S.; Zhang, Y.; Nguyen, L.H.; Tickle, T.L.; Weingart, G.; Ren, B.; Schwager, E.H.; et al. Multivariable association discovery in population-scale meta-omics studies. *PLOS Comput. Biol.* **2021**, *17*, e1009442. [CrossRef]
32. Wehbe, Z.; Nasser, S.A.; El-Yazbi, A.; Nasreddine, S.; Eid, A.H. Estrogen and Bisphenol A in Hypertension. *Curr. Hypertens. Rep.* **2020**, *22*, 23. [CrossRef]
33. Pérez-Bermejo, M.; Mas-Pérez, I.; Murillo-Llorente, M.T. The Role of the Bisphenol A in Diabetes and Obesity. *Biomedicines* **2021**, *9*, 666. [CrossRef] [PubMed]
34. Akash, M.S.H.; Rasheed, S.; Rehman, K.; Imran, M.; Assiri, M.A. Toxicological evaluation of bisphenol analogues: Preventive measures and therapeutic interventions. *RSC Adv.* **2023**, *13*, 21613–21628. [CrossRef]
35. Duar, R.M.; Lin, X.B.; Zheng, J.; Martino, M.E.; Grenier, T.; Pérez-Muñoz, M.E.; Leulier, F.; Gänzle, M.; Walter, J. Lifestyles in transition: Evolution and natural history of the genus Lactobacillus. *FEMS Microbiol. Rev.* **2017**, *41*, S27–S48. [CrossRef] [PubMed]
36. Severi, E.; Hood, D.W.; Thomas, G.H. Sialic acid utilization by bacterial pathogens. *Microbiology* **2007**, *153*, 2817–2822. [CrossRef] [PubMed]
37. Li, Y.; Zhang, H.; Rashid, A.; Hu, A.; Xin, K.; Li, H.; Adyari, B.; Wang, Y.; Yu, C.-P.; Sun, Q. Bisphenol A attenuation in natural microcosm: Contribution of ecological components and identification of transformation pathways through stable isotope tracing. *J. Hazard. Mater.* **2020**, *385*, 121584. [CrossRef]
38. Li, S.; Tian, K.; Qiu, Q.; Yu, Y.; Li, H.; Chang, M.; Sun, X.; Gu, J.; Zhang, F.; Wang, Y.; et al. Study on Genomics of the Bisphenol A-Degrading Bacterium *Pseudomonas* sp. P1. *Water* **2023**, *15*, 830. [CrossRef]
39. Kyrila, G.; Katsoulas, A.; Schoretsaniti, V.; Rigopoulos, A.; Rizou, E.; Doulgeridou, S.; Sarli, V.; Samanidou, V.; Touraki, M. Bisphenol A removal and degradation pathways in microorganisms with probiotic properties. *J. Hazard. Mater.* **2021**, *413*, 125363. [CrossRef]
40. Mukherjee, A.; Lordan, C.; Ross, R.P.; Cotter, P.D. Gut microbes from the phylogenetically diverse genus Eubacterium and their various contributions to gut health. *Gut Microbes* **2020**, *12*, 1802866. [CrossRef] [PubMed]

41. Vacca, M.; Celano, G.; Calabrese, F.M.; Portincasa, P.; Gobbetti, M.; De Angelis, M. The Controversial Role of Human Gut Lachnospiraceae. *Microorganisms* **2020**, *8*, 573. [CrossRef] [PubMed]
42. Lopetuso, L.R.; Scaldaferri, F.; Petito, V.; Gasbarrini, A. Commensal Clostridia: Leading players in the maintenance of gut homeostasis. *Gut Pathog.* **2013**, *5*, 23. [CrossRef] [PubMed]
43. Russell, A.; Copio, J.N.; Shi, Y.; Kang, S.; Franklin, C.L.; Ericsson, A.C. Reduced housing density improves statistical power of murine gut microbiota studies. *Cell Rep.* **2022**, *39*, 110783. [CrossRef]
44. Hildebrand, F.; Nguyen, T.L.A.; Brinkman, B.; Yunta, R.G.; Cauwe, B.; Vandenabeele, P.; Liston, A.; Raes, J. Inflammation-associated enterotypes, host genotype, cage and inter-individual effects drive gut microbiota variation in common laboratory mice. *Genome Biol.* **2013**, *14*, R4. [CrossRef] [PubMed]

metabolites MDPI

Article

Impact of Microplastic Exposure on Blood Glucose Levels and Gut Microbiota: Differential Effects under Normal or High-Fat Diet Conditions

Manjin Xu, Huixia Niu, Lizhi Wu, Mingluan Xing, Zhe Mo, Zhijian Chen, Xueqing Li * and Xiaoming Lou *

Department of Environmental Health, Zhejiang Provincial Center for Disease Control and Prevention, Hangzhou 310051, China; margexmj@outlook.com (M.X.); niuhuixia1118@163.com (H.N.); lzhwu@cdc.zj.cn (L.W.); mlxing@cdc.zj.cn (M.X.); zhmo@cdc.zj.cn (Z.M.); zhjchen@cdc.zj.cn (Z.C.)
* Correspondence: xqli@cdc.zj.cn (X.L.); xmlou@cdc.zj.cn (X.L.)

Abstract: Microplastics are emerging pollutants that have garnered significant attention, with evidence suggesting their association with the pathogenesis of type 2 diabetes mellitus. In order to assess the impact of polystyrene microplastic exposure on alterations in the gut microbiota and the subsequent implications for glucose dysregulation under different dietary conditions in mice, we investigated the effects and disparities in the blood glucose levels induced by polystyrene microplastic exposure in mice fed a high-fat diet versus those fed a normal diet. Using 16S rRNA sequencing and bioinformatics analyses, we explored the dynamic changes and discrepancies in the gut microbiota stability induced by polystyrene microplastic exposure under varied dietary conditions, and we screened for gut genera associated with the potential of polystyrene microplastics to disrupt glucose homeostasis. Our findings indicate that a high-fat diet resulted in abnormal mouse body weight, energy intake, blood glucose levels and related metabolic parameters. Additionally, polystyrene microplastic exposure exacerbated the glucose metabolism disorders induced by a high-fat diet. Furthermore, the composition and diversity of the mouse gut microbiota were significantly altered following microplastic exposure, with 11 gut genera exhibiting a differential presence between mice fed a high-fat diet combined with microplastic exposure compared to those fed a normal diet with microplastic exposure. Moreover, *Ucg-009* played an intermediary role in the association between a high-fat diet and the fasting blood glucose. Hence, our study demonstrates that polystyrene microplastic exposure exacerbates high-fat diet-induced glucose metabolism disorders, whereas its impact on the blood glucose under normal dietary conditions is not significant, highlighting the differential influence attributable to distinct alterations in characteristic gut genera.

Keywords: microplastics; polystyrene; high-fat diet; blood glucose; gut microbiota

1. Introduction

Microplastics (MPs) are plastic fragments and particles with diameters smaller than 5 mm, which are formed by physical and chemical processes [1]. They have become a global environmental pollutant, and in recent years, microplastics have been further categorized into micro-sized plastics (MPs) and nano-sized plastics (NPs) [2]. Despite being conceptualized only two decades ago, extensive scientific attention is now focused on their distribution, migration, and transformation in the environment, as well as their impact on human health. As a novel pollutant, the potential threats posed by microplastics to ecological environments and human health warrant systematic investigation. Humans may ingest up to 5 g of microplastics per week, and due to their small size, these particles can circulate through the bloodstream, accumulating in various tissues and organs, including the liver, kidneys, spleen, lungs, and placenta, potentially exerting localized cellular and tissue toxicity [3,4]. Long-term exposure to microplastics may disrupt normal immune system function [5], induce chronic inflammatory responses [6], and cause cellular damage [7]. Existing research indicates that microplastic exposure may increase the risk of chronic diseases such as cardiovascular diseases and diabetes [8,9]. The exposure pathways include ingestion via food, intake through drinking water, and inhalation through air contact. Among these routes, ingestion via the digestive system is one of the primary ways microplastics enter organisms, where they can colonize, particularly in the gastrointestinal tract, affecting physiological processes [10]. Microplastics have been detected in human feces, confirming their entry into the human gut and emphasizing the potential impact of microplastics on health through the gut microbiota [11].

Diabetes mellitus is a chronic disease characterized by high blood sugar levels, directly or indirectly caused by insulin deficiency. Globally, the number and proportion of diabetes patients continue to rise. According to the International Diabetes Federation (IDF), as of 2021, there were approximately 537 million diabetes patients worldwide, with China alone having 141 million, equating to a prevalence of 12.8%, predominantly type 2 diabetes [12]. Diabetes has been identified by the United Nations and the World Health Organization as one of the five priority non-communicable diseases to address in global health action plans. The increasing prevalence of diabetes and the substantial rise in healthcare costs place significant burdens on societies, economies, and healthcare systems worldwide [13]. A high-fat diet is recognized as a risk factor contributing to the onset and exacerbation of diabetes, influencing blood glucose metabolism levels. Therefore, it is crucial to deepen understanding of the development of diabetes induced by a high-fat diet and propose new treatment and prevention strategies for comprehensive disease management.

There is substantial evidence suggesting a close association between microplastic exposure and the onset and progression of diabetes. Wang et al. found that mice exhibited impaired glucose tolerance and hepatic lipid deposition following high−dose microplastic administration [14]. Insulin resistance in mice induced by microplastic exposure may also be related to metabolic reprogramming of the gut-liver axis [9]. Importantly, numerous studies indicate that disruptions to glucose metabolism caused by microplastic exposure are linked to alterations in the gut microbiota. We hypothesize that microplastics may impact glucose metabolism through changes in the gut microbiota, exacerbated by a high-fat diet. In this study, we established a mouse model of a high-fat diet combined with polystyrene microplastic exposure, measured relevant indicators of blood glucose metabolism, and performed 16S rRNA sequencing and bioinformatics analysis on collected fecal samples at the phylum and genus levels. Subsequently, we conducted statistical analyses to identify gut genera associated with the role of polystyrene microplastics in glucose dysregulation under different dietary patterns. Finally, we analyzed the mediating effects of relevant gut genera. This research will contribute to a comprehensive understanding of the impact of the gut microbiota on glucose dysregulation induced by microplastic exposure under different dietary patterns, providing insights into the toxicological effects of microplastics and raising concerns about their harm to high-risk populations exposed to high-fat diets.

2. Materials and Methods

2.1. Chemicals

The 5 μm monodisperse polystyrene microspheres used in this experiment were purchased from the Tianjin BaseLine ChromTech Research Center (Tianjin, China). The samples were prepared as a 2.5% w/v milky solution, with an initial water suspension concentration of 250 mg/10 mL. The microplastic solution was diluted to 16 mg/mL using deionized water and stored at 4 °C. Particle characterization of the microplastics used in the experiment was conducted using a Kurt particle size analyzer. As shown in Figure S1, the polystyrene microspheres exhibited good mass quality, with no significant differences in the particle size, all averaging around 5 μm.

2.2. Experimental Animals and Exposure Protocol

Prior to commencement, the study protocol was approved by the Zhejiang Chinese Medical University Laboratory Animal Research Center (No. 20230213-07), ensuring compliance with ethical standards in animal research. SPF−grade male C57BL/6 mice, 4–6 weeks old, with uniform genetic backgrounds, were obtained from SLAC Laboratory Animal Co., Ltd. (Shanghai, China). The mice were fed diets from Hangzhou Hangsi Biotechnology Co., Ltd. (Hangzhou, China): a 10 kcal% normal diet and a 60 kcal% high-fat diet, with energy densities of 3.85 kcal/g and 5.24 kcal/g, respectively. The experimental design is shown in Figure 1. After acclimatization for seven days in an animal facility with the temperature maintained at 22 °C, humidity at 55 ± 15%, and a 12 h light-dark cycle, 48 mice were randomly divided into four groups (n = 12/group): normal diet control (ND + water), normal diet microplastic (ND + MP), high-fat diet control (HFD + water), and high-fat diet microplastic (HFD + MP). During the experimental period, mice in the ND + MP and HFD + MP groups were orally gavaged daily with 80 mg/kg body weight of microplastic suspension, following dosage references from previous studies [15]. Mice in the ND + water and HFD + water groups were administered an equivalent volume of high-purity water by oral gavage. The experiment lasted for 14 weeks, during which biweekly measurements were taken for the body weight, food intake, water consumption, and fasting blood glucose. Prior to euthanasia at week 14, fecal samples were collected from each mouse, immediately frozen in liquid nitrogen, and stored at −80 °C. After a 12-h fast, final measurements of the body weight and fasting blood glucose were taken, blood was collected via retro−orbital bleeding, and the mice were euthanized via cervical dislocation. Tissues, including from the pancreas, colon, cecum, small intestine, liver, and epididymal fat pads, were collected and weighed. The experimental procedures were conducted at the Zhejiang Chinese Medical University Laboratory Animal Research Center.

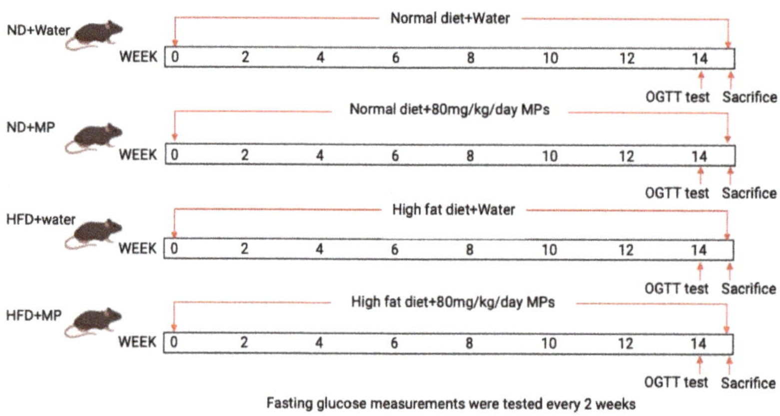

Figure 1. Experimental design.

2.3. Oral Glucose Tolerance Test

At week 14 of the experiment, after a 12−hour fast, the mice were orally administered a glucose solution at a dose of 2 g/kg body weight. Blood samples were collected via tail vein puncture at 0, 30, 60, 90, 120, and 150 min post−glucose administration, and the blood glucose levels were measured using a glucometer. GraphPad was used to plot the blood glucose curve and calculate the area under the curve.

2.4. 16S rRNA High-Throughput Sequencing

Genomic DNA of the gut microbiota from the fecal samples was extracted using the Mag—bind Soil DNA kit. The DNA purity and concentration were assessed by 1% agarose gel electrophoresis. Barcoded primers were used to amplify the diluted genomic DNA following the standard protocol of the Illumina MiSeq platform using the TruSeq® DNA PCR—Free Sample Preparation Kit. Quantification was performed with QuantiFluor™—ST, and the sequences were subjected to quality filtering and base trimming. The sequencing was performed by Shanghai Applied Protein Technology Co., Ltd. (Shanghai, China).

2.5. Gut Microbiota Analysis

The operational taxonomic units (OTUs) were assigned using the Uparse algorithm (version 7.0.1090), with a sequence similarity threshold of 97%. Alpha diversity indices, including Shannon, Simpson, and Chao1, were calculated to assess the community diversity and richness. Principal coordinate analysis (PCoA) based on the weighted and unweighted UniFrac distances was conducted to assess the beta diversity among the study groups. The composition and abundance of the gut microbiota at the phylum and genus levels were analyzed, visualized using bar graphs, and statistically tested for significance using the linear discriminant analysis effect size (LEfSe) and other appropriate statistical methods.

2.6. Statistical Analysis

Statistical analyses and graphing of the experimental data were performed using R version 4.3.0, SPSS version 26.0.0.0, and GraphPad Prism version 10.2.1. The results are presented as means ± standard deviations ($\overline{x} \pm SD$). A one-way ANOVA was used to test for statistical differences among the four groups, and t-tests were employed for comparisons between two groups. Bootstrap methods were used to validate the mediating effects of the gut microbiota. The direct effects of high-fat diet exposure on the fasting blood glucose outcomes and the indirect effects mediated by the gut microbiota were estimated, with the proportion of mediation calculated as the "indirect effect/total effect". Statistical significance was set at $\alpha = 0.05$.

3. Results

3.1. Effects of Polystyrene Microplastic Exposure on Mouse Body Weight Gain, Energy Intake, and Organ Weights

The biweekly changes in the mouse body weight are depicted in Figure 2a. The bar graphs showing the mouse body weights at week 0 and week 14 are presented in Figures 2b and 2c, respectively. The bar graph illustrating the weight gain during the experimental period is shown in Figure 2d. The results indicate that the high-fat diet led to significant weight gain. However, there were no significant differences in body weight between the control and microplastic-exposed mice under both dietary conditions, suggesting that polystyrene microplastic exposure did not significantly affect the mouse body weight.

Details of the average food intake and energy consumption per group are summarized in Tables S1 and S2, respectively. The results show that the high-fat diet led to increased organ weights ($p < 0.05$). However, there were no significant differences in the food intake and energy consumption between the control and microplastic-exposed mice under both dietary conditions, indicating that polystyrene microplastic exposure did not significantly influence the energy intake in the mice.

The absolute and relative weights of various organs and tissues are presented in Supplementary Tables S3 and S4. The absolute weight results indicate that the high-fat diet significantly increased the pancreatic, adipose, and colonic weights in the mice ($p < 0.05$). However, there were no significant differences in the absolute organ weights between the control and microplastic-exposed mice under the different dietary conditions. The relative weight results show that the high-fat diet significantly increased the relative weights of the pancreas and epididymal fat pads, while decreasing the relative weight of the cecal tissue ($p < 0.05$). Again, there were no significant differences in the relative organ weights between the control and microplastic-exposed mice under the different dietary conditions, indicating that microplastic exposure had no significant impact on the organ weights between the dietary groups.

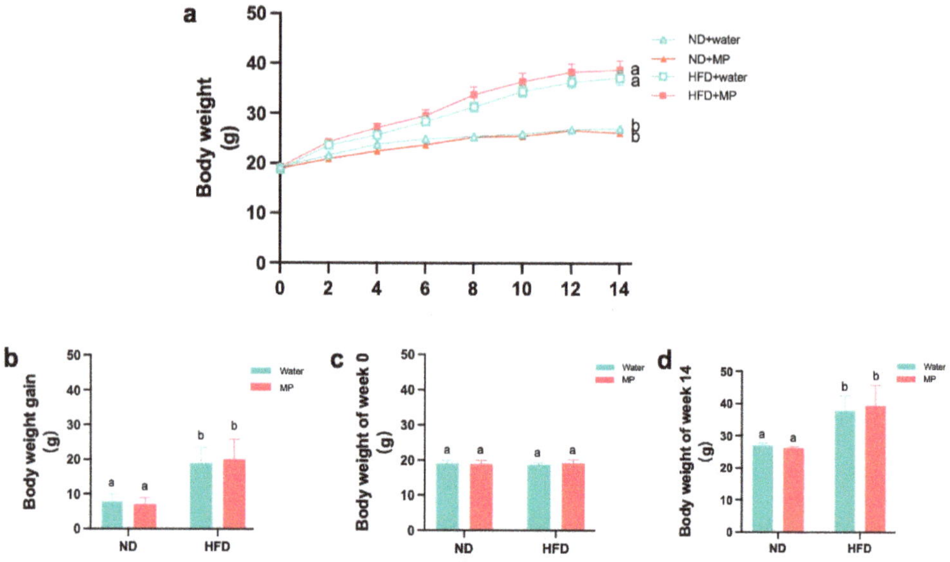

Figure 2. Effects of polystyrene microplastics on the body weight of mice: (**a**) the weight change of the mice at 14 weeks; (**b**) the weight gain of the mice at 14 weeks; (**c**) the weight of the mice at the 0th week; and (**d**) the weight of the mice at the 14th week. Different letters indicate that there is statistical significance in the comparison between groups, $p < 0.05$.

3.2. Polystyrene Microplastic Induces Abnormal Fasting Blood Glucose in Mice

The fasting blood glucose levels were measured biweekly throughout the experiment, and the changes were plotted in line and bar graphs, as shown in Figure 3a. Statistical analysis of the data revealed no significant differences in the fasting blood glucose among the groups from week 0 to week 12. However, starting from week 8, significant differences in the fasting blood glucose between the high-fat diet control group and the normal diet control group mice were observed ($p < 0.05$). Additionally, under the normal diet regimen, there were no significant differences in the fasting blood glucose between the microplastic-exposed mice and the control mice. In contrast, under the high-fat diet regimen, microplastic exposure did not significantly affect the fasting blood glucose levels at 12 weeks but resulted in a significant increase at 14 weeks ($p < 0.05$). These findings indicate that while the high-fat diet induced changes in the mouse blood glucose, polystyrene microplastic exposure alone did not affect the fasting blood glucose significantly. However, co-exposure with a high-fat diet exacerbated the effects of the diet on the mouse fasting blood glucose.

3.3. Polystyrene Microplastic Induces Abnormal Glucose Tolerance in Mice

A few days before the end of the 14th week of exposure, mice fasted for 12 h were subjected to an OGTT (oral glucose tolerance test). The blood glucose levels were measured and recorded during the test, and the results were statistically analyzed, as shown in Figure 3d. According to the experimental results, after glucose solution administration, the blood glucose levels in the mice rose rapidly, with no significant differences among the groups at 30 min. Subsequently, from 30 min onward, the blood glucose began to decline. The mice fed a normal diet exhibited a faster decline in the blood glucose levels compared to those fed a high-fat diet, and from 60 min onward, significant statistical differences in the blood glucose changes emerged between the high-fat diet control group and the normal diet control group ($p < 0.05$). Under the different dietary conditions, there were no significant differences in glucose tolerance between the microplastic-exposed and control mice under the normal diet. However, under the high-fat diet, the mice exposed to microplastics showed slower declines in the blood glucose levels, with significant statistical differences emerging from 60 min onward ($p < 0.05$). Calculations of the area under the OGTT curve revealed significant differences between the high-fat diet control group and the normal diet control group ($p < 0.05$). In the normal diet model, there were no significant differences in the area under the curve between the microplastic-exposed and control mice. In contrast, in the high-fat diet model, the area under the curve was significantly larger in the microplastic-exposed mice compared to the control mice ($p < 0.05$).

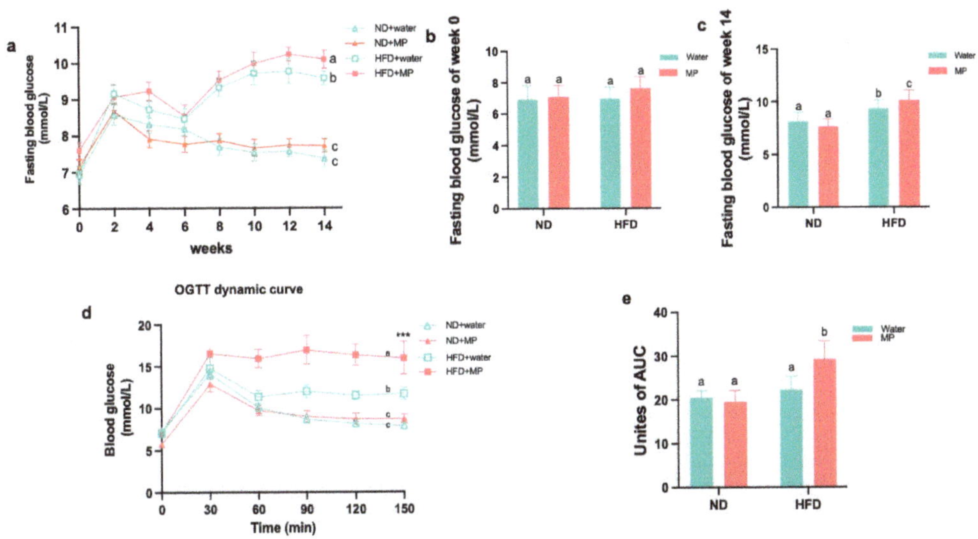

Figure 3. Effects of polystyrene microplastics on the fasting blood glucose and glucose tolerance in mice: (**a**) fasting blood glucose changes of the mice at 14 weeks; (**b**) fasting blood glucose of the mice at 0 week; (**c**) fasting blood glucose of the mice at 14 weeks; (**d**) blood glucose changes of the mice during OGTT test; and (**e**) area under OGTT curve. Different letters indicate that there is statistical significance in the comparison between groups, $p < 0.05$.

3.4. Differences in Gut Microbiota Diversity

The Ace and Chao1 indices estimated that under the normal diet regimen, there were no significant differences in the Ace and Chao1 indices between the microplastic-exposed mice and the control mice. Under the high-fat diet regimen, the microplastic-exposed mice showed slightly lower Ace and Chao1 indices compared to the control mice, but the differences were not statistically significant. Overall, these results indicate that polystyrene microplastics had no significant impact on the species diversity of the gut microbiota. The results of the Shannon and Simpson indices are shown in Figure 4c and 4d, respectively. In the normal diet regimen, there were no statistically significant differences in the Shannon and Simpson indices between the microplastic−exposed mice and the control mice. However, under the high-fat diet regimen, the microplastic−exposed mice exhibited significantly lower Shannon and Simpson indices compared to the control mice ($p < 0.05$), indicating a significant effect. Taken together, these results suggest that polystyrene microplastic exposure affects the richness and evenness of the gut microbiota in mice fed a high-fat diet. As shown in Figure 4e, all the groups had a Coverage index >0.9995, indicating high sample authenticity and representativeness. These sequencing results demonstrate that polystyrene microplastic exposure alters the α-diversity of the gut microbiota in mice under the high-fat diet regimen, with distinct differences compared to the normal diet regimen.

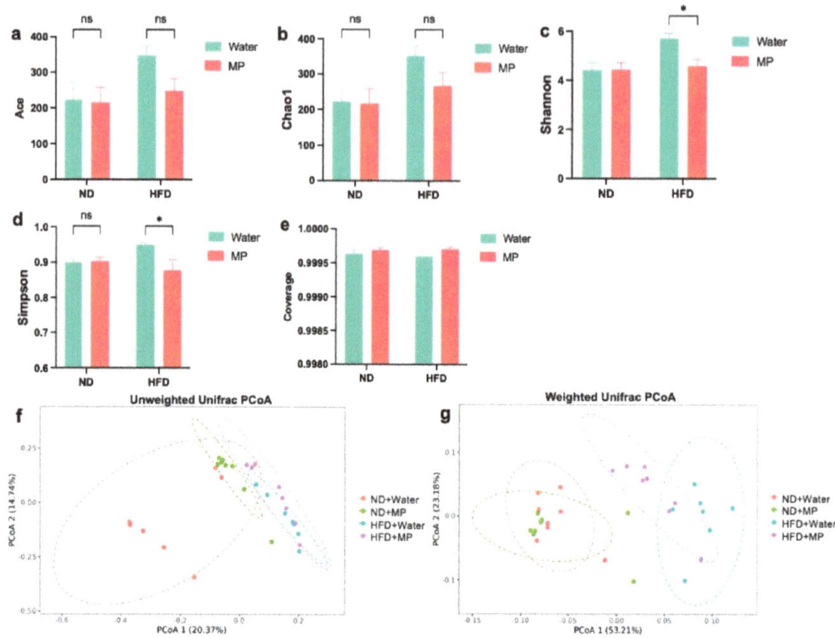

Figure 4. Effects of polystyrene microplastics on the diversity of the intestinal microflora in mice: (**a**) mouse Ace index; (**b**) mouse Chao1 index; (**c**) mouse Shannon index; (**d**) mouse Simpson index; (**e**) mouse Coverage index; (**f**) analysis based on unweighted UniFrac PcoA; and (**g**) analysis based on weighted UniFrac PcoA; * $p < 0.05$.

The β-diversity assessed by unweighted UniFrac is depicted in Figure 4f. The results show significant differences at the OTU level between the mice fed a high-fat diet and those fed a normal diet. In both dietary models, there were differences in the OTU levels between the microplastic–exposed and control groups. The weighted UniFrac results in Figure 4g indicate significant differences in the OTU levels between the mice fed a high-fat diet and those fed a normal diet. Moreover, under both dietary models, there were differences in the OTU levels between the microplastic-exposed and control groups, with a more pronounced difference in the high-fat diet model, characterized by a reduced overlap. These results demonstrate that polystyrene microplastic exposure induces changes in the β-diversity of the gut microbiota in mice, with more significant changes observed under the high-fat diet compared to the normal diet.

3.5. Identification of Characteristic Gut Microbiota Differences

Based on the 16S rRNA sequencing results, the gut microbiota composition and relative abundance at the phylum level were analyzed, as shown in Figure 5a. The results indicate that *Firmicutes*, *Bacteroidota*, *Actinobacteriota*, *Desulfobacterota*, *Campilobacterota*, and *Deferribacterota* were the most abundant phyla in the mouse gut microbiota, with *Firmicutes* and *Bacteroidota* being predominant. Comparing the different treatment groups, significant changes were observed in the gut microbiota at the phylum level in the mice exposed to polystyrene microplastics. Under the normal diet regimen, the microplastic-exposed mice showed increased relative abundance of *Firmicutes*, *Bacteroidota*, *Desulfobacterota*, *Campilobacterota*, and *Deferribacterota* compared to the control mice, while *Actinobacteriota* exhibited a decreasing trend. Under the high-fat diet regimen, the microplastic-exposed mice exhibited increased relative abundance of *Firmicutes*, *Actinobacteriota*, and *Desulfobacterota*, and decreased relative abundance of *Bacteroidota*, *Campilobacterota*, and *Deferribacterota*, compared to the control mice.

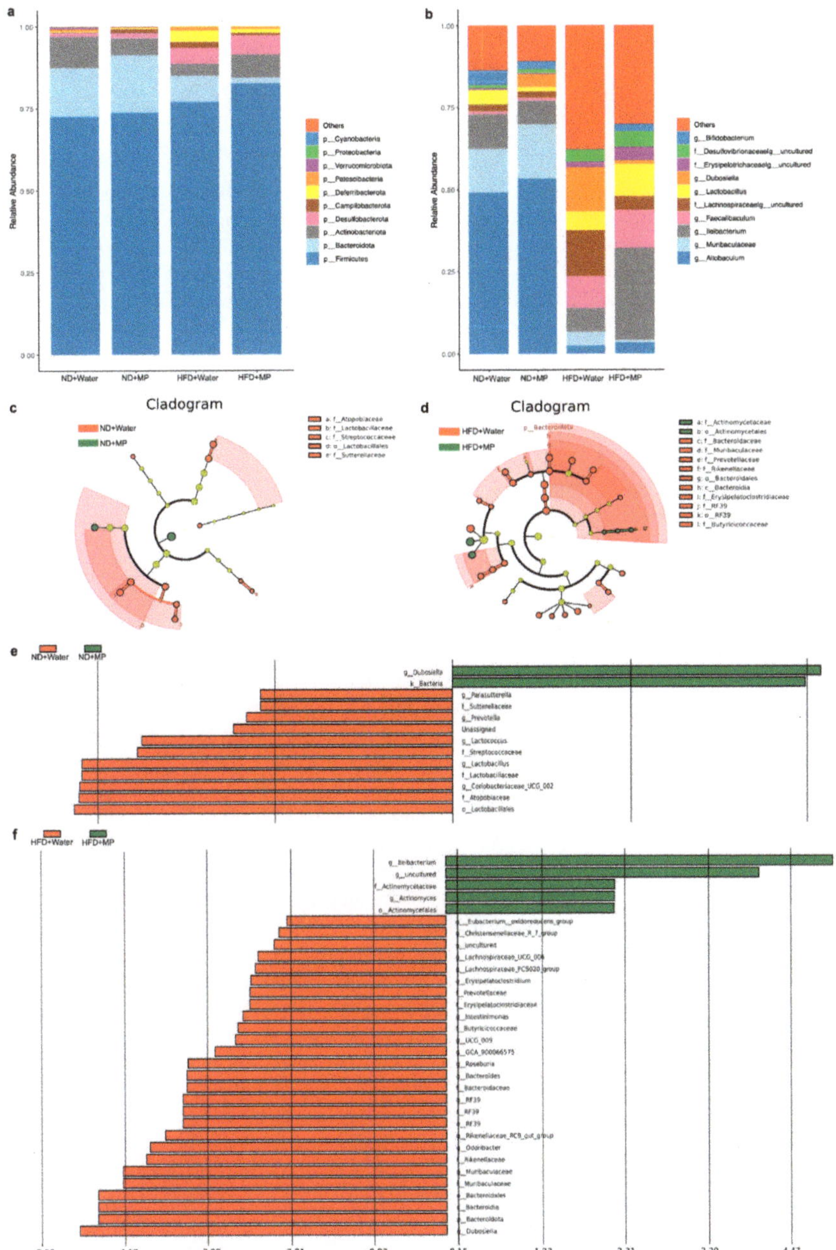

Figure 5. Effects of polystyrene microplastic exposure on the gut microbiota composition in mice: (**a**) changes in the gut microbiota composition at the phyla level; (**b**) changes in the gut microbiota composition at the genera level; (**c**,**e**) cladogram analysis map and bar chart of the LDA score (log10)

distribution of the gut microbiota in the normal diet control group and microplastic group, where red represents microbial communities that play an important role in the (ND + Water) group, green represents microbial communities that play an important role in the (ND + MP) group, yellow represents microbial communities that did not play an important role in either group; and (**d**,**f**) cladogram analysis map and bar chart of the LDA score (log10) distribution of the gut microbiota in the high-fat diet control group and microplastic group, where red represents microbial communities that play an important role in the (HFD + Water) group, green represents microbial communities that play an important role in the (HFD + MP) group, and yellow represents microbial communities that did not play an important role in either group. The current LDA threshold score is over 2.

Further analysis at the genus level was conducted to explore the impact of polystyrene microplastic exposure on the gut microbiota's relative abundance, as shown in Figure 5b. The results reveal that the mouse gut bacteria at the genus level include *Allobaculum*, *Muribaculaceae*, *Ileibacterium*, *Faecalibaculum*, *Lachnospiraceae*, *Lactobacillus*, and *Dubosiella*, among others. Under the normal diet regimen, the microplastic−exposed mice showed increased relative abundance of *Ileibacterium* and *Lactobacillus*, while *Allobaculum*, *Muribaculaceae*, *Faecalibaculum*, *Lachnospiraceae*, and *Dubosiella* exhibited decreased relative abundance compared to the control mice. Under the high-fat diet regimen, the microplastic-exposed mice exhibited increased relative abundance of *Allobaculum*, *Muribaculaceae*, *Ileibacterium*, *Faecalibaculum*, and *Lactobacillus*, while *Lachnospiraceae* and *Dubosiella* exhibited decreased relative abundance compared to the control mice.

The LEfSe analysis further compared the significant differences in the gut microbiota composition between the microplastic-exposed and control groups, identifying characteristic gut microbiota. The LEfSe evolutionary branch diagram and LDA value distribution histogram for the intergroup comparisons are shown in Figure 5c–f. The results indicate that under the normal diet regimen, the microplastic−exposed mice exhibited increased relative abundance of *g__Dubosiella* and *k__Bacteria* compared to the control mice. Additionally, under the high-fat diet regimen, the microplastic−exposed mice showed significantly increased relative abundance of *g__Ileibacterium*, *g__uncultured*, *g__Actinomycetales*, *g__Actinomyces*, and *g__Actinomycetales* compared to the control mice. These results underscore the significant changes in the gut microbiota composition following polystyrene microplastic exposure in mice.

3.6. Changes in Gut Microbiota Genera upon Microplastic Exposure under Different Dietary Models

At the genus level, differential analysis of the gut microbiota was conducted to explore the impact of polystyrene microplastic exposure under different dietary models, as illustrated in Figure 6. The results indicate significant changes in the relative abundance of the gut microbiota genera between the microplastic-exposed and control groups under both the normal diet and high-fat diet regimens. Under the normal diet regimen, the microplastic-exposed mice showed significant changes in relative abundance of seven genera compared to the control mice. In contrast, when comparing the microplastic-exposed mice under the high-fat diet regimen to the control mice, 12 genera exhibited significant changes. Specifically, 11 genera showed significant differences ($p < 0.05$) following microplastic exposure under the high-fat diet regimen compared to the control group. These genera include *Ileibacterium*, *Bifidobacterium*, *Dubosiella*, *Rikenellaceae Rc9 Gut Group*, *Streptococcus*, *Lachnospiraceae Ucg-006*, *Ucg-009*, *Lachnospiraceae Fcs020 Group*, *Christensenellaceae R-7 Group*, *Erysipelatoclostridium*, and one unclassified genus. Among these, the first two genera exhibited significantly increased relative abundance following microplastic exposure under the high-fat diet regimen, while the latter nine genera showed significantly decreased relative abundance. These findings underscore the differential impact of polystyrene microplastic exposure on the relative abundance of the gut microbiota genera depending on the dietary context, highlighting notable shifts, particularly under conditions of high-fat diet consumption.

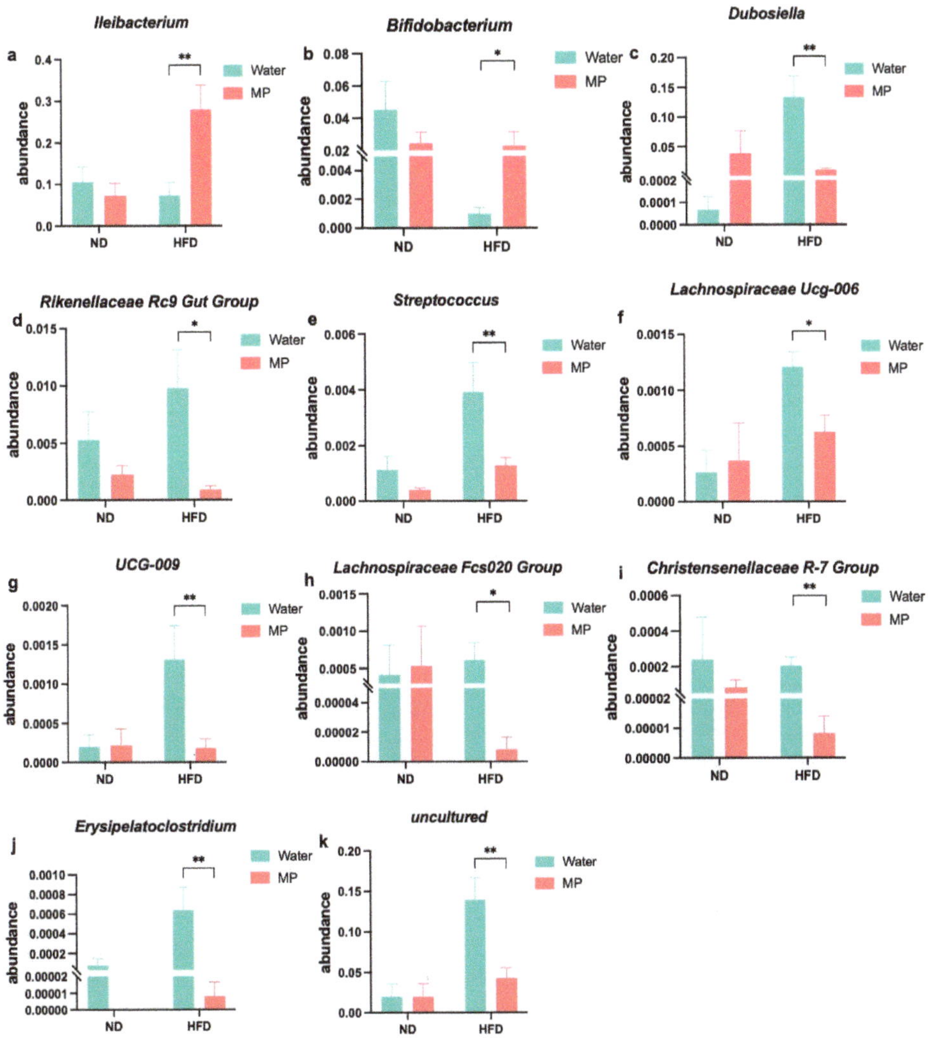

Figure 6. The difference in the intestinal bacteria at different levels caused by exposure to polystyrene microplastics in mice fed with different dietary patterns: (**a**) the change in relative abundance of *Ileibacterium*; (**b**) the change in relative abundance of *Bifidobacterium*; (**c**) the change in relative abundance of *Dubosiella*; (**d**) the change in relative abundance of *Rikenellaceae_RC9_gut_group*; (**e**) the change in relative abundance of *Streptococcus*; (**f**) the relative abundance of *Lachnospiraceae_UCG-006*; (**g**) the change in relative abundance of *UCG-009*; (**h**) change in relative abundance of *Lachnospiraceae Fcs020 Group*; (**i**) the change in relative abundance of *Christensenellaceae_R-7_group*; (**j**) the change in relative abundance of *Erysipelatoclostridium*; and (**k**) the change in relative abundance of *uncultured*. * $p < 0.05$, ** $p < 0.01$.

3.7. Gut Microbiota-Mediated Indirect Effects

The above results suggest that polystyrene microplastic exposure alone does not affect glucose tolerance in mice. However, when exposed simultaneously with a high-fat diet, microplastics exacerbate the diet's impact on glucose tolerance in mice. The gut microbiota may play a key role in the different effects of microplastic exposure under different dietary conditions. To further explore the role of differential changes in specific genera in relation to the outcome of the fasting blood glucose, we conducted mediation analysis to assess the specific gut microbiota−mediated indirect effects and proportions.

Among the 11 genera showing differential changes, we found that *Ucg-009* exhibited a significant mediation effect in the association between a high-fat diet and the fasting blood glucose outcome. The results of the mediating analysis are shown in Table 1. The proportion mediated by *Ucg-009* was 0.6308 ($p < 0.05$).

Table 1. Analysis of the mediating effect of *Ucg-009*.

	Effect Size	Bootstrap 95% CI		p	Proportion Mediated
		Lower	Upper		
Direct effect	0.5361	0.0353	1.2	0.026	0.3692
Indirect effect	0.3139	−0.465	1.08	0.444	0.6308
Total effect	0.85	0.0632	1.68	0.024	1

4. Discussion

As a new environmental contaminant, microplastics have garnered widespread attention. Increasing evidence shows that microplastics can be ingested orally and exert toxic effects on various target organs, including inducing inflammation, neurotoxicity, and oxidative stress, and impairing immune and circulatory systems. However, due to their relatively low toxicity or less pronounced effects, their harmfulness is often overlooked. A high-fat diet has been widely recognized as a risk factor for diabetes [16]. Previous studies found that oral administration of fPS MP to high-fat diet-induced obese mice exacerbates impaired glucose metabolism and insulin resistance, and it promotes systemic inflammation [17]. Leaky gut syndrome (LGS) may be caused by HFD, leading to MP deposition in the intestinal mucosa, causing inflammation of the inner layer of the intestinal mucosa, exacerbating insulin resistance and affecting insulin secretion, thereby affecting the blood glucose levels [18]. Yet, research on the toxic effects and mechanisms of microplastic exposure under high-fat diet conditions remains limited. In our study, based on observations of metabolic markers related to blood glucose and changes in the gut microbiota under different dietary conditions, we explored the association between microplastic exposure and blood glucose impacts in mice.

We designed four exposure groups: normal diet control (ND + water), normal diet microplastic (ND + MP), high-fat diet control (HFD + water), and high-fat diet microplastic (HFD + MP). By comparing the phenotypic differences among these groups, we found that microplastic exposure exacerbated the blood glucose metabolism disruption in the mice fed a high-fat diet while showing no significant impact on the blood glucose in the mice fed a normal diet. Okamura et al. also evaluated the gut outcomes associated with a high-fat diet and MP exposure, and some of the conclusions are similar to our study. They found that a microplastic-induced disruption to blood glucose metabolism only occurred in mice fed a high-fat diet [18]. However, Okamura et al.'s article focuses on the endocrine and metabolic systems, with a particular emphasis on evaluating the gut pathophysiology, while our article focuses on differential changes in the gut microbiota and conducts mediation analysis of these differential changes, with a focus on statistical methods. Comparing the research results, although *Firmicutes* and *Bacteroideta* were both the phyla with the highest content in both studies, our results showed that *Firmicutes* had the highest relative abundance, while Okamura et al.'s study exhibited a higher relative abundance of *Bacteroideta*. *Firmicutes* and *Bacteroidetes* are known as the main bacteria in the gut microbiota, and their relative abundance changes are related to disease, health, obesity, and dietary habits. This difference may be explained by differences in the dietary energy design (3.85 kcal/g and 5.24 kcal/g in our study, and 345 kcal/100 g and 459 kcal/100 g in Okamura et al.'s study, respectively), or methodological differences in the sample processing and DNA analysis [19]. Moreover, in Okamura et al.'s study, microplastics were provided ad libitum. Our study used the method of gavage administration for microplastic exposure, and the concentrations of the microplastics in the two studies were also different. These differences in experimental conditions may lead to different bacterial abundances. Although there are differences in the abundance of the gut microbiota, combining the results of the two studies, we hypothesize that microplastic exposure aggravates the blood glucose metabolism disruption under a high-fat diet, and the gut microbiota may play a role in this process.

At the genus level, our study identified differential changes in 11 genera, such as *Ileibacterium*, *Bifidobacterium*, and *Dubosiella*, following microplastic exposure under different dietary conditions. Among these, *Ucg-009* was found to mediate the association between a high-fat diet and the fasting blood glucose outcome. The gut microbiota plays critical roles in regulating various pathways in human metabolism and physiology, including immune, energy, lipid, and glucose metabolism [20]. Dysbiosis of the gut microbiota has been implicated in the pathogenesis of obesity and type 2 diabetes [21]. *Ileibacterium*, considered a harmful bacterium, has been positively correlated with blood lipid levels and metabolic disorders [22]. It has also been implicated in type 2 diabetes and diabetic nephropathy [23]. Additionally, Xiao et al. demonstrated that decreased abundance of *Ileibacterium* and *Bifidobacterium* positively regulated the gut microbiota and suppressed blood glucose disorders [24]. Similarly, our experimental results showed significant changes in *Ileibacterium* under different dietary feeding patterns with microplastic exposure, where its relative abundance significantly increased under a high-fat diet but showed no significant change under a normal diet with microplastic exposure. Microplastic exposure may aggravate the blood glucose disruption induced by a high-fat diet by increasing the levels of *Ileibacterium*.

Dubosiella has been linked to the development of certain diseases, such as diabetes [25], and is widely used as a probiotic in the medical field [26]. *Christensenellaceae R-7 Group* is considered beneficial in regulating diabetes-related gut types [27]. In our study, we found significant reductions in *Dubosiella* and *Christensenellaceae R-7 Group* following microplastic exposure under a high-fat diet, which differed from the changes observed under a normal diet with exposure to microplastics. *Rikenellaceae RC9 Gut Group* has been reported to promote the expression of fat synthesis genes and increase obesity [28]. Keisiii et al. found that the *Lachnospiraceae* family is associated with metabolic dysfunction and the development of obesity and diabetes in mice [29]. However, in our experimental results, the relative abundances of *Rikenellaceae RC Gut Group* and various genera within *Lachnospiraceae* were decreased under microplastic exposure in a high-fat diet model. Further research is needed to determine the role of *Rikenellaceae RC Gut Group* and the *Lachnospiraceae* family in exacerbating the blood glucose disruption induced by microplastics. It is reported that *Ucg-009* is negatively correlated with HOMA-IR, HDL-C, ALT, AST, TC, and lipopolysaccharide (LPS), indicating that *Ucg-009* is beneficial to hypoglycemia and hypolipidemia in diabetic mice [30]. However, we found that *Ucg-009* significantly decreased after exposure to microplastics under high-fat diet feeding, and this change in the relative abundance was validated to play a mediating role in the association between high-fat diet microplastic exposure and blood glucose changes. This result indicated that exposure to high-fat dietary microplastics has adverse effects on blood glucose by reducing *Ucg-009*. Thus, the differential effects of microplastics on blood glucose in mice under different dietary conditions could be attributed to differential changes in characteristic gut genera.

To the best of our knowledge, there was a limited understanding of the exacerbation of glucose metabolism disorders under high-fat diet conditions due to microplastic exposure and the role of the gut microbiota therein. Our findings reveal significant changes in the blood glucose and distinct gut microbiota in mice exposed to microplastics under a high-fat diet pattern. Statistical methods were applied to validate the mediating role of the gut microbiota, providing evidence for further exploration of characteristic gut microbiota in the context of microplastic-induced glucose changes under high-fat diet conditions. This study offers new insights into the toxicological effects and health risks associated with microplastic exposure. Further mechanistic studies and clinical trials are necessary to comprehensively explore the roles and specific mechanisms of microplastics in blood glucose metabolism in high-fat diet populations, which could provide valuable insights for preventing and treating diabetes in high-risk populations.

5. Conclusions

This study investigated the differential effects of microplastic exposure on the blood glucose levels and gut microbiota homeostasis in mice fed normal and high-fat diets. The results demonstrate that microplastic exposure exacerbates the blood glucose disruption in mice fed a high-fat diet while showing no significant impact on those fed a normal diet. Additionally, the differential effects of microplastics on the blood glucose in mice under different dietary conditions may be associated with differential changes in characteristic gut genera. This study contributes to a better understanding of the health risks posed by microplastics and their potential impact on exacerbating glucose metabolic disruption in populations with a high-fat diet.

Supplementary Materials: The following supporting information can be downloaded at https://www.mdpi.com/article/10.3390/metabo14090504/s1, Figure S1: Kurt particle size analysis of polystyrene microplastic microspheres (data provided by Tianjin Bessler chromatographic Technology Development Center); Table S1: Average food intake of the mice; Table S2. Average energy intake of the mice; Table S3. Absolute weight of the mouse viscera and tissue; Table S4. Relative weight of the organs and tissues in the mice.

Author Contributions: Conceptualization, M.X. (Manjin Xu) and H.N.; writing—original draft preparation, M.X. (Manjin Xu); writing—review and editing, M.X. (Manjin Xu), H.N, L.W., M.X. (Mingluan Xing), Z.M. and Z.C.; project administration, X.L. (Xueqing Li); funding acquisition, X.L. (Xiaoming Lou). All authors have read and agreed to the published version of the manuscript.

Funding: This research was funded by the Central Guiding Local Science and Technology Development Fund Projects (No. 2023ZY1024).

Institutional Review Board Statement: The animal study protocol was approved by the Zhejiang Chinese Medical University Laboratory Animal Research Center (approval code: No. 20230213-07 and approval date: 13 February 2023).

Informed Consent Statement: Not applicable.

Data Availability Statement: The original contributions presented in the study are included in the article/Supplementary Mmaterial, further inquiries can be directed to the corresponding authors.

Conflicts of Interest: The authors declare no conflicts of interest.

References

1. Thompson, R.; Olsen, Y.; Mitchell, R.; Davis, A.; Rowland, S.; John, A.; McGonigle, D.; Russell, A. Lost at Sea: Where Is All the Plastic? *Science* **2004**, *304*, 838. [CrossRef] [PubMed]
2. Ter Halle, A.; Jeanneau, L.; Martignac, M.; Jardé, E.; Pedrono, B.; Brach, L.; Gigault, J. Nanoplastic in the North Atlantic Subtropical Gyre. *Environ. Sci. Technol.* **2017**, *51*, 13689–13697. [CrossRef] [PubMed]
3. Wright, S.; Kelly, F. Plastic and Human Health: A Micro Issue? *Environ. Sci. Technol.* **2017**, *51*, 6634–6647. [CrossRef] [PubMed]
4. Jeong, C.; Won, E.; Kang, H.; Lee, M.; Hwang, D.; Hwang, U.; Zhou, B.; Souissi, S.; Lee, S.; Lee, J. Microplastic Size-Dependent Toxicity, Oxidative Stress Induction, and p-JNK and p-P38 Activation in the Monogonont Rotifer (Brachionus Koreanus). *Environ. Sci. Technol.* **2016**, *50*, 8849–8857. [CrossRef] [PubMed]
5. Tang, Y.; Rong, J.; Guan, X.; Zha, S.; Shi, W.; Han, Y.; Du, X.; Wu, F.; Huang, W.; Liu, G. Immunotoxicity of Microplastics and Two Persistent Organic Pollutants Alone or in Combination to a Bivalve Species. *Environ. Pollut.* **2020**, *258*, 113845. [CrossRef]
6. Li, B.; Ding, Y.; Cheng, X.; Sheng, D.; Xu, Z.; Rong, Q.; Wu, Y.; Zhao, H.; Ji, X.; Zhang, Y. Polyethylene Microplastics Affect the Distribution of Gut Microbiota and Inflammation Development in Mice. *Chemosphere* **2020**, *244*, 125492. [CrossRef]
7. Liu, T.; Hou, B.; Wang, Z.; Yang, Y. Polystyrene Microplastics Induce Mitochondrial Damage in Mouse GC-2 Cells. *Ecotoxicol. Environ. Saf.* **2022**, *237*, 113520. [CrossRef]
8. Marfella, R.; Prattichizzo, F.; Sardu, C.; Fulgenzi, G.; Graciotti, L.; Spadoni, T.; D'Onofrio, N.; Scisciola, L.; La Grotta, R.; Frigé, C.; et al. Microplastics and Nanoplastics in Atheromas and Cardiovascular Events. *N. Engl. J. Med.* **2024**, *390*, 900–910. [CrossRef]
9. Shi, C.; Han, X.; Guo, W.; Wu, Q.; Yang, X.; Wang, Y.; Tang, G.; Wang, S.; Wang, Z.; Liu, Y.; et al. Disturbed Gut–Liver Axis Indicating Oral Exposure to Polystyrene Microplastic Potentially Increases the Risk of Insulin Resistance. *Environ. Int.* **2022**, *164*, 107273. [CrossRef]

10. Li, W.; Chen, X.; Li, M.; Cai, Z.; Gong, H.; Yan, M. Microplastics as an Aquatic Pollutant Affect Gut Microbiota within Aquatic Animals. *J. Hazard. Mater.* **2022**, *423*, 127094. [CrossRef]
11. Liebmann, B.; Köppel, S.; Königshofer, P.; Bucsics, T.; Reiberger, T.; Schwabl, P. Assessment of Microplastic Concentrations in Human Stool Final Results of a Prospective Study. *United Eur. Gastroenterol.* **2018**, *2018* (Suppl. S1), 2884.
12. Wang, Y.; Wei, Z.; Xu, K.; Wang, X.; Gao, X.; Han, Q.; Wang, S.; Chen, M. The Effect and a Mechanistic Evaluation of Polystyrene Nanoplastics on a Mouse Model of Type 2 Diabetes. *Food Chem. Toxicol.* **2023**, *173*, 113642. [CrossRef] [PubMed]
13. Perry, R.; Samuel, V.; Petersen, K.; Shulman, G. The Role of Hepatic Lipids in Hepatic Insulin Resistance and Type 2 Diabetes. *Nature* **2014**, *510*, 84–91. [CrossRef] [PubMed]
14. Wang, Q.; Wu, Y.; Zhang, W.; Shen, T.; Li, H.; Wu, J.; Zhang, L.; Qin, L.; Chen, R.; Gu, W.; et al. Lipidomics and Transcriptomics Insight into Impacts of Microplastics Exposure on Hepatic Lipid Metabolism in Mice. *Chemosphere* **2022**, *308*, 136591. [CrossRef] [PubMed]
15. Huang, D.; Zhang, Y.; Long, J.; Yang, X.; Bao, L.; Yang, Z.; Wu, B.; Si, R.; Zhao, W.; Peng, C.; et al. Polystyrene Microplastic Exposure Induces Insulin Resistance in Mice via Dysbacteriosis and Pro-Inflammation. *Sci. Total Environ.* **2022**, *838*, 155937. [CrossRef] [PubMed]
16. Assmann, G.; Sacks, F.; Awad, A.; Ascherio, A.; Bonanome, A.; Berra, B.; Booyse, F.; Carmena, R.; Dahlan, W.; DeLorgeril, M.; et al. Dietary Fat Consensus Statements. *Am. J. Med.* **2002**, *113*, 5–8.
17. Lee, A.G.; Kang, S.; Yoon, H.J.; Im, S.; Oh, S.J.; Pak, Y.K. Polystyrene Microplastics Exacerbate Systemic Inflammation in High-Fat Diet–Induced Obesity. *Int. J. Mol. Sci.* **2023**, *24*, 12421. [CrossRef]
18. Okamura, T.; Hamaguchi, M.; Hasegawa, Y.; Hashimoto, Y.; Majima, S.; Senmaru, T.; Ushigome, E.; Nakanishi, N.; Asano, M.; Yamazaki, M.; et al. Oral Exposure to Polystyrene Microplastics of Mice on a Normal or High-Fat Diet and Intestinal and Metabolic Outcomes. *Environ. Health Perspect.* **2023**, *131*, 27006. [CrossRef]
19. Magne, F.; Gotteland, M.; Gauthier, L.; Zazueta, A.; Pesoa, S.; Navarrete, P.; Balamurugan, R. The Firmicutes/Bacteroidetes Ratio: A Relevant Marker of Gut Dysbiosis in Obese Patients? *Nutrients* **2020**, *12*, 1474. [CrossRef]
20. de Vos, W.; Tilg, H.; Van Hul, M.; Cani, P. Gut Microbiome and Health: Mechanistic Insights. *Gut* **2022**, *71*, 1020–1032. [CrossRef]
21. Turnbaugh, P.; Ley, R.; Mahowald, M.; Magrini, V.; Mardis, E.; Gordon, J. An Obesity-Associated Gut Microbiome with Increased Capacity for Energy Harvest. *Nature* **2006**, *444*, 1027–1031. [CrossRef] [PubMed]
22. Xia, F.; Xiang, S.; Chen, Z.; Song, L.; Li, Y.; Liao, Z.; Ge, B.; Zhou, B. The Probiotic Effects of AB23A on High-Fat-Diet-Induced Non-Alcoholic Fatty Liver Disease in Mice May Be Associated with Suppressing the Serum Levels of Lipopolysaccharides and Branched-Chain Amino Acids. *Arch. Biochem. Biophys.* **2021**, *714*, 109080. [CrossRef] [PubMed]
23. Wu, C.; Fei, J.; Xu, Q.; Tao, Y.; Zhou, Z.; Wang, Y.; Wu, J.; Gu, H.F. Interaction between Plasma Metabolomics and Intestinal Microbiome in Db/Db Mouse, an Animal Model for Study of Type 2 Diabetes and Diabetic Kidney Disease. *Metabolites* **2022**, *12*, 775. [CrossRef] [PubMed]
24. Xiao, N.; Ruan, S.; Mo, Q.; Zhao, M.; Feng, F. The Effect of Sodium Benzoate on Host Health: Insight into Physiological Indexes and Gut Microbiota. *Foods* **2023**, *12*, 4081. [CrossRef] [PubMed]
25. Ai, X.; Wu, C.; Yin, T.; Zhur, O.; Liu, C.; Yan, X.; Yi, C.; Liu, D.; Xiao, L.; Li, W.; et al. Antidiabetic Function of Lactobacillus Fermentum MF423-Fermented Rice Bran and Its Effect on Gut Microbiota Structure in Type 2 Diabetic Mice. *Front. Microbiol.* **2021**, *12*, 682290. [CrossRef]
26. Liu, T.; Chen, M.; Tu, W.; Liang, Q.; Tao, W.; Jin, Z.; Xiao, Y.; Chen, L. Network and 16S rRNA Sequencing-Combined Approach Provides Insightal Evidence of Vitamin K2 for Salt-Sensitive Hypertension. *Front. Nutr.* **2021**, *8*, 639467. [CrossRef]
27. Hu, R.; Zeng, F.; Wu, L.; Wan, X.; Chen, Y.; Zhang, J.; Liu, B. Fermented Carrot Juice Attenuates Type 2 Diabetes by Mediating Gut Microbiota in Rats. *Food Funct.* **2019**, *10*, 2935–2946. [CrossRef]
28. Kang, Y.; Li, Y.; Du, Y.; Guo, L.; Chen, M.; Huang, X.; Yang, F.; Hong, J.; Kong, X. Konjaku Flour Reduces Obesity in Mice by Modulating the Composition of the Gut Microbiota. *Int. J. Obes.* **2019**, *43*, 1631–1643. [CrossRef]
29. Kameyama, K.; Itoii, K. Intestinal Colonization by a Lachnospiraceae Bacterium Contributes to the Development of Diabetes in Obese Mice. *Microbes Environ.* **2014**, *29*, 427–430. [CrossRef]
30. Ma, Q.; Zhai, R.; Xie, X.; Chen, T.; Zhang, Z.; Liu, H.; Nie, C.; Yuan, X.; Tu, A.; Tian, B.; et al. Hypoglycemic Effects of Lycium Barbarum Polysaccharide in Type 2 Diabetes Mellitus Mice via Modulating Gut Microbiota. *Front. Nutr.* **2022**, *9*, 916271. [CrossRef]

metabolites

MDPI

Article

The Dose–Response Effect of Fluoride Exposure on the Gut Microbiome and Its Functional Pathways in Rats

Zhe Mo [1,2,†], Jian Wang [1,†], Xinyue Meng [1], Ailin Li [1], Zhe Li [1], Wenjun Que [1], Tuo Wang [1], Korto Fatti Tarnue [1], Xu Ma [1], Ying Liu [1], Shirui Yan [1], Lei Wu [1], Rui Zhang [1], Junrui Pei [1,*] and Xiaofeng Wang [2,*]

[1] Key Laboratory of Etiology and Epidemiology, Chinese Center for Disease Control and Prevention, Center for Endemic Disease Control, Education Bureau of Heilongjiang Province & National Health Commission (23618504), Institute for Fluorosis Disease Control, Harbin Medical University, Harbin 150081, China; zhmo@cdc.zj.cn (Z.M.); 102690@hrbmu.edu.cn (J.W.); 2019020210@hrbmu.edu.cn (X.M.); liailin@hrbmu.edu.cn (A.L.); 2021020138@hrbmu.edu.cn (Z.L.); 2021020163@hrbmu.edu.cn (W.Q.); 2021020139@hrbmu.edu.cn (T.W.); 20210200005@hrbmu.edu.cn (K.F.T.); 2022020080@hrbmu.edu.cn (X.M.); 2022020107@hrbmu.edu.cn (Y.L.); 2023020208@hrbmu.edu.cn (S.Y.); 2023020210@hrbmu.edu.cn (L.W.); 2023020215@hrbmu.edu.cn (R.Z.)

[2] Department of Environmental Health, Zhejiang Provincial Center for Disease Control and Prevention, Hangzhou 310051, China

* Correspondence: peijunrui@ems.hrbmu.edu.cn (J.P.); xfwang@cdc.zj.cn (X.W.)

† These authors contributed equally to this work.

Abstract: Metabolic activities within the gut microbiome are intimately linked to human health and disease, especially within the context of environmental exposure and its potential ramifications. Perturbations within this microbiome, termed "gut microbiome perturbations", have emerged as plausible intermediaries in the onset or exacerbation of diseases following environmental chemical exposures, with fluoride being a compound of particular concern. Despite the well-documented adverse impacts of excessive fluoride on various human physiological systems—ranging from skeletal to neurological—the nuanced dynamics between fluoride exposure, the gut microbiome, and the resulting dose–response relationship remains a scientific enigma. Leveraging the precision of 16S rRNA high-throughput sequencing, this study meticulously examines the ramifications of diverse fluoride concentrations on the gut microbiome's composition and functional capabilities within Wistar rats. Our findings indicate a profound shift in the intestinal microbial composition following fluoride exposure, marked by a dose-dependent modulation in the abundance of key genera, including *Pelagibacterium*, *Bilophila*, *Turicibacter*, and *Roseburia*. Moreover, discernible alterations were observed in critical functional and metabolic pathways of the microbiome, such as D-lyxose ketol-isomerase and DNA polymerase III subunit gamma/tau, underscoring the broad-reaching implications of fluoride exposure. Intriguingly, correlation analyses elucidated strong associations between specific bacterial co-abundance groups (CAGs) and these shifted metabolic pathways. In essence, fluoride exposure not only perturbs the compositional equilibrium of the gut microbiota but also instigates profound shifts in its metabolic landscape. These intricate alterations may provide a mechanistic foundation for understanding fluoride's potential toxicological effects mediated via gut microbiome modulation.

Keywords: fluoride exposure; gut microbiome; functional pathways; toxicity

1. Introduction

Fluorine is an essential element for the normal growth and development of organisms. However, excessive exposure can lead to fluorosis [1]. Due to the uneven distribution of chemical elements on the Earth's crust, some regions contain excessive fluoride, leading to endemic fluorosis among the local population primarily manifested as dental and skeletal fluorosis [2]. Additionally, it negatively affects human reproductive development, the liver and kidney, and the endocrine, nervous, and genetic systems [2–4]. Drinking water is the primary source of fluoride exposure. Over 20 countries worldwide have excessive fluoride concentrations in their groundwater [5]. In China, more than 70 million people are still at risk of high fluoride exposure [6].

Several mechanisms have been proposed for fluoride-induced diseases, which include stress pathways, signaling routes, cell cycle dysregulation, cell apoptosis, and epigenetic changes [7–9]. Increasing evidence suggests that disturbances in the gut microbiota and their subsequent effects on metabolism and physiological functions play a significant role in disease development [10]. Given fluoride's toxicity, it is crucial to elucidate the impact of high fluoride exposure on the gut microbiota and its metabolism and functions. Excessive fluoride intake through drinking water induces an immune response, damages the structure of the caecum and rectum, inhibits the proliferation of intestinal epithelial cells, restricts glycoprotein secretion, decreases the distribution of goblet cells and hypertrophic cells, and alters the diversity and composition of the mouse gut microbiome [11–13]. An imbalance in the gut microbiota might compromise the integrity of the intestinal barrier and directly lead to diseases [14].

Studies on the effects of fluoride exposure on the gut microbiota have yielded inconsistent results. Research indicates that exposure to 100 mg/L fluoride increases the diversity and richness of the gut microbiome in Kunming and ICR mice [11,12]. However, other studies have found no significant changes in the composition and function of the gut microbiota in BALB/c mice exposed to a 4 ppm fluoride dose [7]. Population studies have shown that children with dental fluorosis in high-fluoride-exposure areas have slightly lower bacterial diversity and richness compared to normal groups [15]. Most current studies are mouse-based, although some research has confirmed that rats have a gut microbiota more similar to humans [16]. Moreover, the majority of studies focus on the effects of fluoride exposure on microbiota diversity and richness, with fewer investigating the dose–response relationship between fluoride exposure and the gut microbiota.

Building on preliminary research on the effects of drinking water fluoride on rat gut microbiota [17], this study aims to analyze the dose–response relationship of fluoride exposure in drinking water to the rat gut microbiota. Initially, 16S rRNA amplicon sequencing will be employed to observe changes in the gut microbiota and explore dose–response relationships. Subsequently, predictions on the metagenome's functions and metabolic pathways [18] will be made to observe changes in these areas. Lastly, a comprehensive analysis of the gut microbiota's composition and function will be conducted to determine the correlation between the microbial community and its functionality.

2. Materials and Methods

2.1. Animal Experiment

Specific-pathogen-free (SPF) male Wistar rats (4 weeks of age, body weight 127.96 ± 14.01 g) were purchased from Vital River Laboratories (Beijing, China). Rats were fed with a pelleted rodent diet (Beijing Keao Xieli Feed Co., Ltd., Beijing, China) and filtered water ad libitum and were maintained in the SPF animal facility of Harbin Medical University. A total of twenty-five rats were housed under maintained conditions of a 12:12 h light/dark cycle, constant environmental temperature of 23 ± 3 °C, and 40–70% relative humidity. All experiments were approved by the Ethics Committee of the Endemic Disease Control Center of Harbin Medical University. After 1 week of acclimation, rats were randomized into five groups (5 rats/group): control rats received distilled water alone, and distilled water with 25 mg/L, 50 mg/L, 100 mg/L, and 150 mg/L of NaF was administered to rats in 4 groups for 12 weeks, respectively. The dosages were chosen based on previous studies concerning environmental fluoride levels, the toxicities of fluorides, and the exposure level and duration in rodents (see Section 4). The feces of each rat were regularly obtained in a sterile tube after animal model realization and stored at −80 °C.

2.2. DNA Extraction and 16S rRNA Gene Sequencing

Total bacterial DNA was extracted from fecal samples using the environmental sample DNA extraction kit based on the manufacturer's instructions. Its concentration was monitored on the Qubit PicoGreen fluorescence quantification system (Thermo Fisher Scientific Inc., Shanghai, China) and purity was detected using 1% agarose gels. The variable region 3 and 4 (V3/V4) of the 16S RNA gene was PCR-amplified and then detected using 2% agarose gels. The amplifications were purified, quantified, and pooled, and then sequenced on an Illumina MiSeq sequencer (Genergy Biotechnology, Shanghai, China).

2.3. Sequencing Data Analysis

The raw mate-paired fastq files were merged, quality-filtered, and demultiplexed to obtain the optimized sequences. USEARCH software verson 11 with UPARSE-OTU algorithm (http://www.drive5.com/usearch/manual/uparseotu_algo.html (accessed on 5 October 2023)) was applied to cluster the operational taxonomical units (OTUs) with a threshold of 97% sequence similarity, and the Ribosomal Database Project classifier was used for taxonomic identification.

2.4. Microbiome Co-Abundance Groups (CAGs)

The abundant genus or species were applied to construct co-abundance groups (CAGs). The function "cor" was used to calculate their Kendall correlations, which were visualized with hierarchical Ward clustering with a Spearman correlation distance metric to define CAGs with function "hcut" in the "factoextra" package. The best number of clusters was determined according to the number of significance using a pairwise Adonis test among each group of the Kendall correlation matrix using the "pairwise.adonis" function in the "pairwiseAdonis" package. The expression levels of CAGs were calculated based on the sum of relative abundance of the same CAGs.

2.5. Functional Prediction via Tax4fun2 Package

Alterations in the intestinal microbiome do not necessarily signify functional changes [19]. Hence, studying the functional changes resulting from fluoride exposure, rather than merely the taxonomic composition of the intestinal microbiota, becomes imperative for understanding its health impacts. We employed the "Tax4fun2" package in R 4.2.3 software to predict alterations in the functions of the intestinal microbiota [18]. Tax4Fun2, an upgraded version of the Tax4Fun package, enables rapid functional spectrum predictions of prokaryotes based on 16S rRNA gene sequences. By integrating user-defined genome information, it significantly enhances the accuracy and robustness of predictive functional profiles [18]. Firstly, user-supplied 16S rRNA gene sequences are searched against the 16S rRNA reference sequences via BLAST using the runRefBlast function [18]. 16S rRNA gene sequences are aligned with reference sequences provided by Tax4Fun2 to identify closely related sequences. Secondly, functional predictions are subsequently calculated using the makeFunctionalPrediction function. During this step, the OTUs table supplied by the user is summarized based on the results of the next-neighbor search. Based on these search results, the abundance of operational taxonomic units (OTUs) for each sample is summarized, creating an association matrix (AM) that includes the functional profiles of identified reference sequences from the 16S rRNA search. Predicted profiles are later summarized based on KEGG pathways. Only OTUs passing a defined similarity threshold (default = 97%) are considered in the functional prediction [18]. In brief, the amplicon sequence variants were applied to predict the abundance of gene families (function) and MetaCyc pathways.

2.6. Statistical Analysis

The differences of abundances in the gut microbiome of phylum among groups were detected using an ANOVA test, and the pairwise differences between groups were identified via Tukey correction. Linear regression was applied to explore the relationship between fluoride exposures and gut microbiome/CAGs/functions, and then expressed as Pearson correlations. Spearman correlations between abundance in CAGs and MetaCyc pathways were performed to identify the association. All statistical analysis was conducted in R 4.2.3 software (R Foundation for Statistical Computing, Vienna, Austria). A two-tailed p value < 0.05 was defined as statistically significant.

3. Results

3.1. Distribution of the Gut Microbiome with Fluoride Exposure

The gut microbiomes among the control and different NaF groups were dominated by taxa belonging to Firmicutes, Bacteroidetes, Proteobacteria, and Actinobacteria in descending order (Figure 1 and Table 1). A significant difference was only observed in the abundances of Proteobacteria among groups (p value = 0.043), and the result of the pairwise test showed that the 150 mg/L F group may have higher abundances than the 50 mg/L F group (adjusted p value = 0.046).

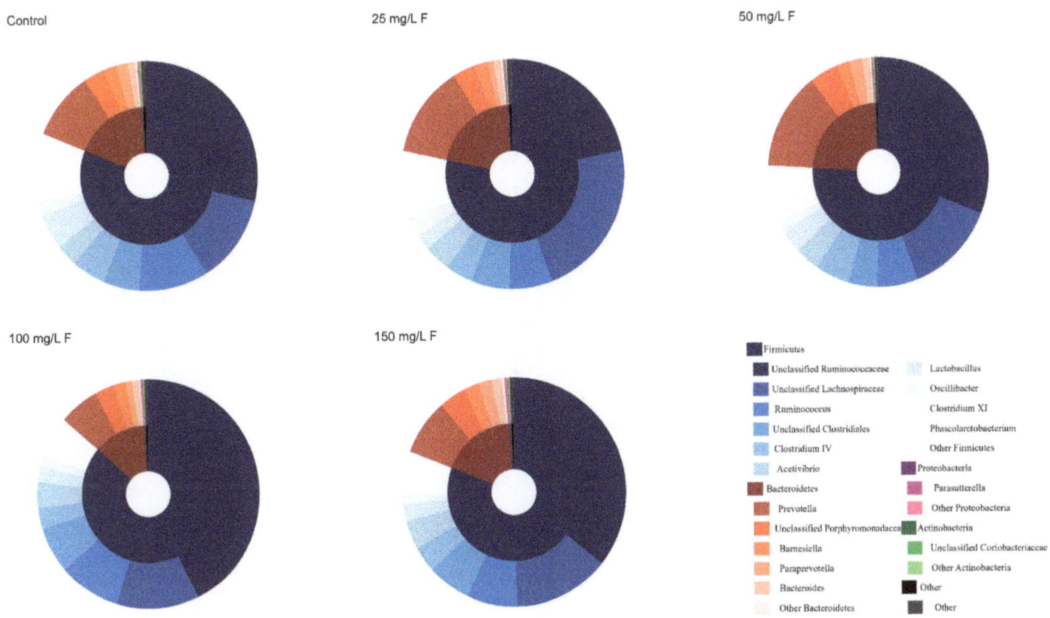

Figure 1. Mean relative abundance of the most abundant phylum and genus in the rats' gut microbiome composition.

Table 1. Relative abundance of the most abundant phyla in the rats' gut microbiome composition (mean ± SD).

Phylum	Control	25 mg/L F	50 mg/L F	100 mg/L F	150 mg/L F	F	p Value
Firmicutes	81.12 ± 18.73	78.33 ± 9.95	75.97 ± 13.54	86.36 ± 6.66	80.84 ± 9.48	0.490	0.743
Bacteroidetes	17.71 ± 18.35	20.55 ± 9.75	23.09 ± 13.51	12.62 ± 6.76	17.94 ± 9.75	0.502	0.735
Proteobacteria	0.25 ± 0.11	0.35 ± 0.19	0.24 ± 0.12	0.35 ± 0.13	0.51 ± 0.12	2.995	0.043
Actinobacteria	0.31 ± 0.24	0.27 ± 0.12	0.24 ± 0.12	0.22 ± 0.09	0.29 ± 0.15	0.283	0.885

3.2. Influence of Fluoride Exposure on Gut Microbiome in Genus

To clearly identify differences in gut microbiome structure among fluoride exposure groups, co-abundance correlations were clustered into four groups based on an Adonis test (Figure 2A). Significant associations were not observed between CAG1/CAG2 and fluoride exposure (p value > 0.05). CAG3 has a positive association with fluoride exposure, which means the relative abundance of CAG3 rises with increasing fluoride exposure doses, while CAG4 was negatively associated with fluoride exposure, which means the relative abundance of CAG4 decreases with increasing fluoride exposure doses (Figure 2B). Genera in *Bilophila, Holdemania, Pelagibacterium,* and *Ruminococcaceae* from CAG3 (Figure 2C and Table S1) were also found to be positively correlated with fluoride exposure, while *Corynebacterium, Lachnospiracea incertae sedis, Roseburia,* and *Turicibacter* from CAG4 (Figure 2C and Table S1) were negatively correlated with fluoride exposure.

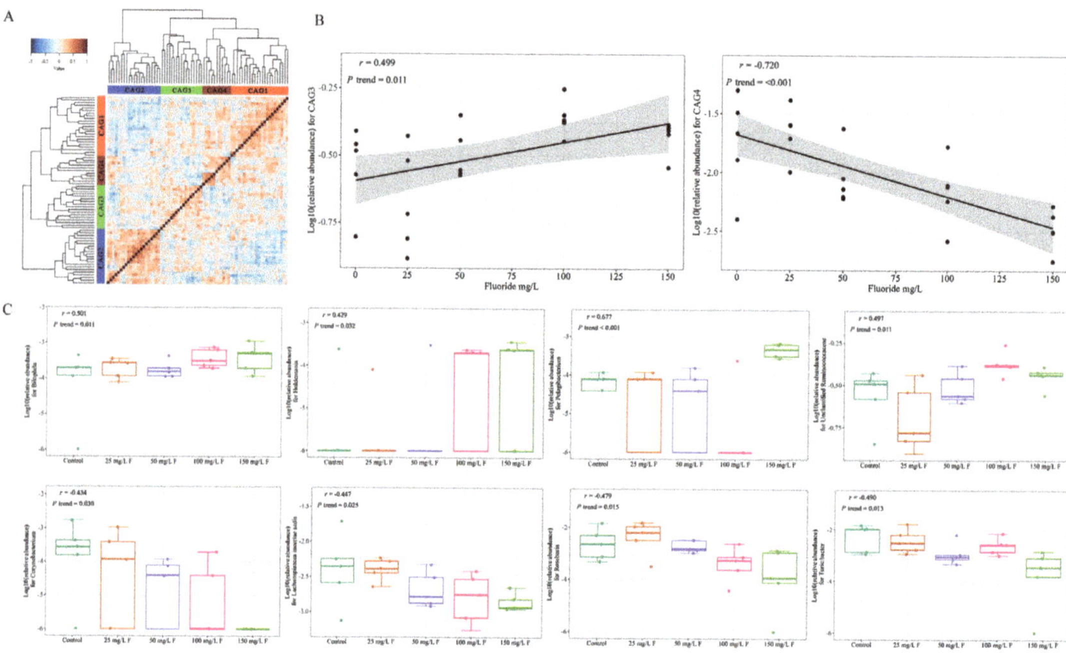

Figure 2. (**A**) Kendall correlations between genera clustered using hierarchical Ward method with a Spearman correlation distance metric. The primary genera attributed to the most CAGs are (proportion > 10%) CAG1 (*Unclassified Lachnospiraceae, Ruminococcus,* and *Unclassified Clostridiales*), CAG2 (*Prevotella* and *Unclassified Porphyromonadaceae*), CAG3 (*Unclassified Ruminococcaceae*), and CAG4 (*Turicibacter, Lachnospiracea incertae sedis, Roseburia,* and *Clostridium sensu stricto*). (**B**) The linear association between fluoride exposure and CAGs. (**C**) Boxplots show the effect of fluoride exposure on the gut microbiome in each genus with the Pearson correlation.

3.3. *Effect of Fluoride Exposure on Gut Microbiome in Species*

Co-abundance correlations were clustered into six groups based on the Adonis test (Figure 3A). With the exception of CAG1, no significant association with fluoride exposure was observed in the other CAGs. CAG1 has a positive association with fluoride exposure, which means the relative abundance of CAG3 rises with increasing fluoride exposure doses (Figure 3B). Species in *Unclassified Pelagibacterium, Candidatus Soleaferrea massiliensis, Parabacteroides distasonis, Uncultured bacterium adhufec108, Unclassified Bdellovibrionales, Unclassified Ruminococcaceae, Bacteroidales bacterium ph8, Bacteroides fragilis (T), Uncultured Kopriimonas* sp., *Unclassified Odoribacter, Parabacteroides goldsteinii, Unclassified Alphaproteobacteria, Bacteroides uniformis, Uncultured Erysipelotrichales bacterium,* and *Uncultured Desulfovibrionaceae bacterium* from CAG1 (Figure 3C and Table S2) were also found to be positively correlated with fluoride exposure.

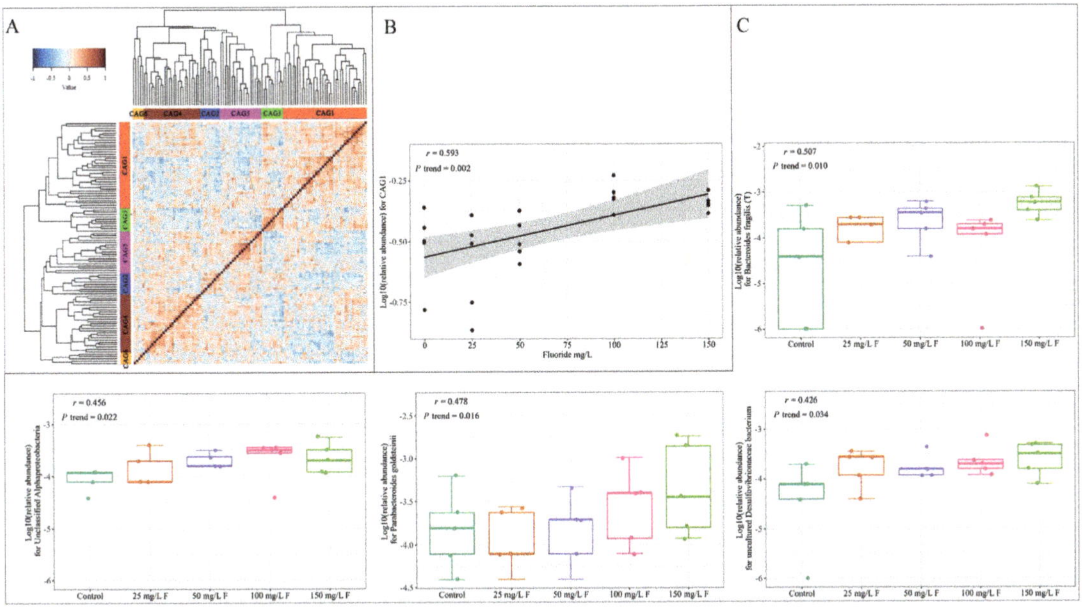

Figure 3. (**A**) Kendall correlations between species clustered using hierarchical Ward method with a Spearman correlation distance metric. The primary species attributed to the most CAGs are (proportion >10%) CAG1 (*Unclassified Ruminococcaceae*), CAG2 (*Uncultured rumen bacterium*), CAG3 (*Unclassified Prevotella*, *Uncultured bacterium* and *Unclassified Porphyromonadaceae*), CAG4 (*Uncultured Firmicutes bacterium*, *Rumen bacterium NK4A214* and *Unclassified Clostridium IV*), CAG5 (*Unclassified Lachnospiraceae*), and CAG6 (*Ruminococcus* sp., *Uncultured bacterium adhufec* and *Unclassified Blautia*). (**B**) The linear association between fluoride exposure and CAGs. (**C**) Boxplots show the effect of fluoride exposure on the gut microbiome in each species with Pearson correlation (due to space limitations, only some figures are shown).

3.4. Functional Changes in the Gut Microbiome with Fluoride Exposure

Figure 4 and Table S3 show that the biological functions of the gut microbiome were significantly associated with fluoride exposure. Metabolism pathways including energy metabolism and metabolism of terpenoids and polyketides were strongly positively correlated with fluoride exposure (Figure 5, Tables 2 and 3) and strongly correlated with CAG3 in genus, CAG4 in genus, and CAG1 in species abundance (Figure 6). Furthermore, Genetic Information Processing pathways including Translation; Folding sorting, and degradation; and Transcription were negatively correlated with fluoride exposure (Figure 5, Tables 2 and 3). Finally, Human Diseases pathways including drug resistance: antimicrobial, drug resistance: antineoplastic, and infectious diseases: viral were negatively correlated with fluoride exposure, while cardiovascular diseases was positively correlated (Figure 5, Tables 2 and 3). Meanwhile, it also was negatively correlated with CAG3 in genus and CAG1 in species abundance (Figure 6).

Table 2. Correlation coefficients between MetaCyc pathways of level 3 and fluoride exposure.

MetaCyc Pathways of Level 3	Correlation Coefficient	*p* Value
Metabolism	0.723	<0.001
Genetic Information Processing	−0.426	0.034
Human Diseases	−0.451	0.024

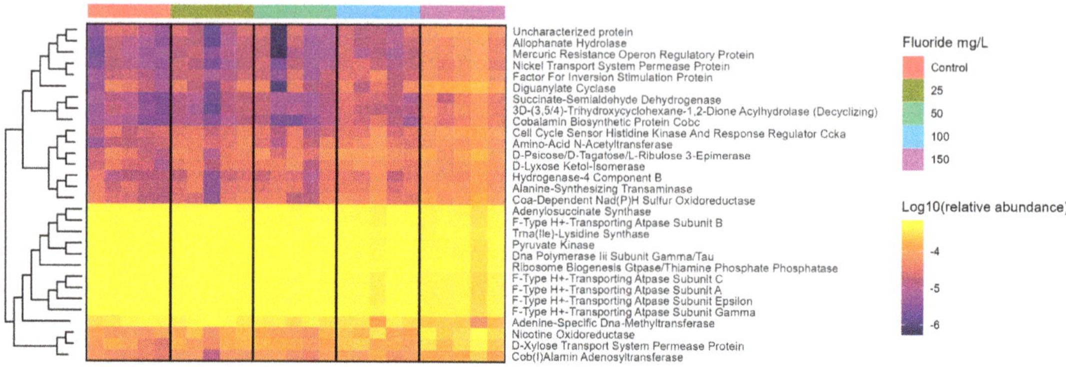

Figure 4. The relative abundance of the most correlated function (the absolute of Spearman correlation coefficient > 0.75 and p-value < 0.05) with fluoride exposure visualized via Heatmap plot.

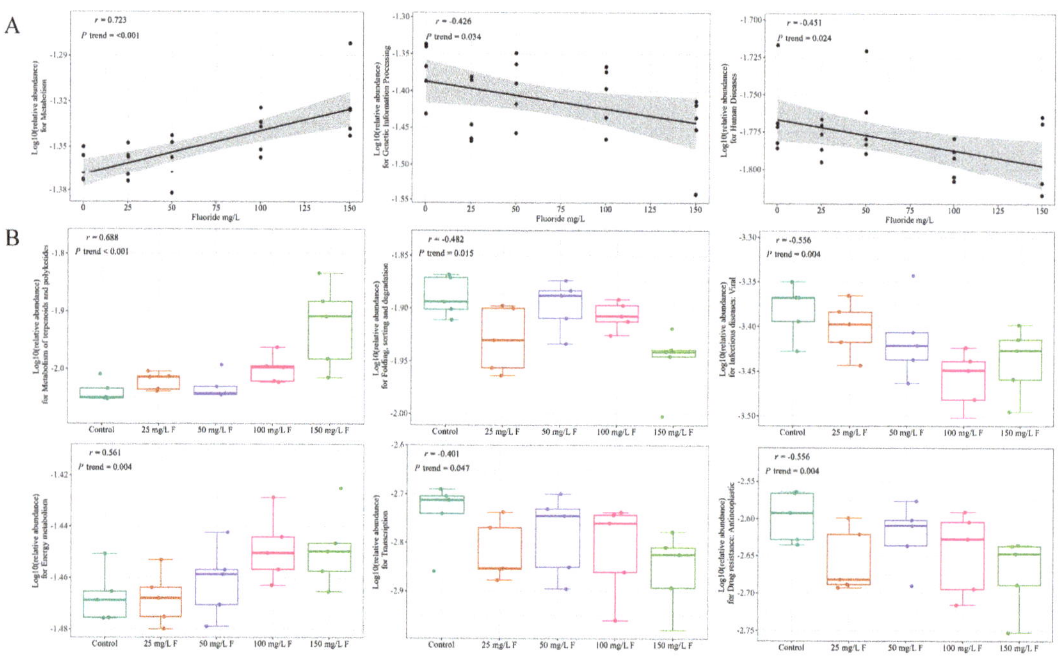

Figure 5. (**A**) The linear association between fluoride exposure and MetaCyc pathways of level 3. (**B**) Boxplots show the effect of fluoride exposure on the MetaCyc pathways of level 2 with Pearson correlation.

Table 3. Correlation coefficients between MetaCyc pathways of level 2 and fluoride exposure.

MetaCyc Pathways of Level 3	MetaCyc Pathways of Level 2	Correlation Coefficient	*p* Value
Genetic Information Processing	Translation	−0.399	0.048
Genetic Information Processing	Folding, sorting, and degradation	−0.482	0.015
Genetic Information Processing	Transcription	−0.401	0.047
Human Diseases	Drug resistance: Antimicrobial	−0.442	0.027
Human Diseases	Drug resistance: Antineoplastic	−0.411	0.041
Human Diseases	Infectious diseases: Viral	−0.556	0.004
Human Diseases	Cardiovascular diseases	0.399	0.048
Metabolism	Energy metabolism	0.561	0.004
Metabolism	Metabolism of terpenoids and polyketides	0.688	<0.001

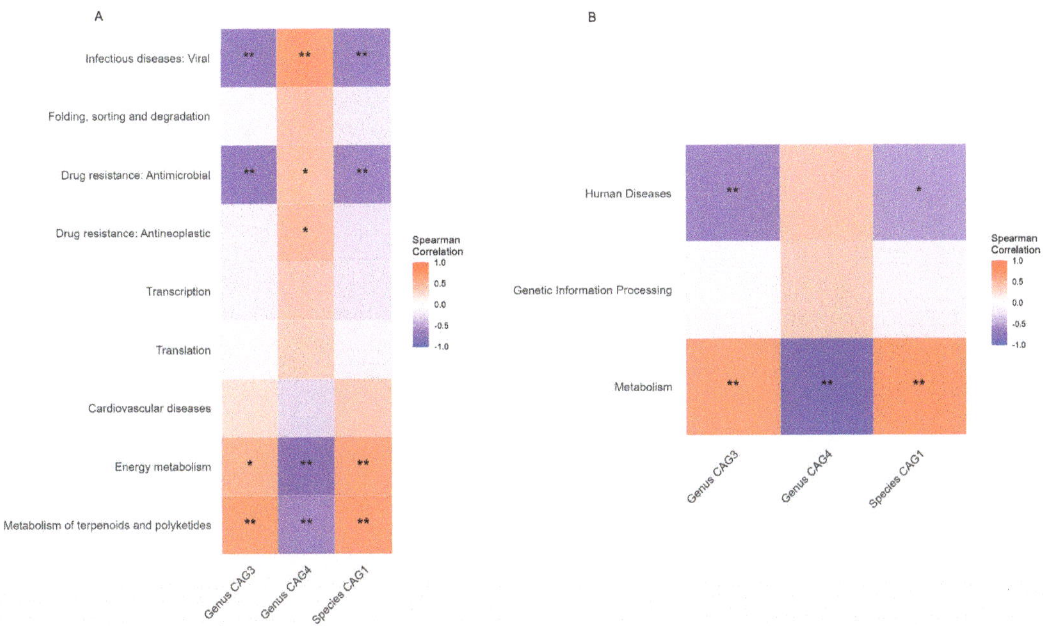

Figure 6. (**A**) Spearman correlations of the differentially abundant MetaCyc pathways of level 2 and CAGs; (**B**) Spearman correlations of the differentially abundant MetaCyc pathways of level 3 and CAGs. * indicates p-value < 0.05, ** indicates p-value < 0.01.

The results of the predictions indicate that fluoride exposure exhibits a strong positive dose–response relationship (up-regulation) with functions like D-lyxose ketol-isomerase, alanine-synthesizing transaminase, Fis family transcriptional regulator, CoA-dependent NAD(P)H sulfur oxidoreductase, 3D-(3,5/4)-trihydroxycyclohexane-1,2-dione acylhydrolase (decyclizing), and nickel transport system permease protein. Conversely, it displays a strong negative dose–response relationship (down-regulation) with functions like DNA polymerase III subunit gamma/tau, F-type H+-transporting ATPase subunit gamma, F-type H+-transporting ATPase subunit a, and F-type H+-transporting ATPase subunit c. Fluoride exposure has a pronounced positive dose–response relation (up-regulation) with metabolic pathways like metabolism of terpenoids and polyketides and energy metabolism, while it has a pronounced negative dose–response relation (down-regulation) with pathways like viral infectious diseases and folding, sorting, and degradation.

3.5. Associations between Functional Alterations and Gut Microbiome

The association analysis between functional alterations and the gut microbiome presented correlations ranging from −0.69 to 0.65. Significant positive correlations were found in Genus CAG3/Species CAG1 with MetaCyc pathways of level 2, such as metabolism of terpenoids and polyketides (Spearman's correlation 0.60/0.62) and energy metabolism (0.49/0.57), and with MetaCyc pathways of level 3, such as metabolism (0.58/0.64). Meanwhile, a negative correlation was observed with MetaCyc pathways of level 2, such as drug resistance: antimicrobial (−0.67/−0.59) and infectious diseases: viral (−0.65/−0.62), and with MetaCyc pathways of level 3, such as human diseases (−0.59/−0.50) (Figure 6).

For the genus CAG4, significant positive correlations were found with MetaCyc pathways of level 2, such as infectious diseases: viral (Spearman's correlation 0.60), drug resistance: antineoplastic (0.41), and drug resistance: antimicrobial (0.40), while a negative correlation was observed with MetaCyc pathways of level 2, such as energy metabolism (−0.69) and metabolism of terpenoids and polyketides (−0.68), and with MetaCyc pathways of level 3, such as metabolism (−0.79) (Figure 6).

In all, the three fluoride exposure-associated CAGs were significantly correlated with the metabolism of terpenoids and polyketides and thallium, energy metabolism, drug resistance: antimicrobial, and infectious diseases: viral (Figure 6A).

4. Discussion

In this study, we investigated the impact of varying concentrations of fluoride exposure on alterations in the gut microbial community and its functional changes. The research demonstrates significant dose–response changes in the composition of the gut microbiome in rats due to fluoride exposure. Moreover, these disturbed intestinal bacteria are closely associated with alterations in many gut microbiota-related functions, indicating that fluoride exposure not only interferes with the bacterial abundance but also significantly alters the functional characteristics of the gut microbiota, leading to a disruption in the host's homeostasis post exposure.

The gut microbiota, including micro-organisms, their genomes, and the surrounding environment in the gut, has received unprecedented attention in the past decade [20]. Increasing evidence suggests that metabolic activities within the gut microbiota are intricately linked to human health and disease [21]. Several crucial functions executed by the gut microbiota are now widely recognized, encompassing polysaccharide digestion, biosynthesis of vitamins and nutrients, colonization resistance, and modulation of the immune system [21–25]. Moreover, the influence of the gut microbiota on the host's metabolism and physiology extends beyond the intestine, impacting distant organs such as the liver, muscles, and brain [21,25]. Different gut bacteria have certain functions. For instance, *Bilophila* is conditionally pathogenic in healthy human hosts, which promotes intestinal inflammatory effects, intestinal barrier defects, systemic inflammation, bile acid dysmetabolism, and changes in the functional profile of the microbiome [26]. *Turicibacter* strains lead to changes in host bile acid profiles, and were positioned as modulators of host fat biology [27]. Roseburia has been proved to be a probiotic that can prevent intestinal inflammation and maintain energy homeostasis by producing metabolites [28]. The composition and function of the gut microbiota can be easily influenced by various intrinsic and extrinsic factors [29], and exposure to exogenous pollutants can lead to functional dysregulation of the gut microbiota [24,25,29]. More research is suggesting that the detrimental functional alterations in the gut microbiota due to exposure to these external pollutants might be associated with increased disease risk [21].

Numerous studies highlight that excessive fluoride exposure has adverse effects on human health, primarily resulting in dental and skeletal fluorosis [2] as well as unfavorable impacts on human reproductive development, the liver, kidney, endocrine system, nervous system, and genetics [2–4]. In recent years, there has been growing attention on soft tissue lesions associated with excessive fluoride exposure [30]. Studies have shown that fluoride metabolism is five times faster in rodents than in humans [31]. The fluoride doses used in the current study (25, 50, 100, and 150 mg/L NaF corresponding to 11.3, 22.6, 45.2, and 67.8 mg/L fluoride ion, respectively) are equal to environmental levels. Although the World Health Organization sets the guideline limit for fluoride in drinking water at 1.5 mg/L [32], the levels of fluoride in drinking water in endemic fluorosis areas are up to 16 mg/L [33]. Moreover, fluoride can be ingested from food and the air. Thus, the doses in this study were supposed to mimic the real human exposure in areas of endemic fluorosis. Research indicates that excessive fluoride exposure can impair the organism's intestinal morphology and ultrastructure, leading to gastrointestinal diseases [12,34,35]. For instance, after exposure to 100 mg/L sodium fluoride for 90 days, C57BL/6J mice exhibited disrupted small intestine tissue structure and ultrastructural disorder, with a marked decrease in the ratio of villus height to crypt depth [36]. Exposure to 50 mg/L and 100 mg/L fluoride (calculated as fluoride ion) for 70 days resulted in severe structural damage to the rats' colon and rectum, with a significant inhibition of epithelial cell proliferation [12,13]. However, the current research on the effects of fluoride exposure on the gut microbiota remains relatively limited [7,11,15]. Studies have found that environmental chemicals can directly impact gut bacteria by interrupting specific metabolic pathways or gene expressions, leading to differential selective pressures. This, in turn, reshapes the gut microbial community due to the unique set of metabolic pathways and genomes possessed by different bacterial species [37]. (Some environmental chemicals can directly affect the gut bacteria by interrupting specific metabolic pathways or gene expression, leading to distinct selection pressures, hence shaping the gut microbial community due to the uniqueness of the set of metabolic pathways and genome possessed by different bacterial species.) The potential mechanism behind the impact of fluoride exposure on the intestinal microbiota is likely related to the changes observed; however, the specific mechanisms remain unclear and warrant further research.

The current study indicates that exposure to fluoride induces significant dose-responsive alterations in the composition of the intestinal microbiome. Notably, different fluoride concentrations showed a positive correlation with CAG3 in different genera, with the genera *Bilophila, Holdemania, Pelagibacterium,* and *Ruminococcaceae* significantly positively correlated with fluoride concentrations. A plethora of studies demonstrate that there is a marked increase in the genus *Bilophila* among patients with diabetic nephropathy [38–40], IgA nephropathy [41], intrahepatic cholestasis of pregnancy [42], autism-related *Bilophila* abundance increase [43], and ovarian premature aging mice [44]. Additionally, an increase in both the *Bilophila* and *Ruminococcaceae* genera affects the reproductive performance of sows [45]. Chronic hepatitis B patients showed a correlation between an increase in the genus *Holdemania* and a heightened risk of chronic hepatitis B [46]. Elevated blood urea nitrogen levels in children with idiopathic nephrotic syndrome positively correlate with the *Ruminococcaceae* genus [47]. Moreover, patients with conditions such as diabetic nephropathy [38], liver–kidney transplantation [48], hepatitis C virus (HCV) infection [49], non-alcoholic fatty liver [50], liver fibrosis [51], polycystic ovary syndrome [52], thyroid cancer [53], Parkinson's disease [54], autism spectrum disorder [55], multiple sclerosis [56], and spinal cord injury [57] all show a significant rise in the relative abundance of the *Ruminococcaceae* genus. After exposure to perfluorooctanesulfonic acid, there was a trend of increased abundance of *Ruminococcaceae,* but only the high-dose group was significantly higher than the control group [58]. This suggests that fluoride exposure-induced augmentation of the aforementioned intestinal microbes might further be associated with pathological changes in the kidneys, liver, and nervous system, reproductive development, and the thyroid.

In contrast, different fluoride concentrations were negatively correlated with CAG4 in different genera. The genera *Corynebacterium, Lachnospiracea incertae sedis, Roseburia,* and *Turicibacter* showed a significant negative correlation with fluoride concentrations. Recent studies revealed a significant decrease in the relative abundance of *Corynebacterium* in depression model rats compared to controls [59]. Patients with conditions such as polycystic ovary syndrome [60], IgA nephropathy [61], Parkinson's disease [54], and non-alcoholic fatty liver [62] all display a marked reduction in the abundance of *Lachnospiracea incertae sedis*. Studies suggest that as the abundance of *Lachnospiracea incertae sedis* increases, levels of indolyl sulfate, formyl sulfate, and phenylacetylglutamine in circulation decrease, correlating with improved renal function [63]. Kidney diseases, including diabetic kidney disease (DKD) [64], chronic kidney disease (CKD) [65], and end-stage renal disease (ESRD) [66], are associated with a notable decrease in the Roseburia genus, potentially linked to its involvement in butyrate production, indole synthesis, and mucin degradation [67]. Autoimmune diseases show a marked reduction in *Roseburia* abundance in autoimmune disease mice [68], and patients with IgA nephropathy also exhibit a significant reduction in *Roseburia* compared to healthy individuals [62]. Neurological conditions such as schizophrenia [69], Parkinson's disease [70], and spinal cord injury [71], as well as reproductive disorders like infertility, show a decrease in the number of *Roseburia* [72]. Compared to controls, benign prostatic hyperplasia rat groups displayed a marked reduction in *Turicibacter* [73], as did unilateral ureteral obstruction (UUO) rats [74]. This implies that fluoride exposure-induced diminishment of the aforementioned intestinal microbes might further correlate with pathologies in autoimmunity, the kidneys, liver, and nervous system, and reproductive development. The current study indicated that fluoride exposure causes a certain up-regulation of metabolic pathways like metabolism of terpenoids and polyketides and energy metabolism, while it causes a certain down-regulation of pathways like viral infectious diseases and folding, sorting, and degradation. Concurrently, strong correlations exist between these metabolic pathways and CAGs. In summation, these functional and metabolic pathway changes may represent potential mechanisms by which fluoride exerts toxicity by affecting the intestinal microbiota. In summation, our results suggest that the bacteria sensitive to fluoride in a fecal sample could be a biomarker of fluoride exposure, and these functional and metabolic pathway changes may be used to further explore the mechanism of bacteria in the pathogenesis of fluorosis.

The concept of intestinal microbiota toxicity has been increasingly recognized. Intestinal microbiota toxicity refers to the structural and functional changes in the gut microbial community caused by exposure to certain environmental chemicals, potentially leading to a series of adverse health consequences [75]. Environmental exposure is a significant risk factor for a range of human diseases that overlap with those associated with gut microbial composition [76,77]. Consequently, microbial community toxicity may represent the missing link between environmental exposure and microbial community-associated human diseases [75]. At present, this has not received adequate attention, and the dose–response relationship of gut microbiota toxicity has largely remained unexplored. In this study, we investigated the toxicity of fluoride on the gut microbiota and further explored the dose-response relationship between them.

5. Conclusions

We analyzed the effects of fluoride exposure at various concentrations on the gut microbial community and its functions in rats. Exposure to different concentrations of fluoride significantly altered the composition of the gut microbiota, showing a dose–response relationship with changes in the abundance levels of multiple bacterial genera. The functions and metabolic pathways of the gut microbiota experienced substantial alterations post fluoride exposure, also displaying a dose–response relationship. Moreover, correlation analyses identified that some gut bacterial CAGs (co-abundance groups) were highly associated with the altered metabolic pathways of the gut microbiota. In summary, fluoride exposure not only disrupted the gut microbiota composition in terms of abundance but also affected its functions and metabolic pathways. This study explored the pathogenesis of fluorosis and provided a clue for the biomarkers of fluoride exposure.

Supplementary Materials: The following supporting information can be downloaded at https://www.mdpi.com/article/10.3390/metabo13111159/s1: Table S1: Correlation coefficient between gut microbiome in genera and fluoride exposure; Table S2: Correlation coefficient between gut microbiome in species and fluoride exposure; Table S3: Correlation coefficient between functional alterations and fluoride exposure.

Author Contributions: Conceptualization: J.P., X.W., Z.M. and J.W.; Experiment: X.M. (Xinyue Meng), A.L., Z.L., W.Q. and T.W.; Data Analysis: Z.M., X.M. (Xu Ma) and Y.L.; Writing—Original Draft: Z.M., S.Y., L.W., R.Z. and K.F.T. All authors have read and agreed to the published version of the manuscript.

Funding: This research was funded by the National Natural Science Foundation of China (82273749 and 81773468), the Natural Science Foundation of Zhejiang Province, China (LTGY23H240001), and the Opening Foundation of NHC Key Laboratory of Etiology and Epidemiology (Harbin Medical University) (NHCKLEE20230908).

Institutional Review Board Statement: This study was approved by the Ethics Committee of the Endemic Disease Control Center of Harbin Medical University (3 March 2017).

Informed Consent Statement: Not applicable.

Data Availability Statement: Data available on request due to restrictions eg privacy or ethica.

Conflicts of Interest: The authors declare no conflict of interest.

References

1. Pramanik, S.; Saha, D. The genetic influence in fluorosis. *Environ. Toxicol. Pharmacol.* **2017**, *56*, 157–162. [CrossRef] [PubMed]
2. Ozsvath, D.L. Fluoride and environmental health: A review. *Rev. Environ. Sci. Bio./Technol.* **2009**, *8*, 59–79. [CrossRef]
3. Petrone, P.; Giordano, M.; Giustino, S.; Guarino, F.M. Enduring fluoride health hazard for the Vesuvius area population: The case of AD 79 Herculaneum. *PLoS ONE* **2011**, *6*, e21085. [CrossRef] [PubMed]
4. Patil, M.M.; Lakhkar, B.B.; Patil, S.S. Curse of Fluorosis. *Indian J. Pediatr.* **2018**, *85*, 375–383. [CrossRef] [PubMed]
5. Meenakshi; Maheshwari, R.C. Fluoride in drinking water and its removal. *J. Hazard. Mater.* **2006**, *137*, 456–463. [CrossRef] [PubMed]
6. Guan, Z.; Wang, L.; Sun, D. Endemic Fluorosis. In *Endemic Disease in China*; Sun, D., Ed.; Springer: Singapore, 2019; pp. 61–96.
7. Yasuda, K.; Hsu, T.; Gallini, C.A.; McLver, L.J.; Schwager, E.; Shi, A.; DuLong, C.R.; Schwager, R.N.; Abu-Ali, G.S.; Franzosa, E.A.; et al. Fluoride Depletes Acidogenic Taxa in Oral but Not Gut Microbial Communities in Mice. *mSystems* **2017**, *2*, e00047-17. [CrossRef]
8. Zuo, H.; Chen, L.; Kong, M.; Qiu, L.; Lü, P.; Wu, P.; Yang, Y.; Chen, K. Toxic effects of fluoride on organisms. *Life Sci.* **2018**, *198*, 18–24. [CrossRef]
9. Wang, F.; Li, C.; Qin, Y.; Han, X.; Gao, J.; Zhang, A.; Luo, P.; Pan, X. Analysis of the microRNA Profile of Coal-Burning Endemic Fluorosis Using Deep Sequencing and Bioinformatic Approaches. *Bull. Environ. Contam. Toxicol.* **2019**, *103*, 56–63. [CrossRef]
10. Lu, K.; Abo, R.P.; Schlieper, K.A.; Graffam, M.E.; Levine, S.; Wishnok, J.S.; Swenberg, J.A.; Tannenbaum, S.R.; Fox, J.G. Arsenic exposure perturbs the gut microbiome and its metabolic profile in mice: An integrated metagenomics and metabolomics analysis. *Environ. Health Perspect.* **2014**, *122*, 284–291. [CrossRef] [PubMed]
11. Liu, J.; Wang, H.W.; Lin, L.; Miao, C.Y.; Zhang, Y.; Zhou, B.H. Intestinal barrier damage involved in intestinal microflora changes in fluoride-induced mice. *Chemosphere* **2019**, *234*, 409–418. [CrossRef]
12. Fu, R.; Niu, R.; Li, R.; Yue, B.; Zhang, X.; Cao, Q.; Wang, J.; Sun, Z. Fluoride-Induced Alteration in the Diversity and Composition of Bacterial Microbiota in Mice Colon. *Biol. Trace Elem. Res.* **2020**, *196*, 537–544. [CrossRef]
13. Wang, H.W.; Miao, C.Y.; Liu, J.; Zhang, Y.; Zhu, S.Q.; Zhou, B.H. Fluoride-induced rectal barrier damage and microflora disorder in mice. *Environ. Sci. Pollut. Res. Int.* **2020**, *27*, 7596–7607. [CrossRef] [PubMed]
14. Chelakkot, C.; Ghim, J.; Ryu, S.H. Mechanisms regulating intestinal barrier integrity and its pathological implications. *Exp. Mol. Med.* **2018**, *50*, 1–9. [CrossRef] [PubMed]
15. Wu, H. *Study on the Characteristics of Gut Microbiota for Children with Dental Fluorosis in Drinking Water-Born Endemic Fluorosis Areas*; Zhengzhou University: Zhengzhou, China, 2019.
16. Nguyen, T.L.; Vieira-Silva, S.; Liston, A.; Raes, J. How informative is the mouse for human gut microbiota research? *Dis. Model Mech.* **2015**, *8*, 1–16. [CrossRef]
17. Zhong, N.; Ma, Y.; Meng, X.; Sowanou, A.; Wu, L.; Huang, W.; Gao, Y.; Pei, J. Effect of Fluoride in Drinking Water on Fecal Microbial Community in Rats. *Biol. Trace Elem. Res.* **2022**, *200*, 238–246. [CrossRef]
18. Wemheuer, F.; Taylor, J.A.; Daniel, R.; Johnston, E.; Meinicke, P.; Thomas, T.; Wemheuer, B. Tax4Fun2: Prediction of habitat-specific functional profiles and functional redundancy based on 16S rRNA gene sequences. *Environ. Microbiome* **2020**, *15*, 11. [CrossRef]
19. Liu, H.; Li, F.Y.; Liu, J.; Shi, C.; Tang, K.; Yang, Q.; Liu, Y.; Fu, Q.; Gao, X.; Wang, N.; et al. The reciprocal changes in dominant species with complete metabolic functions explain the decoupling phenomenon of microbial taxonomic and functional composition in a grassland. *Front. Microbiol.* **2023**, *14*, 1113157. [CrossRef]
20. Marchesi, J.R.; Ravel, J. The vocabulary of microbiome research: A proposal. *Microbiome* **2015**, *3*, 31. [CrossRef]

21. Nicholson, J.K.; Holmes, E.; Kinross, J.; Burcelin, R.; Gibson, G.; Jia, W.; Pettersson, S. Host-gut microbiota metabolic interactions. *Science* **2012**, *336*, 1262–1267. [CrossRef]
22. Sharon, G.; Garg, N.; Debelius, J.; Knight, R.; Dorrestein, P.C.; Mazmanian, S.K. Specialized metabolites from the microbiome in health and disease. *Cell. Metab.* **2014**, *20*, 719–730. [CrossRef] [PubMed]
23. Buffie, C.G.; Bucci, V.; Stein, R.R.; McKenney, P.T.; Ling, L.; Gobourne, A.; No, D.; Liu, H.; Kinnebrew, M.; Viale, A.; et al. Precision microbiome reconstitution restores bile acid mediated resistance to Clostridium difficile. *Nature* **2015**, *517*, 205–208. [CrossRef] [PubMed]
24. Koh, A.; De Vadder, F.; Kovatcheva-Datchary, P.; Bäckhed, F. From Dietary Fiber to Host Physiology: Short-Chain Fatty Acids as Key Bacterial Metabolites. *Cell* **2016**, *165*, 1332–1345. [CrossRef] [PubMed]
25. Schroeder, B.O.; Bäckhed, F. Signals from the gut microbiota to distant organs in physiology and disease. *Nat. Med.* **2016**, *22*, 1079–1089. [CrossRef]
26. Natividad, J.M.; Lamas, B.; Pham, H.P.; Michel, M.L.; Rainteau, D.; Bridonneau, C.; da Costa, G.; van Hylckama Vlieg, J.; Sovran, B.; Chamignon, C.; et al. Bilophila wadsworthia aggravates high fat diet induced metabolic dysfunctions in mice. *Nat. Commun.* **2018**, *9*, 2802. [CrossRef]
27. Lynch, J.B.; Gonzalez, E.L.; Choy, K.; Faull, K.F.; Jewell, T.; Arellano, A.; Liang, J.; Yu, K.B.; Paramo, J.; Hsiao, E.Y. Gut microbiota Turicibacter strains differentially modify bile acids and host lipids. *Nat. Commun.* **2023**, *14*, 3669. [CrossRef]
28. Nie, K.; Ma, K.; Luo, W.; Shen, Z.; Yang, Z.; Xiao, M.; Tong, T.; Yang, Y.; Wang, X. Roseburia intestinalis: A Beneficial Gut Organism From the Discoveries in Genus and Species. *Front. Cell. Infect. Microbiol.* **2021**, *11*, 757718. [CrossRef]
29. Schmidt, T.S.B.; Raes, J.; Bork, P. The Human Gut Microbiome: From Association to Modulation. *Cell* **2018**, *172*, 1198–1215. [CrossRef]
30. Ghosh, A.; Mukherjee, K.; Ghosh, S.K.; Saha, B. Sources and toxicity of fluoride in the environment. *Res. Chem. Intermed.* **2013**, *39*, 2881–2915. [CrossRef]
31. Mumtaz, N.; Pandey, G.; Labhasetwar, P.K. Global Fluoride Occurrence, Available Technologies for Fluoride Removal and Electrolytic Defluoridation: A Review. *Crit. Rev. Environ. Sci. Technol.* **2015**, *45*, 2357–2389. [CrossRef]
32. JK, F. *Fluoride in Drinking-Water*; World Health Organization: Geneva, Switzerland, 2006.
33. Godebo, T.R.; Jeuland, M.; Tekle-Haimanot, R.; Shankar, A.; Alemayehu, B.; Assefa, G.; Whitford, G.; Wolfe, A. Bone quality in fluoride-exposed populations: A novel application of the ultrasonic method. *Bone Rep.* **2020**, *12*, 100235. [CrossRef]
34. Miao, L.; Gong, Y.; Li, H.; Xie, C.; Xu, Q.; Dong, X.; Elwan, H.A.M.; Zou, X. Alterations in cecal microbiota and intestinal barrier function of laying hens fed on fluoride supplemented diets. *Ecotoxicol. Environ. Saf.* **2020**, *193*, 110372. [CrossRef]
35. Ferreira, M.K.M.; Aragão, W.A.B.; Bittencourt, L.O.; Puty, B.; Dionizio, A.; Souza, M.P.C.; Buzalaf, M.A.R.; de Oliveira, E.H.; Crespo-Lopez, M.E.; Lima, R.R. Fluoride exposure during pregnancy and lactation triggers oxidative stress and molecular changes in hippocampus of offspring rats. *Ecotoxicol. Environ. Saf.* **2021**, *208*, 111437. [CrossRef]
36. Li, M.; Wang, J.; Wu, P.; Manthari, R.K.; Zhao, Y.; Li, W.; Wang, J. Self-recovery study of the adverse effects of fluoride on small intestine: Involvement of pyroptosis induced inflammation. *Sci. Total Environ.* **2020**, *742*, 140533. [CrossRef]
37. Koppel, N.; Maini Rekdal, V.; Balskus, E.P. Chemical transformation of xenobiotics by the human gut microbiota. *Science* **2017**, *356*, eaag2770. [CrossRef]
38. Chen, R.; Zhu, D.; Yang, R.; Wu, Z.; Xu, N.; Chen, F.; Zhang, S.; Chen, H.; Li, M.; Hou, K. Gut microbiota diversity in middle-aged and elderly patients with end-stage diabetic kidney disease. *Ann. Transl. Med.* **2022**, *10*, 750. [CrossRef]
39. Han, S.; Chen, M.; Cheng, P.; Zhang, Z.; Lu, Y.; Xu, Y.; Wang, Y. A systematic review and meta-analysis of gut microbiota in diabetic kidney disease: Comparisons with diabetes mellitus, non-diabetic kidney disease, and healthy individuals. *Front. Endocrinol.* **2022**, *13*, 1018093. [CrossRef]
40. He, X.; Sun, J.; Liu, C.; Yu, X.; Li, H.; Zhang, W.; Li, Y.; Geng, Y.; Wang, Z. Compositional Alterations of Gut Microbiota in Patients with Diabetic Kidney Disease and Type 2 Diabetes Mellitus. *Diabetes Metab. Syndr. Obes.* **2022**, *15*, 755–765. [CrossRef] [PubMed]
41. Dong, R.; Bai, M.; Zhao, J.; Wang, D.; Ning, X.; Sun, S. A Comparative Study of the Gut Microbiota Associated With Immunoglobulin a Nephropathy and Membranous Nephropathy. *Front. Cell. Infect. Microbiol.* **2020**, *10*, 557368. [CrossRef] [PubMed]
42. Li, G.H.; Huang, S.J.; Li, X.; Liu, X.S.; Du, Q.L. Response of gut microbiota to serum metabolome changes in intrahepatic cholestasis of pregnant patients. *World J. Gastroenterol.* **2020**, *26*, 7338–7351. [CrossRef] [PubMed]
43. Alamoudi, M.U.; Hosie, S.; Shindler, A.E.; Wood, J.L.; Franks, A.E.; Hill-Yardin, E.L. Comparing the Gut Microbiome in Autism and Preclinical Models: A Systematic Review. *Front. Cell. Infect. Microbiol.* **2022**, *12*, 905841. [CrossRef] [PubMed]
44. Zheng, H.; Liang, X.; Zhou, H.; Zhou, T.; Liu, X.; Duan, J.; Duan, J.A.; Zhu, Y. Integrated gut microbiota and fecal metabolome analyses of the effect of Lycium barbarum polysaccharide on D-galactose-induced premature ovarian insufficiency. *Food Funct.* **2023**, *14*, 7209–7221. [CrossRef] [PubMed]
45. Yuan, X.; Yan, J.; Hu, R.; Li, Y.; Wang, Y.; Chen, H.; Hou, D.X.; He, J.; Wu, S. Modulation of Gut Microbiota and Oxidative Status by β-Carotene in Late Pregnant Sows. *Front. Nutr.* **2020**, *7*, 612875. [CrossRef] [PubMed]
46. Zhang, Q.; Zhou, J.; Zhang, X.; Mao, R.; Zhang, C. Mendelian randomization supports causality between gut microbiota and chronic hepatitis B. *Front. Microbiol.* **2023**, *14*, 1243811. [CrossRef] [PubMed]
47. He, H.; Lin, M.; You, L.; Chen, T.; Liang, Z.; Li, D.; Xie, C.; Xiao, G.; Ye, P.; Kong, Y.; et al. Gut Microbiota Profile in Adult Patients with Idiopathic Nephrotic Syndrome. *Biomed. Res. Int.* **2021**, *2021*, 8854969. [CrossRef] [PubMed]

48. Sivaraj, S.; Chan, A.; Pasini, E.; Chen, E.; Lawendy, B.; Verna, E.; Watt, K.; Bhat, M. Enteric dysbiosis in liver and kidney transplant recipients: A systematic review. *Transpl. Int.* **2020**, *33*, 1163–1176. [CrossRef] [PubMed]
49. Sultan, S.; El-Mowafy, M.; Elgaml, A.; El-Mesery, M.; El Shabrawi, A.; Elegezy, M.; Hammami, R.; Mottawea, W. Alterations of the Treatment-Naive Gut Microbiome in Newly Diagnosed Hepatitis C Virus Infection. *ACS Infect. Dis.* **2021**, *7*, 1059–1068. [CrossRef] [PubMed]
50. Lee, G.; You, H.J.; Bajaj, J.S.; Joo, S.K.; Yu, J.; Park, S.; Kang, H.; Park, J.H.; Kim, J.H.; Lee, D.H.; et al. Distinct signatures of gut microbiome and metabolites associated with significant fibrosis in non-obese NAFLD. *Nat. Commun.* **2020**, *11*, 4982. [CrossRef]
51. Martin, H.R.; Sales Martinez, S.; Stebliankin, V.; Tamargo, J.A.; Campa, A.; Narasimhan, G.; Hernandez, J.; Rodriguez, J.A.B.; Teeman, C.; Johnson, A.; et al. Diet Quality and Liver Health in People Living with HIV in the MASH Cohort: A Multi-Omic Analysis of the Fecal Microbiome and Metabolome. *Metabolites* **2023**, *13*, 271. [CrossRef]
52. Eyupoglu, N.D.; Ergunay, K.; Acikgoz, A.; Akyon, Y.; Yilmaz, E.; Yildiz, B.O. Gut Microbiota and Oral Contraceptive Use in Overweight and Obese Patients with Polycystic Ovary Syndrome. *J. Clin. Endocrinol. Metab.* **2020**, *105*, e4792–e4800. [CrossRef]
53. Ishaq, H.M.; Mohammad, I.S.; Hussain, R.; Parveen, R.; Shirazi, J.H.; Fan, Y.; Shahzad, M.; Hayat, K.; Li, H.; Ihsan, A.; et al. Gut-Thyroid axis: How gut microbial dysbiosis associated with euthyroid thyroid cancer. *J. Cancer* **2022**, *13*, 2014–2028. [CrossRef]
54. Shen, T.; Yue, Y.; He, T.; Huang, C.; Qu, B.; Lv, W.; Lai, H.Y. The Association Between the Gut Microbiota and Parkinson's Disease, a Meta-Analysis. *Front. Aging Neurosci.* **2021**, *13*, 636545. [CrossRef] [PubMed]
55. Sun, H.; You, Z.; Jia, L.; Wang, F. Autism spectrum disorder is associated with gut microbiota disorder in children. *BMC Pediatr.* **2019**, *19*, 516. [CrossRef]
56. Galluzzo, P.; Capri, F.C.; Vecchioni, L.; Realmuto, S.; Scalisi, L.; Cottone, S.; Nuzzo, D.; Alduina, R. Comparison of the Intestinal Microbiome of Italian Patients with Multiple Sclerosis and Their Household Relatives. *Life* **2021**, *11*, 620. [CrossRef] [PubMed]
57. Pang, R.; Wang, J.; Xiong, Y.; Liu, J.; Ma, X.; Gou, X.; He, X.; Cheng, C.; Wang, W.; Zheng, J.; et al. Relationship between gut microbiota and lymphocyte subsets in Chinese Han patients with spinal cord injury. *Front. Microbiol.* **2022**, *13*, 986480. [CrossRef] [PubMed]
58. Wang, G.; Sun, S.; Wu, X.; Yang, S.; Wu, Y.; Zhao, J.; Zhang, H.; Chen, W. Intestinal environmental disorders associate with the tissue damages induced by perfluorooctane sulfonate exposure. *Ecotoxicol. Environ. Saf.* **2020**, *197*, 110590. [CrossRef]
59. Yu, M.; Jia, H.; Zhou, C.; Yang, Y.; Zhao, Y.; Yang, M.; Zou, Z. Variations in gut microbiota and fecal metabolic phenotype associated with depression by 16S rRNA gene sequencing and LC/MS-based metabolomics. *J. Pharm. Biomed. Anal.* **2017**, *138*, 231–239. [CrossRef]
60. Wang, Q.; Wang, Q.; Zhao, L.; Bin, Y.; Wang, L.; Wang, L.; Zhang, K.; Li, Q. Blood Bacterial 16S rRNA Gene Alterations in Women With Polycystic Ovary Syndrome. *Front. Endocrinol.* **2022**, *13*, 814520. [CrossRef]
61. Han, S.; Shang, L.; Lu, Y.; Wang, Y. Gut Microbiome Characteristics in IgA Nephropathy: Qualitative and Quantitative Analysis from Observational Studies. *Front. Cell. Infect. Microbiol.* **2022**, *12*, 904401. [CrossRef]
62. Kim, H.N.; Joo, E.J.; Cheong, H.S.; Kim, Y.; Kim, H.L.; Shin, H.; Chang, Y.; Ryu, S. Gut Microbiota and Risk of Persistent Nonalcoholic Fatty Liver Diseases. *J. Clin. Med.* **2019**, *8*, 1089. [CrossRef]
63. Barrios, C.; Beaumont, M.; Pallister, T.; Villar, J.; Goodrich, J.K.; Clark, A.; Pascual, J.; Ley, R.E.; Spector, T.D.; Bell, J.T.; et al. Gut-Microbiota-Metabolite Axis in Early Renal Function Decline. *PLoS ONE* **2015**, *10*, e0134311. [CrossRef]
64. Wang, Y.; Zhao, J.; Qin, Y.; Yu, Z.; Zhang, Y.; Ning, X.; Sun, S. The Specific Alteration of Gut Microbiota in Diabetic Kidney Diseases-A Systematic Review and Meta-Analysis. *Front. Immunol.* **2022**, *13*, 908219. [CrossRef] [PubMed]
65. Voroneanu, L.; Burlacu, A.; Brinza, C.; Covic, A.; Balan, G.G.; Nistor, I.; Popa, C.; Hogas, S.; Covic, A. Gut Microbiota in Chronic Kidney Disease: From Composition to Modulation towards Better Outcomes-A Systematic Review. *J. Clin. Med.* **2023**, *12*, 1948. [CrossRef]
66. Jiang, S.; Xie, S.; Lv, D.; Wang, P.; He, H.; Zhang, T.; Zhou, Y.; Lin, Q.; Zhou, H.; Jiang, J.; et al. Alteration of the gut microbiota in Chinese population with chronic kidney disease. *Sci. Rep.* **2017**, *7*, 2870. [CrossRef]
67. Lohia, S.; Vlahou, A.; Zoidakis, J. Microbiome in Chronic Kidney Disease (CKD): An Omics Perspective. *Toxins* **2022**, *14*, 176. [CrossRef] [PubMed]
68. Wang, T.; Sternes, P.R.; Guo, X.K.; Zhao, H.; Xu, C.; Xu, H. Autoimmune diseases exhibit shared alterations in the gut microbiota. *Rheumatology* **2023**, kead364. [CrossRef] [PubMed]
69. Li, S.; Song, J.; Ke, P.; Kong, L.; Lei, B.; Zhou, J.; Huang, Y.; Li, H.; Li, G.; Chen, J.; et al. The gut microbiome is associated with brain structure and function in schizophrenia. *Sci. Rep.* **2021**, *11*, 9743. [CrossRef]
70. Kleine Bardenhorst, S.; Cereda, E.; Severgnini, M.; Barichella, M.; Pezzoli, G.; Keshavarzian, A.; Desideri, A.; Pietrucci, D.; Aho, V.T.E.; Scheperjans, F.; et al. Gut microbiota dysbiosis in Parkinson disease: A systematic review and pooled analysis. *Eur. J. Neurol.* **2023**, *30*, 3581–3594. [CrossRef]
71. Gungor, B.; Adiguzel, E.; Gursel, I.; Yilmaz, B.; Gursel, M. Intestinal Microbiota in Patients with Spinal Cord Injury. *PLoS ONE* **2016**, *11*, e0145878. [CrossRef]
72. Komiya, S.; Naito, Y.; Okada, H.; Matsuo, Y.; Hirota, K.; Takagi, T.; Mizushima, K.; Inoue, R.; Abe, A.; Morimoto, Y. Characterizing the gut microbiota in females with infertility and preliminary results of a water-soluble dietary fiber intervention study. *J. Clin. Biochem. Nutr.* **2020**, *67*, 105–111. [CrossRef]
73. Li, L.Y.; Han, J.; Wu, L.; Fang, C.; Li, W.G.; Gu, J.M.; Deng, T.; Qin, C.J.; Nie, J.Y.; Zeng, X.T. Alterations of gut microbiota diversity, composition and metabonomics in testosterone-induced benign prostatic hyperplasia rats. *Mil. Med. Res.* **2022**, *9*, 12. [CrossRef]

74. Chen, L.; Chen, D.Q.; Liu, J.R.; Zhang, J.; Vaziri, N.D.; Zhuang, S.; Chen, H.; Feng, Y.L.; Guo, Y.; Zhao, Y.Y. Unilateral ureteral obstruction causes gut microbial dysbiosis and metabolome disorders contributing to tubulointerstitial fibrosis. *Exp. Mol. Med.* **2019**, *51*, 1–18. [CrossRef] [PubMed]
75. Tu, P.; Chi, L.; Bodnar, W.; Zhang, Z.; Gao, B.; Bian, X.; Stewart, J.; Fry, R.; Lu, K. Gut Microbiome Toxicity: Connecting the Environment and Gut Microbiome-Associated Diseases. *Toxics* **2020**, *8*, 19. [CrossRef] [PubMed]
76. Thayer, K.A.; Heindel, J.J.; Bucher, J.R.; Gallo, M.A. Role of environmental chemicals in diabetes and obesity: A National Toxicology Program workshop review. *Environ. Health Perspect.* **2012**, *120*, 779–789. [CrossRef] [PubMed]
77. Ananthakrishnan, A.N. Epidemiology and risk factors for IBD. *Nat. Rev. Gastroenterol. Hepatol.* **2015**, *12*, 205–217. [CrossRef]

metabolites

MDPI

Article

Deciphering Gut Microbiome Responses upon Microplastic Exposure via Integrating Metagenomics and Activity-Based Metabolomics

Pengcheng Tu [1,†], Jingchuan Xue [2,†], Huixia Niu [1,3], Qiong Tang [4], Zhe Mo [1], Xiaodong Zheng [5], Lizhi Wu [1], Zhijian Chen [1], Yanpeng Cai [2] and Xiaofeng Wang [1,*]

1 Department of Environmental Health, Zhejiang Provincial Center for Disease Control and Prevention, 3399 Binsheng Road, Hangzhou 310051, China
2 Guangdong Provincial Key Laboratory of Water Quality Improvement and Ecological Restoration for Watersheds, Institute of Environmental and Ecological Engineering, Guangdong University of Technology, Guangzhou 510006, China
3 School of Medicine, Ningbo University, Ningbo 315000, China
4 College of Standardization, China Jiliang University, Hangzhou 310018, China
5 Department of Food Science and Nutrition, Zhejiang University, Hangzhou 310058, China
* Correspondence: xfwang@cdc.zj.cn
† These authors contributed equally to this work.

Abstract: Perturbations of the gut microbiome are often intertwined with the onset and development of diverse metabolic diseases. It has been suggested that gut microbiome perturbation could be a potential mechanism through which environmental chemical exposure induces or exacerbates human diseases. Microplastic pollution, an emerging environmental issue, has received ever increasing attention in recent years. However, interactions between microplastic exposure and the gut microbiota remain elusive. This study aimed to decipher the responses of the gut microbiome upon microplastic polystyrene (MP) exposure by integrating 16S rRNA high-throughput sequencing with metabolomic profiling techniques using a C57BL/6 mouse model. The results indicated that MP exposure significantly perturbed aspects of the gut microbiota, including its composition, diversity, and functional pathways that are involved in xenobiotic metabolism. A distinct metabolite profile was observed in mice with MP exposure, which probably resulted from changes in gut bacterial composition. Specifically, untargeted metabolomics revealed that levels of metabolites associated with cholesterol metabolism, primary and secondary bile acid biosynthesis, and taurine and hypotaurine metabolism were changed significantly. Targeted approaches indicated significant perturbation with respect to the levels of short-chain fatty acids derived from the gut microbiota. This study can provide evidence for the missing link in understanding the mechanisms behind the toxic effects of microplastics.

Keywords: microplastic exposure; gut microbiome; metabolomics; toxicity

1. Introduction

It has been well demonstrated that the gut microbiota plays a key role in immune response [1,2], metabolic processes [3], epithelial homeostasis [4], etc. Mounting evidence has established significant associations between an imbalanced gut microbiota and various adverse health outcomes such as inflammation [5], obesity [6], diabetes [7,8], and cancer [9]. The gut microbiota evolves through several transitions during infancy and relatively stabilizes thereafter if no significant perturbations occur. However, it has been well recognized that exposure to numerous environmental chemicals, including heavy metals [10–14], pesticides [14,15], artificial sweeteners [15–17], and others [18], is able to alter the gut microbiome. In particular, environmentally driven perturbations in the gut microbiome may lead to adverse effects on the host. For example, arsenic exposure perturbs the gut microbiome and induces alterations of diverse microbiota-related metabolites, which is suggested as a novel mechanism underlying arsenic toxicity [13,19,20]. Likewise, nicotine-induced changes in gut microbial metabolic pathways and metabolites related to neurotransmitters might contribute to the neurotoxicity of nicotine [21]. The functional damage driven by environmental exposure in the gut microbiome has recently been proposed as gut microbiome toxicity [18].

Recently, microplastic pollution and its potential health effects have become an emerging global environmental health issue. Plastic particles with a size of less than 5 mm are usually called microplastics [2]. Widespread occurrence of microplastics has been reported in a variety of environmental matrices, including oceans, rivers, soil, and even table salt [1]. Microplastics can easily accumulate in the environment due to their ubiquity and persistence. Studies have shown that microplastics can be ingested by marine organisms and passed through the food chain [7,8]. Microplastics have been detected in a variety of food products such as bottled water, honey, beer, and canned fish [22]. Thus, similar to how humans can be exposed to organic contaminants such as bisphenol A, they can also be exposed to microplastics via a variety of exposure routes [23,24]. The resulting accumulation of microplastics in tissues may cause a variety of toxic effects, such as growth inhibition, energy deficiency, inflammation, oxidative stress, and metabolic abnormalities [7,8]. However, the potential mechanisms remain elusive. In addition, it has been recently reported that the gut microbiota can be perturbed by microplastic exposure. For instance, microplastics, including polystyrene and polyethylene particles, could affect gut microbial composition and induce adverse effects such as inflammation [25,26]. Likewise, the profile of metabolites such as bile acids in the gut microbiota could also be impacted upon exposure to microplastics [27,28]. The gut microbiota not only directly impacts intestinal homeostasis locally through microbial metabolic products, but it also triggers systemic effects on remote tissues/organs such as liver, adipose, or brain by producing metabolites that can act as signaling molecules [29,30]. Moreover, the role of the gut microbiota in chemical toxicity has been well recognized [18]. Given the profound role of the gut microbiome in human health coupled with extensive and constant exposure to microplastics, it is of necessity to elucidate effects of microplastic exposure on the gut microbiome.

The main goal of the present study is to elucidate the effect of MP exposure on the gut microbiota. Microplastics comprise a variety of particles, of which polystyrene is the most classic example of a microplastic that exists widely throughout the environment. Provided the ubiquitous existence of polystyrene coupled with the use of polystyrene as a representative microplastic in a number of studies [26–28,31,32], polystyrene (5 μm, 0.1 mg/day) microplastic particles were used in the present study as representative microplastics for exposure to the gut microbiota. Polystyrene is widely used as a raw material in a variety of consumer products, including food packaging, leading to its ubiquitous environmental occurrence [33]. Thus, microplastic polystyrene is often used in a number of studies to investigate microplastic toxicity [26,27,34]. MP at 5 μm is the smallest diameter of plastic debris found in marine habitats [35], and it is within the smaller size range of particles that have been known to be ingested by aquatic organisms [36]. The dose is chosen based on environmentally relevant concentrations of MP in the environment [37]. Two complementary omic approaches have been employed to achieve a comprehensive understanding of how the gut microbiome is impacted by MP exposure at an environmentally relevant level. The 16S rRNA gene sequencing technique has been used to identify bacteria at the species level perturbed by MP exposure, which has been used as a mainstay of sequence-based bacterial analysis for decades. The other technique, activity-based metabolomics, including both untargeted and targeted approaches, allows the identification of metabolites with significant change upon exposure. An untargeted approach enables the comprehensive comparison of metabolomes under different conditions, while a targeted approach can enable the highly sensitive analysis of specific gut-microbiome-derived metabolites, which is critical in understanding drivers of physiological activities related to the gut microbiome.

2. Materials and Methods

2.1. Chemicals

Five-micrometer green, fluorescent polystyrene microplastic particles were purchased from Tianjin BaseLine ChromTech Research Center (Tianjin, China), with an excitation wavelength of 488 nm and an emission wavelength of 520 nm.

2.2. Animals and Experimental Design

Twenty male specific-pathogen-free (SPF) C57BL/6 mice (~4 weeks old) were purchased from SLAC Laboratory Animal Co., Ltd (Shanghai, China). After 1 week of acclimation, mice were randomly assigned into 6 cages with 3 or 4 mice per cage. A total of 10 mice were marked as controls (Control group, $n = 10$), and the rest of them were marked as MP-treated mice (treatment group, $n = 10$). The mice were housed under a temperature of 22 °C and 40–70% humidity with a 12/12 h light/dark cycle. Mice from the treatment group were given MP particles (0.1 mg/day [27,37]) by oral gavage; the control mice were given equivalent volume of water also by oral gavage. After 6-week MP treatment, fecal samples were collected individually, and these were put into liquid nitrogen immediately and stored at −80 °C for further experiments. Before sacrifice, mice were euthanized in a carbon dioxide chamber after 12 h of fasting. Mouse tissues were collected and quickly frozen in liquid nitrogen and stored at −80 °C. The animal experiment was conducted at Laboratory Animal Research Center of Zhejiang Chinese Medical University with approval by the Animal Care and Use Committee of Zhejiang Chinese Medical University (No. 20201103-08). All experiments were in accordance with relevant guidelines. All mice were treated humanely with regard for alleviation of suffering.

2.3. 16S rRNA Gene Sequencing

Microbial DNA was extracted from mouse fecal samples ($n = 8$ per group, ~50 mg per mouse) using QIAamp DNA Stool Mini Kit (Qiagen, Germany), as per the manufacturer's instructions. Purified amplicons were pooled in equimolar and then paired-end sequenced using an Illumina MiSeq platform (Illumina, San Diego, CA, USA), as per the standard protocols by Majorbio Bio-Pharm Technology Co. Ltd. (Shanghai, China). The V4 region of the 16S rRNA gene was amplified by PCR with primers of 515 (5′-GTGCCAGCMGCCGCGGTAA) and 806 (5′- GGACTACHVGGGTWTCTAAT). PCR reaction conditions were 3 min at 95 °C, followed by 30 cycles of 45 s at 95 °C, 60 s at 50 °C and 90 s at 72 °C. Raw fastq files were quality-filtered by Trimmomatic and merged by FLASH. The taxonomy of each 16S rRNA gene sequence was analyzed by the RDP Classifier algorithm. OTUs were clustered using a 97% similarity cutoff with UPARSE (version 7.1). Tax4Fun, an open-source R package for pathway prediction, was used to profile functional genes of the gut microbiome, based on marker genes from the 16S rRNA gene sequencing data combined with the KEGG database of reference genomes [38].

2.4. Quantification of Fecal Short-Chain Fatty Acids (SCFAs)

Briefly, 50 mg of fecal samples ($n = 6$ per group) were mixed with an NaOH solution containing internal standards of SCFAs (acetate, butyrate, and propionate), and then homogenized for 2 min. The homogenates were centrifuged, and the supernatants were mixed with 1-propanol/pyridine (3:2, v/v) and propyl chloroformate for derivatization. Derivatives were extracted using hexane and transferred to an autosampler vial for injection. An Agilent 7820A GC-5977B MS system (Agilent Technologies, Santa Clara, CA, USA) using an HP-5 ms capillary column was applied for GC/MS analysis. The initial oven temperature was held at 50 °C for 2 min, then ramped to 70 °C at a rate of 10 °C min^{-1}, to 85 °C at a rate of 3 °C min^{-1}, to 110 °C at a rate of 5 °C min^{-1}, to 290 °C at a rate of 30 °C min^{-1}, and finally held at 290 °C for 8 min. In addition, the temperatures of the front inlet, transfer line, and electron impact (EI) ion source were set to 260, 290, and 230 °C, respectively [39].

2.5. Untargeted Metabolomic Analysis

To extract fecal metabolites, 50 mg of mouse fecal samples ($n = 6$ per group) were mixed with pre-cooled solution (methanol:acetonitrile:water = 2:2:1, v/v), which was added to 20 µL of L-2-chlorophenylalanine as the internal standard. Samples were extracted ultrasonically for 30 min at 5 °C. After 10 min of incubation at −20 °C, the sample was then centrifuged at 14,000× g/4 °C for 20 min. The supernatant was dried up using nitrogen. Before injection, the samples were dissolved with 100 µL of acetonitrile solution (acetonitrile:water = 1:1, v/v), followed by vortexing for 30 s and ultrasonic extraction for 5 min at 4 °C. The supernatant was used for injection after centrifuging at 14,000× g/4 °C for 15 min. Fecal metabolite profiles were analyzed using an Agilent 1290 Infinity LC Ultra High-Performance Liquid Chromatography System (UPLC) with a Hydrophilic Interaction Liquid Chromatography (HILIC) column in ShangHai Applied Protein Technology Co. Ltd. (Shanghai, China). A total of 20 µL of supernatant from each sample was mixed together as the quality-control sample. The mobile phase composition of A was as follows: water + 25 mM ammonium acetate + 25 mM ammonia. For B, it was as follows: Acetonitrile. The solvent gradients were as follows: at 0–0.5 min, B was 95%; at 0.5–7 min, B changed linearly from 95% to 65%; at 7–8 min, B changed linearly from 65% to 40%; at 8–9 min, B was maintained at 40%; at 9–9.1 min, B changed linearly from 40% to 95%; at 9.1–12 min, B was maintained at 95%. Mass spectrometry settings were as follows: ion source gas1: 60; ion source gas2: 60; curtain gas: 30; source temperature: 600 °C; ion sapary voltage floating: ±5500 V; TOF MS scan m/z range: 60–1000 Da; product ion scan m/z range: 25–1000 Da; TOF MS scan accumulation time: 0.20 s/spectra; product ion scan accumulation time: 0.05 s/spectra; and information-dependent acquisition (IDA) was applied for MS/MS with a collision energy of 35 ± 15 eV. Raw data, converted from Wiff format into mzXML format by ProteoWizard, were imported into XCMS software for baseline filtering, peak identification, retention time correction, and peak alignment. Identification of metabolites was performed based on the accuracy m/z value (<25 ppm) and the comparison of MS/MS spectra with an in-house database established by ShangHai Applied Protein Technology Co. Ltd. (Shanghai, China), which was built using available authentic standards.

2.6. Statistical Analysis

Differences in gut bacterial abundances were assessed by a nonparametric test via Metastats. Two-tailed Welch's t-test was used to analyze metabolites that differed in abundance between groups corrected for the FDR. In addition, alpha rarefaction and principal co-ordinates analysis (PCoA) were used to assess diversities in the gut microbial communities. Principal components analysis (PCA) and a hierarchical clustering algorithm were used to visualize the comparison of metabolite profiles and pathways. The correlation matrix between gut bacterial species and metabolites was generated using Spearman's correlation coefficient. Unless otherwise indicated, all results are expressed as mean values with standard deviation (** $p < 0.01$; * $p < 0.05$). The symbol * represents statistically significant difference.

3. Results

3.1. MP-Induced Alterations in Diversity and Composition of the Gut Microbial Community

After 6-week exposure of MP, there was no significant change in final body weight, weight gain, or food intake observed between the control and the MP-treated groups. Details regarding body weight, food intake, and organ indices are provided in the supplementary information (Figure S1). To investigate the impact of MP on the gut microbiome, fecal samples of mice were collected for taxonomic characterization and metabolite profiling after 6-week exposure to MP (Figure 1A). The overall composition of the gut microbial community was assessed by examining taxonomic similarity between the sequencing samples in the control and the MP-treated groups. As shown in Figure 1B, the sequences assigned to Firmicutes were more enriched in fecal samples from the MP-treated mice, whereas reads assigned to Bacteroidetes were slightly lower in these samples. Overall, the MP-treated group and the control group shared 170 bacterial species (Figure 1C). Interestingly, there were 12 and 19 unique genera in the gut microbiota of the control and MP-treated mice, respectively. Bacterial communities were clustered using Principal Coordinate Analysis (PCoA), which revealed that the gut microbiota of the MP-treated mice was distinctly clustered compared to that of the controls (Figure 1D).

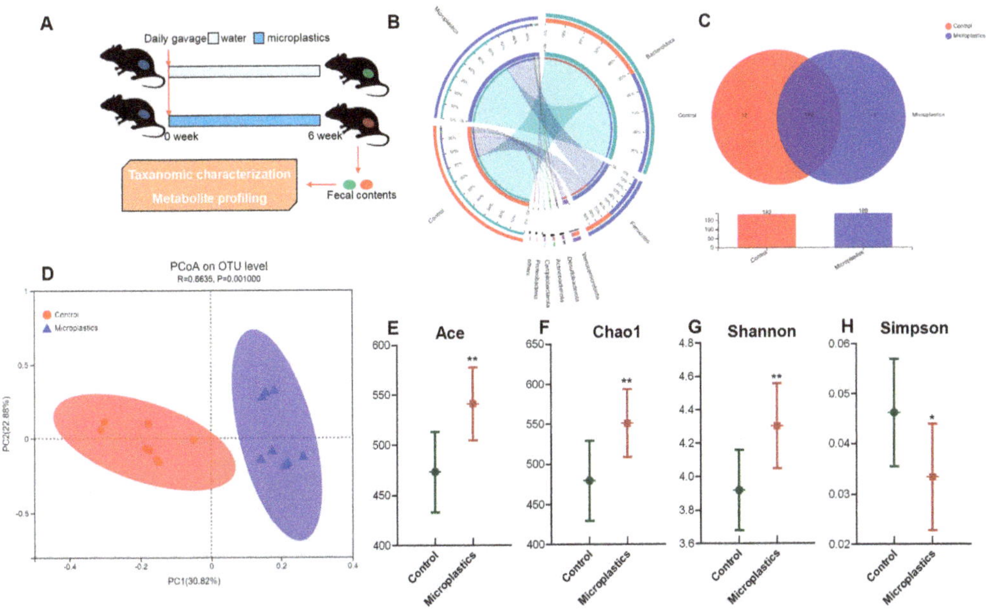

Figure 1. MP-induced alterations in the mouse gut microbial community. (**A**) Experimental design. (**B**) Circos diagram for component profiles of the gut microbial community at phylum level (n = 8). (**C**) Venn diagram for component profiles of the gut microbial community at species level (n = 8). (**D**) Component profiles analyzed by PcoA model (n = 8). (**E–H**) Alpha diversity indices of Ace, Chao1, Shannon, and Simpson (n = 8, * $p < 0.05$, ** $p < 0.01$).

Ace and Chao1 are indices estimating the number of OTUs in the community. A larger value of Ace or Chao1 indicates more OTUs of the microbiota community. Significantly increased values of Ace and Chao1 were observed in the MP-treated group compared to the controls (Figure 1E,F). Shannon and Simpson are indicators reflecting the diversity of the microbial community. Increased diversity comes with a larger Shannon value or a smaller Simpson value. MP exposure significantly increased the diversity of the gut microbiota community (Figure 1G). Taken together, we found that MP treatment increased the number of OTUs and diversity in the gut microbiota of mice. In addition, the phylogenetic tree shown in Figure 2A indicated that the gut microbiota were significantly modified by MP treatment. The red color represents bacteria that are abundant in the control mice, and the blue color represents abundant bacteria in the MP-treated group. As shown, phyla of Firmicutes and Campilobacterota were enriched in the mice upon microplastic exposure. LEfSe analysis disclosed that MP treatment is associated with an expansion of Clostridia and Lachnospiraceae (o_Lachnospirales) (Figure 2B). Particularly, at a species level, MP treatment is associated with an increase in a species from f_Muribaculaceae and a species from g_Anaerotruncus but a decrease in a species from g_Dubosiella (Figure 2C).

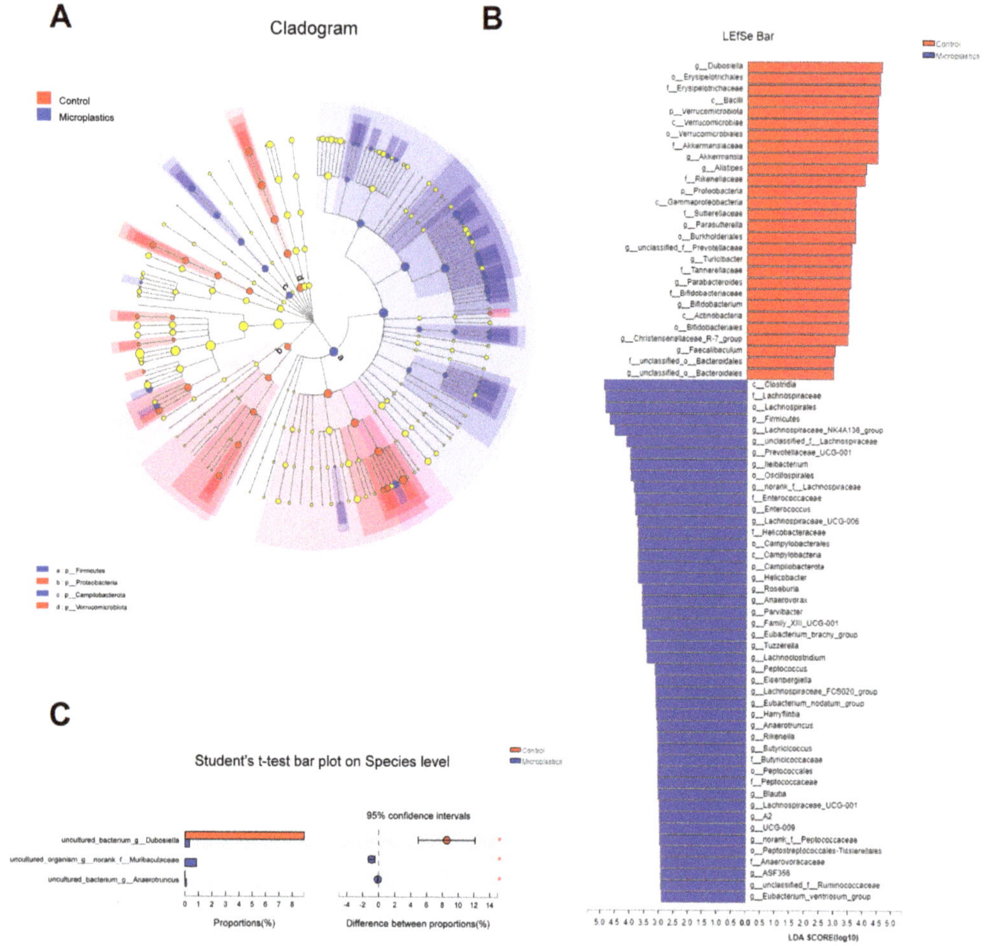

Figure 2. (**A**) Cladogram (phylum to genus) by discriminant analysis of Lefse with different colors representing control (Red) or microplastics (Blue) groups (n = 8). (**B**) Discriminant analysis of Lefse multi-level species difference (n = 8). (**C**) Community barplot analysis at species level (n = 8, * $p < 0.05$).

3.2. MP-Induced Changes in Functional Pathways of the Gut Microbiome

We next investigated changes in functional pathways of the gut microbiome in response to MP exposure. As shown in Figure 3A, the heatmap representing distribution of gene abundances shows strong comparison of functional pathways between the control and MP-treated groups. In particular, abundances of bacterial genes that were involved in pathways including xenobiotic biodegradation and metabolism, bacterial infectious disease, and drug resistance were significantly enriched in the MP-treated group compared to the controls (Figure 3B,D). In addition, abundances of bacterial genes encoding xenobiotic-related genes including ABC transporters, the bacterial secretion system, lysosomes, flagellar assembly, and quorum sensing were significantly perturbed with the administration of MP (Figure 3E–I).

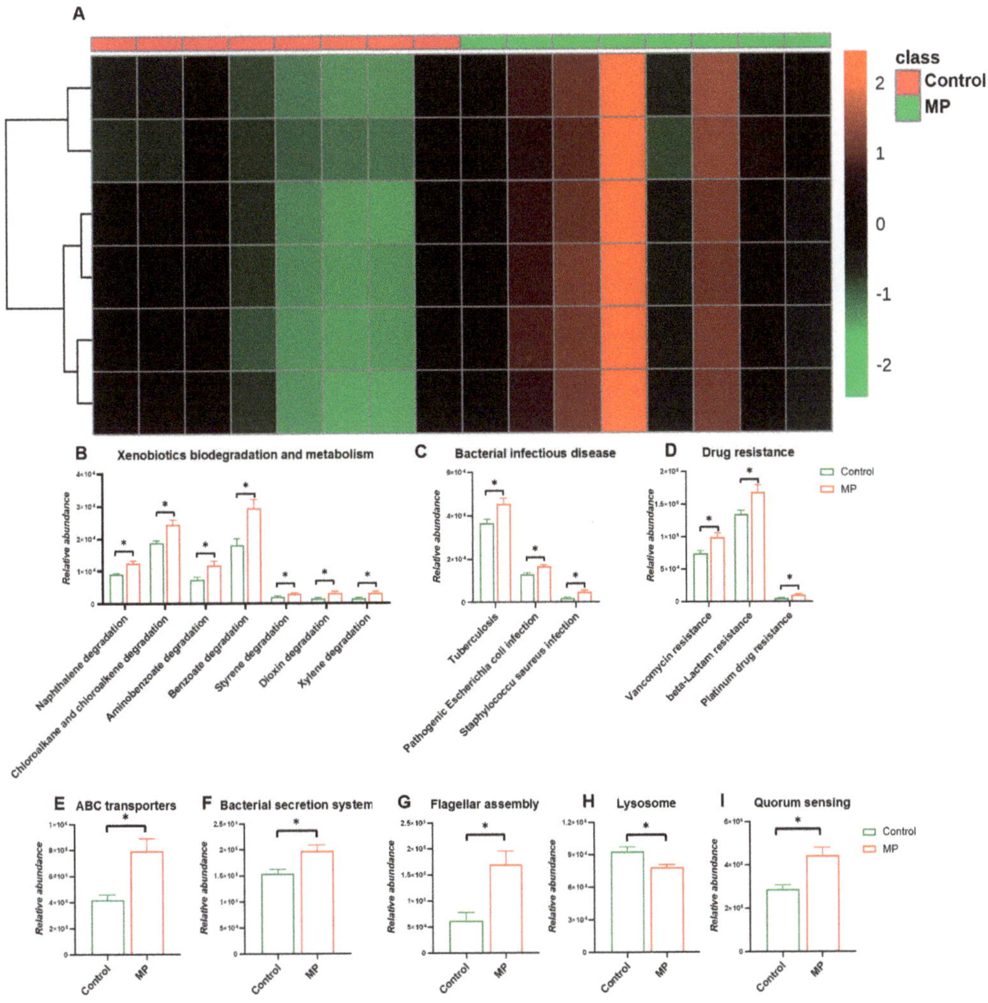

Figure 3. Comparisons of Functional pathways (n = 8). (**A**) Heatmap constructed by abundances of bacterial pathways at KEGG Level 1. The abundance of pathways related to xenobiotic biodegradation and metabolism (**B**), bacterial infectious disease (**C**), and drug resistance (**D**). The abundance of pathways related to ABC transporters I, bacterial secretion system (**F**), flagellar assembly (**G**), lysosome (**H**), and quorum sensing (**I**). *, $p < 0.05$.

3.3. MP Alters Metabolite Profiles of the Gut Microbiome

To further assess the impact of MP on metabolic products of the gut bacteria, an untargeted metabolomics experiment was performed on mouse fecal samples. Table 1 lists annotated dysregulated features with fold changes of $\geq$1.5 or $\leq$0.67 ($p < 0.05$) for microplastic challenges, including numerous metabolites that were related to gut microbial activities. PCA analysis clearly separated the two cultivars under both positive and negative ion modes (Figure 4A,C). The distribution of metabolic profiles of the MP-treated group was well-separated from the control group. Differential metabolites were screened with criteria of fold change >1.5 or <0.67 and p-Value < 0.05, which were visually represented with volcano plots (Figure 4B,D). In addition, enrichment analysis indicated that the pathways of cholesterol metabolism, primary and secondary bile acid biosynthesis, and taurine and hypotaurine metabolism, which play a key role in bile acid metabolism, were significantly enriched in the MP-treated mice (Figure S2). Fold change and classification of identified metabolites under positive and negative ion modes were visually presented in Figure 4E,F, respectively.

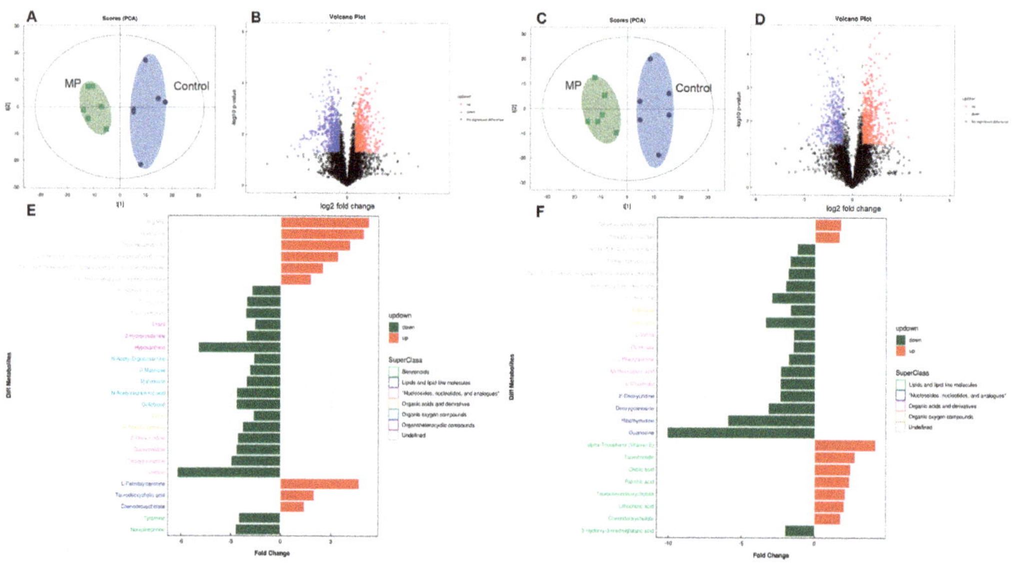

Figure 4. Comparisons of metabolite profiles ($n = 6$). Volcano plots represent distribution of metabolite fingerprints in positive (**A**) and negative (**C**) ion modes. PCA score charts of metabolite profiles in positive (**B**) and negative (**D**) ion modes. Fold change and classification of identified metabolites under positive (**E**) and negative (**F**) ion modes.

3.4. Key Metabolites That Are Associated with MP Exposure

The structures of altered metabolites are diverse, with a number of the metabolites being either directly generated or modified by the gut microbes. As shown in Figure 5, several typical gut-microbiota-related metabolites were differentiated upon MP exposure. Specifically, significantly increased levels of bile acids were observed in the gut microbiota (Figure 5A). Accordingly, the abundance of the bile acid metabolism pathway also showed significant increase upon MP treatment (Figure 5B). Moreover, modulation of purine and pyrimidine nucleoside abundances, in addition to pathways of nucleotide metabolism, was found in the gut microbiome of mice (Figure 5C,D). Given the essential role of SCFAs in metabolic activities of the gut microbiota, we additionally measured fecal levels of SCFAs. Fecal levels of SCFAs, including acetic acid, propionic acid, butyric acid, and isobutyric acid, were significantly decreased upon MP exposure (Figure 5F). Taken together, significant changes of key metabolites derived from the gut microbiota upon MP exposure were observed, which could contribute to the toxicity of MP.

Table 1. Significantly dysregulated ($p < 0.05$) metabolomic classes and the corresponding metabolites in mice exposed to microplastic.

Annotated Feature	Adduct	m/z	rt(s) [a]	VIP [b]	Fold Change [c]	*p*-Value	HMDB ID
Nucleosides, nucleotides, and analogues							
2′-Deoxyuridine	(M − H)−	227.0671	112.8	10.7	0.4	6.80×10^{-4}	12
Deoxycytidine	(M + H)+	228.0964	206.6	1	0.4	7.24×10^{-3}	14
Deoxyguanosine	(M − H)−	266.0890	230.3	2.1	0.3	6.30×10^{-3}	85
Guanosine	(M − H)−	282.0840	262.8	1.3	0.1	1.20×10^{-2}	133
Uridine	(M + H)+	245.0757	159.2	1.7	0.2	3.44×10^{-4}	296
Ribothymidine	(M − H)−	257.0780	142	4	0.2	3.20×10^{-3}	884
Lipids and lipid-like molecules							
3-Hydorxy−3-methylglutaric acid	(M − H)−	161.0452	373.8	1.1	0.5	9.00×10^{-4}	355
alpha-Tocopherol (Vitamin E)	(M − H)−	429.3724	31.5	1.9	4.2	3.10×10^{-2}	1893
Cholic acid	(M − H)−	407.2802	227.2	14	2.5	1.70×10^{-4}	619
Palmitic acid	(M − H)−	255.2327	46.7	9.4	2.4	4.60×10^{-2}	220
Taurochenodeoxycholate	(M − H)−	498.2886	140.7	3	2.1	1.60×10^{-2}	951
Taurocholate	(M − H)−	514.2840	200.3	2.4	2.8	4.40×10^{-2}	36
Chenodeoxycholate	(M + CH_3COO)−	451.3053	160.3	4.4	1.8	8.00×10^{-3}	518
Lithocholic acid	(M + CH_3COO)−	435.3105	82.6	3.3	2.1	2.10×10^{-2}	761
L-Palmitoylcarnitine	(M + H)+	400.3401	172.9	2.4	4.7	1.41×10^{-2}	222
Taurodeoxycholic acid	(M + NH4)+	517.3270	140.8	1.7	2	1.01×10^{-2}	896
Organic acids and derivatives							
DL-lactate	(M − H)−	89.0243	304.1	1.8	0.7	4.00×10^{-2}	1311
L-Glutamate	(M − H)−	146.0458	398.3	2	0.4	4.70×10^{-2}	148
L-Phenylalanine	(M − H)−	164.0719	261.5	3.1	0.6	5.00×10^{-3}	159
L-Valine	(M − H)−	116.0714	304.9	2.3	0.7	2.50×10^{-2}	883
Methylmalonic acid	(M − H)−	117.0188	104.9	1.2	0.4	1.10×10^{-2}	202
L-Arabinose	(M − H)−	149.0449	133	2	0.6	1.50×10^{-2}	646
Galactinol	(M + CH_3COO)−	401.1292	391.2	1.1	0.3	1.20×10^{-2}	5826
N-Acetyl-L-tyrosine	(M − H + 2Na)+	268.0606	242.4	1.3	0.4	3.45×10^{-2}	866
N-Acetyl-D-glucosamine	(M + H)+	222.0966	256.4	3	0.6	2.76×10^{-2}	215
N-Acetylneuraminic acid	(M + H)+	310.1121	373.6	2.2	0.4	4.82×10^{-5}	230
Cellobiose	(M + NH_4)+	360.1487	389.6	4.4	0.4	3.05×10^{-2}	55
D-Mannose	(M + NH_4)+	198.0958	302.8	2	0.5	4.46×10^{-3}	169
Maltotriose	(M + NH_4)+	522.2001	449.6	1.6	0.5	2.10×10^{-2}	1262
Benzenoids							
Tyramine	(M + H)+	138.0900	218.2	1.2	0.4	2.46×10^{-3}	306
Norepinephrine	(M + H − H_2O)+	152.0691	105.2	2.7	0.4	3.43×10^{-4}	216
Organoheterocyclic compounds							
2-Hydroxyadenine	(M + H)+	152.0559	262.4	1.5	0.5	3.03×10^{-2}	403
Hypoxanthine	(M + H)+	137.0448	217.3	4.2	0.2	4.17×10^{-2}	157
Uracil	(M + H)+	113.0334	84.7	1.9	0.6	3.31×10^{-2}	300
Others							
1-Palmitoyl−2-hydroxy-sn-glycero−3-phosphoethanolamine	(M − H)−	452.2773	200.6	4.6	0.5	2.00×10^{-2}	n.a. [d]
2-Methylbenzoic acid	(M − H)−	135.0445	133.3	1.2	0.6	1.20×10^{-2}	2340
3b-Hydroxy−5-cholenoic acid	(M − H)−	373.2733	61.2	2.5	0.5	3.00×10^{-2}	308
gamma-Linolenic acid	(M − H)−	277.2174	46.1	6.4	0.8	4.20×10^{-2}	3073
Cholesterol 3-sulfate	(M − H)−	465.3042	26.3	15.8	1.7	1.40×10^{-2}	653
Isobutyric acid	(M − H)−	87.0452	132.6	6	0.3	1.30×10^{-4}	1873
Tetrahydrocorticosterone	(M − H)−	349.2373	67.6	1.4	1.8	7.50×10^{-3}	268
Isoetharine	(M + CH_3CN + Na)+	303.1689	197.6	1.6	4.9	2.00×10^{-2}	14366
1-Palmitoyl-sn-glycero−3-phosphocholine	(M + H)+	496.3396	194.5	10.3	1.8	3.93×10^{-3}	n.a.
1-Stearoyl−2-hydroxy-sn-glycero−3-phosphoethanolamine	(M + H)+	482.3230	195.7	3.1	3.4	1.73×10^{-5}	n.a.
2′-Deoxyinosine	(M + H)+	253.0924	179.6	3.5	0.5	3.21×10^{-2}	71
Arg-Phe	(M + H)+	322.1851	340.8	1.2	5.3	1.28×10^{-3}	n.a.
DL-Indole−3-lactic acid	(M + H − H_2O)+	188.0693	259.1	1.3	0.6	3.25×10^{-2}	671
Thioetheramide-PC	(M + Na)+	758.5646	129.4	2.1	4.1	5.77×10^{-3}	n.a.
D-(-)-Lyxose	(M + NH_4)+	168.0853	152.6	1.3	0.5	9.26×10^{-4}	n.a.
1-O-(cis−9-Octadecenyl)−2-O-acetyl-sn-glycero−3-phosphocholine	(M + H)+	550.3832	188.7	1.7	2.5	4.49×10^{-4}	n.a.

[a] rt—retention time; [b] VIP—variable importance in projection; [c] the value of fold change higher than or equal to 1.5 indicates that the corresponding metabolite was upregulated, and the value of fold change lower than or equal to 0.7 indicates that the corresponding metabolite was downregulated; [d] n.a., the HMDB ID is not available for the corresponding metabolite.

3.5. Correlation between the Gut Microbiome and Metabolites

To explore the functional correlation between the gut microbiome changes and metabolite perturbations induced by MP treatment, we performed functional correlation analysis between the gut microbial species and metabolites with significant changes. Strong correlations were identified between the relative abundances of gut bacterial species and altered metabolite profiles (rho > 0.7 or <−0.7; $p < 0.05$). Figure 6 lists several typical gut-microbiota-related metabolites that are highly correlated with specific gut bacteria. For example, taurocholate, a key bile acid, negatively correlates with B1 (s__uncultured_bacterium_g__Dubosiella) but positively correlates with B2 (s__uncultured_bacterium_g__Anaerotruncus).

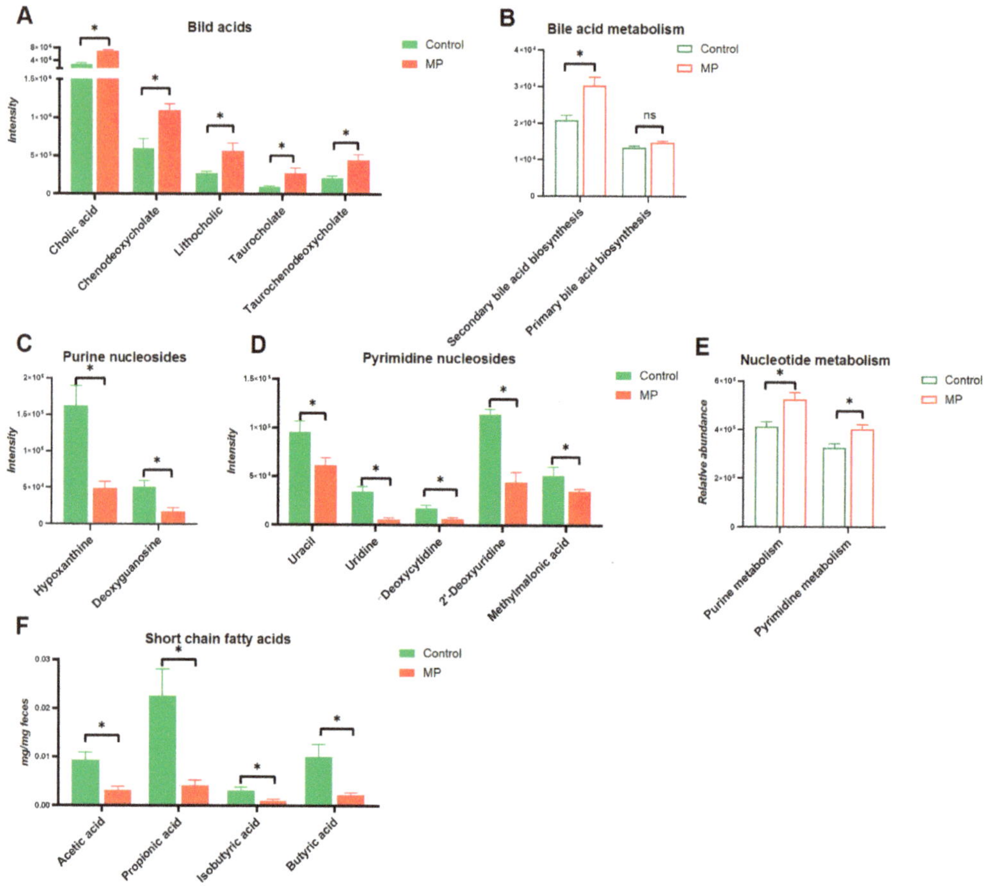

Figure 5. Key metabolites in the gut microbiota associated with MP exposure. (**A**) bile acids ($n = 6$). (**B**) Pathways involved in bile acid metabolism ($n = 8$). (**C**) Purine nucleosides ($n = 6$). (**D**) Pyrimidine nucleosides ($n = 6$). (**E**) Pathways involved in nucleotide metabolism ($n = 8$). (**F**) Short-chain fatty acids ($n = 6$). *, $p < 0.05$.

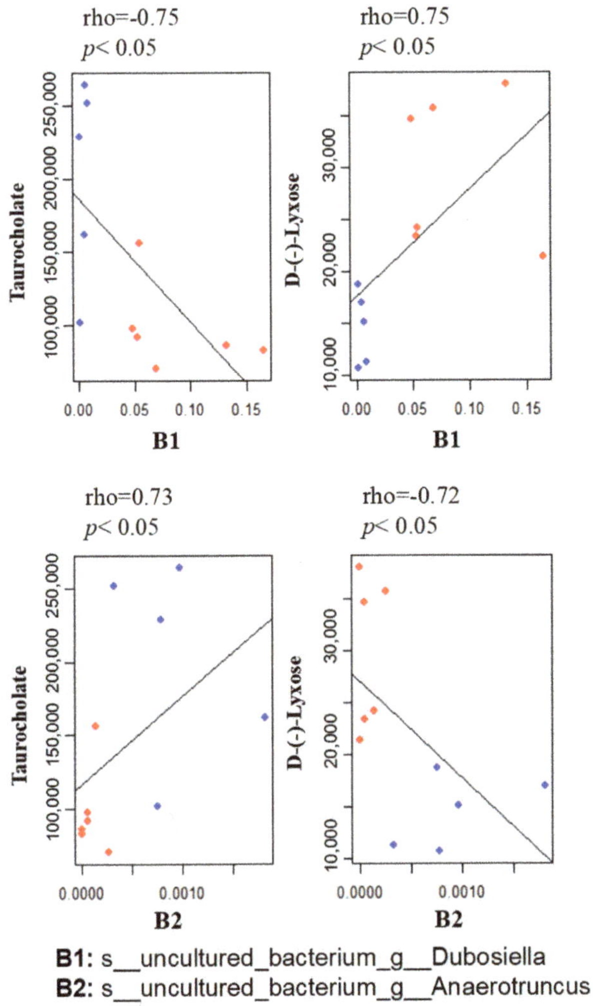

Figure 6. Scatter plots illustrating statistical associations between key metabolites and gut bacterial species (Blue dot: MP; Red dot: Control; rho > 0.7 or <−0.7; $p < 0.05$).

4. Discussion

We used high-throughput 16S rRNA gene sequencing and metabolomics profiling to investigate the impact of MP exposure on the gut microbiota. The results clearly showed that MP exposure induced a significant alteration in the gut microbial composition of mice. In addition, perturbations in gut bacterial composition were associated with changes in a variety of gut-microbiota-related metabolic products, suggesting that MP exposure not only perturbs the gut microbiota at the abundance level but also essentially changes the metabolite profile. Specifically, MP exposure significantly perturbed aspects of the gut microbiota, including its composition, diversity, and functional pathways that are involved in xenobiotic metabolism. A distinct metabolite profile was observed, which probably resulted from changes in gut bacterial composition and metabolic pathways induced by MP exposure. A number of key metabolites, including bile acids, purine and pyrimidine nucleosides, lipids, and SCFAs that are associated with MP exposure, may contribute to the toxic effects of MP.

The underlying mechanisms by which MP may exert toxic effects are still elusive. Mounting evidence suggests that metabolic changes associated with gut microbiome perturbations are key risk factors for the onset and development of various adverse outcomes [40]. The gut microbiome could not only directly impact intestinal homeostasis locally through bacterial products, but it could also trigger systemic effects on remote tissues including liver, adipose, or brain by producing metabolites that serve as signaling molecules [29,30]. Therefore, metabolic changes, especially perturbations in microbiota-generated metabolites, play an essential role in the disease development.

The gut microliome is critical to the energy metabolism of the host. Dysbiosis of the gut microbiome may be associated with obesity and diabetes [41,42]. In the present study, we observed an increase in the proportion of Firmicutes and a slight decrease in the proportion of Bacteroidetes in the MP-treated mice compared to the controls, which is a typical characteristic of obesity-driven dysbiosis and is consistent with previous reports [6]. Similar changes regarding Firmicutes and Bacteroidetes were also reported in a previous study [27]. However, this result is in contradiction with another previous study, which reported a decrease in the proportion of Firmicutes and a slight increase in the proportion of Bacteroidetes in the MP-treated mice compared to the controls [28]. In addition, we observed an increase in microbial diversity in the MP-treated mice compared to the controls. Our follow-up results shown in Figures 3 and S3 indicate that MP may lead to the appearance of potentially pathogenic bacteria, providing a possible explanation for the increase of microbial diversity. Moreover, abundances of bacterial genes that are involved in pathways including bacterial infectious disease and drug resistance were significantly enriched in the MP-treated group compared to the controls (Figure 3C,D), further supporting the appearance of potentially pathogenic bacteria in the gut of the MP-treated mice. It is worth noting that the microbes in MP dilutes may also impact the composition of the gut microbiota of the MP-treated mice, and verification of sterility is able to exclude the possible effects of this factor.

The gut microbiome has profound roles in xenobiotic metabolism and evolved in the toxicity of environmental agents. Our results show enrichment of pathways involved in xenobiotic biodegradation and metabolism (Figure 3B). Particularly, abundances of bacterial genes encoding xenobiotic-related genes including ABC transporters, the bacterial secretion system, and lysosomes were significantly altered in the gut microbiome of the MP-treated mice. It is reported that ABC transporters are capable of exporting microplastic particles [43]. Consistently, we also observed alterations in metabolites that are related to ABC transporters (Figure S4). Therefore, MP treatment would induce functional changes in xenobiotic-metabolism-related pathways of the gut microbiome, providing additional evidence regarding the interactions between the gut microbiome and xenobiotic metabolism and toxicity.

Recent evidence suggested that perturbations of the gut microbiome and its functions may be a potential mechanism underlying the toxic effects of environmental agents [18]. The gut bacteria could directly communicate with the host through the production of a variety of endogenous metabolites. Bile acids are cholesterol derivatives that are synthesized in the liver before undergoing extensive enterohepatic recycling as well as modification by gut bacteria. It is established that not only are bile acids involved in digestion and absorption, but also they act as signaling molecules affecting diverse pathways by activating a number of nuclear receptors [44]. Bile acids and intermediates, in addition to bile-acid-related pathways, were significantly perturbed in the MP-treated mice, indicating that MP exposure affects the homeostasis of bile acids. The underlying mechanisms remain elusive; however, MP-induced gut-microbiome perturbations may be involved. Previous reports demonstrated that the gut microbiota would affect primary and secondary bile acid profiles in the tissues of antibiotic-treated rats [45]. Moreover, it is reported that bile-acid signaling via the relevant receptors is associated with the regulation of the immune system and inflammatory response [46]. It is particularly interesting that taurocholate, a key bile acid molecule, negatively correlates with abundances of s__uncultured_bacterium_g__Dubosiella, whereas it positively correlates with abundances of s__uncultured_bacterium_g__Anaerotruncus. Therefore, alterations in the bile-acid profile in the MP-treated mice could be associated with species changes in the gut microbiota, suggesting the involvement of the bacterial metabolic activities of bile acids in chemical toxicity. MP-induced alterations in bile-acid-related metabolites were reported in a previous study; their results showed that exposure to 5 μm MP impacted bile acid metabolism of mice via measurement of the total bile acids in serum and the liver [27]. This is further confirmed in a more recent study in which plasma metabolites were profiled in mice upon MP exposure. Perturbations of bile acids and derivatives were also observed in mouse plasma such as taurochenodeoxycholate (TUDCA) [47], which increased with a fold change of two in fecal samples of the MP-treated mice compared to that of the controls, according to the results of the present study. It is worth noting that we cannot exclude an effect on host metabolism due to styrene monomers derived by MP degradation, which could also exert an impact on biliary and taurine metabolism. Likewise, an important class of gut-microbiota-derived metabolites, the functions of SCFAs have been extensively studied. For example, it is well documented that butyrate has anti-inflammatory effects, mainly via suppression of nuclear factor kappa β (NF-κB) activation in macrophages and histone deacetylation (HDAc) in acute myeloid leukemia [48,49]. It has recently been reported that propionate and butyrate have a potential role in regulatory T-cell production and function at the whole-animal level through inhibition of HDAc [50,51]. Thus, perturbations of metabolic profiles, especially bile acids and SCFAs in the gut microbiome, may be linked to MP-induced toxic effects. Taken together, these results support the hypothesis that perturbations in the gut microbial composition and key metabolites could be one of the underlying mechanisms of MP toxicity.

5. Conclusions

In summary, MP exposure perturbs the gut microbial composition and key metabolites in mice. This finding represents an important step toward understanding how MP exposure affects the gut microbiome and its functions. Future studies are warranted to address more intriguing issues. For instance, the dose-, time-, and particle-size-dependent effects of MP exposure on the gut microbiota need to be defined. Nevertheless, our results indicate that MP exposure not only changes the gut microbiota at the abundance level but also substantially alters the metabolic profiles with perturbations in key metabolites. Key metabolites and bacterial species were identified. Perturbations in the gut microbiota and these gut-microbiota-related metabolites by MP exposure support the hypothesis that environmental chemicals including MP could lead to toxic effects via perturbation of the gut microbiome and its metabolic profiles.

Supplementary Materials: The following supporting information can be downloaded at: https://www.mdpi.com/article/10.3390/metabo13040530/s1. Figure S1. Metabolic impact of MP exposure. (A) Body weight curve. (B) Total weight gain. (C) Food intake. Organ indices of liver (D), kidney (D), and cecum (F). Student's *t*-test, $p < 0.05$ was considered statistically significant. Figure S2. KEGG enrichment analysis of differential metabolites. Figure S3. Wilcoxonrank-sum test on phenotype. Figure S4. Alterations in metabolites that are related to ABC transporters.

Author Contributions: Conceptualization: P.T., X.W. and X.Z.; Experiment: Q.T., H.N. and P.T.; Data analysis: Z.M., L.W., Z.C. and P.T.; Writing-original draft: P.T., Y.C. and J.X.; Writing-review & editing: all authors. All authors have read and agreed to the published version of the manuscript.

Funding: This research was funded by the Program for Guangdong Introducing Innovative and Entrepreneurial Teams (2019ZT08L213), the Key-Area Research and Development Program of Guangdong Province (2020B1111380003), the Guangdong Provincial Key Laboratory Project (2019B121203011), and the Natural Science Foundation of China (42007374).

Institutional Review Board Statement: The study was conducted in accordance with the Declaration of Helsinki, and approved by the Animal Care and Use Committee of Zhejiang Chinese Medical University (No. 20201103-08).

Informed Consent Statement: Not applicable.

Data Availability Statement: Not applicable.

Conflicts of Interest: The authors declare no conflict of interest.

References

1. Pickard, J.M.; Zeng, M.Y.; Caruso, R.; Núñez, G. Gut microbiota: Role in pathogen colonization, immune responses, and inflammatory disease. *Immunol. Rev.* **2017**, *279*, 70–89. [CrossRef] [PubMed]
2. Round, J.L.; Mazmanian, S.K. The gut microbiota shapes intestinal immune responses during health and disease. *Nat. Rev. Immunol.* **2009**, *9*, 313–323. [CrossRef] [PubMed]
3. Tremaroli, V.; Bäckhed, F. Functional interactions between the gut microbiota and host metabolism. *Nature* **2012**, *489*, 242–249. [CrossRef] [PubMed]
4. Shi, N.; Li, N.; Duan, X.; Niu, H. Interaction between the gut microbiome and mucosal immune system. *Mil. Med. Res.* **2017**, *4*, 14. [CrossRef]
5. Clemente, J.C.; Manasson, J.; Scher, J.U. The role of the gut microbiome in systemic inflammatory disease. *BMJ* **2018**, *360*, j5145. [CrossRef]
6. Turnbaugh, P.J.; Ley, R.E.; Mahowald, M.A.; Magrini, V.; Mardis, E.R.; Gordon, J.I. An obesity-associated gut microbiome with increased capacity for energy harvest. *Nature* **2006**, *444*, 1027–1031. [CrossRef]
7. Gurung, M.; Li, Z.; You, H.; Rodrigues, R.; Jump, D.B.; Morgun, A.; Shulzhenko, N. Role of gut microbiota in type 2 diabetes pathophysiology. *EBioMedicine* **2020**, *51*, 102590. [CrossRef]
8. Zheng, P.; Li, Z.; Zhou, Z. Gut microbiome in type 1 diabetes: A comprehensive review. *Diabetes Metab. Res. Rev.* **2018**, *34*, e3043. [CrossRef]
9. Louis, P.; Hold, G.L.; Flint, H.J. The gut microbiota, bacterial metabolites and colorectal cancer. *Nat. Rev. Microbiol.* **2014**, *12*, 661–672. [CrossRef]
10. Chi, L.; Bian, X.; Gao, B.; Ru, H.; Tu, P.; Lu, K. Sex-Specific Effects of Arsenic Exposure on the Trajectory and Function of the Gut Microbiome. *Chem. Res. Toxicol.* **2016**, *29*, 949–951. [CrossRef]
11. Chi, L.; Gao, B.; Bian, X.; Tu, P.; Ru, H.; Lu, K. Manganese-induced sex-specific gut microbiome perturbations in C57BL/6 mice. *Toxicol. Appl. Pharmacol.* **2017**, *331*, 142–153. [CrossRef] [PubMed]
12. Gao, B.; Chi, L.; Mahbub, R.; Bian, X.; Tu, P.; Ru, H.; Lu, K. Multi-Omics Reveals that Lead Exposure Disturbs Gut Microbiome Development, Key Metabolites, and Metabolic Pathways. *Chem. Res. Toxicol.* **2017**, *30*, 996–1005. [CrossRef] [PubMed]
13. Lu, K.; Abo, R.P.; Schlieper, K.A.; Graffam, M.E.; Levine, S.; Wishnok, J.S.; Swenberg, J.A.; Tannenbaum, S.R.; Fox, J.G. Arsenic exposure perturbs the gut microbiome and its metabolic profile in mice: An integrated metagenomics and metabolomics analysis. *Environ. Health Perspect* **2014**, *122*, 284–291. [CrossRef] [PubMed]
14. Gao, B.; Bian, X.; Chi, L.; Tu, P.; Ru, H.; Lu, K. Editor's Highlight: OrganophosphateDiazinon Altered Quorum Sensing, Cell Motility, Stress Response, and Carbohydrate Metabolism of Gut Microbiome. *Toxicol. Sci.* **2017**, *157*, 354–364. [CrossRef] [PubMed]
15. Bian, X.; Tu, P.; Chi, L.; Gao, B.; Ru, H.; Lu, K. Saccharin induced liver inflammation in mice by altering the gut microbiota and its metabolic functions. *Food Chem. Toxicol.* **2017**, *107*, 530–539. [CrossRef]
16. Bian, X.; Chi, L.; Gao, B.; Tu, P.; Ru, H.; Lu, K. The artificial sweetener acesulfame potassium affects the gut microbiome and body weight gain in CD-1 mice. *PLoS ONE* **2017**, *12*, e0178426. [CrossRef]
17. Bian, X.; Chi, L.; Gao, B.; Tu, P.; Ru, H.; Lu, K. Gut Microbiome Response to Sucralose and Its Potential Role in Inducing Liver Inflammation in Mice. *Front. Physiol.* **2017**, *8*, 487. [CrossRef]
18. Tu, P.; Chi, L.; Bodnar, W.; Zhang, Z.; Gao, B.; Bian, X.; Stewart, J.; Fry, R.; Lu, K. Gut Microbiome Toxicity: Connecting the Environment and Gut Microbiome-Associated Diseases. *Toxics* **2020**, *8*, 19. [CrossRef]
19. Chi, L.; Bian, X.; Gao, B.; Tu, P.; Ru, H.; Lu, K. The Effects of an Environmentally Relevant Level of Arsenic on the Gut Microbiome and Its Functional Metagenome. *Toxicol. Sci.* **2017**, *160*, 193–204. [CrossRef]
20. Chi, L.; Gao, B.; Tu, P.; Liu, C.-W.; Xue, J.; Lai, Y.; Ru, H.; Lu, K. Individual susceptibility to arsenic-induced diseases: The role of host genetics, nutritional status, and the gut microbiome. *Mamm. Genome* **2018**, *29*, 63–79. [CrossRef]
21. Chi, L.; Mahbub, R.; Gao, B.; Bian, X.; Tu, P.; Ru, H.; Lu, K. Nicotine Alters the Gut Microbiome and Metabolites of Gut-Brain Interactions in a Sex-Specific Manner. *Chem. Res. Toxicol.* **2017**, *30*, 2110–2119. [CrossRef] [PubMed]
22. Toussaint, B.; Raffael, B.; Angers-Loustau, A.; Gilliland, D.; Kestens, V.; Petrillo, M.; Rio-Echevarria, I.M.; Van den Eede, G. Review of micro- and nanoplastic contamination in the food chain. *Food Addit. Contam. Part A* **2019**, *36*, 639–673. [CrossRef] [PubMed]

23. Prata, J.C.; da Costa, J.P.; Lopes, I.; Duarte, A.C.; Rocha-Santos, T. Environmental exposure to microplastics: An overview on possible human health effects. *Sci. Total Environ.* **2020**, *702*, 134455. [CrossRef] [PubMed]
24. Vandenberg, L.N.; Hauser, R.; Marcus, M.; Olea, N.; Welshons, W.V. Human exposure to bisphenol A (BPA). *Reprod. Toxicol.* **2007**, *24*, 139–177. [CrossRef]
25. Li, B.; Ding, Y.; Cheng, X.; Sheng, D.; Xu, Z.; Rong, Q.; Wu, Y.; Zhao, H.; Ji, X.; Zhang, Y. Polyethylene microplastics affect the distribution of gut microbiota and inflammation development in mice. *Chemosphere* **2020**, *244*, 125492. [CrossRef]
26. Lu, L.; Wan, Z.; Luo, T.; Fu, Z.; Jin, Y. Polystyrene microplastics induce gut microbiota dysbiosis and hepatic lipid metabolism disorder in mice. *Sci. Total Environ.* **2018**, *631–632*, 449–458. [CrossRef]
27. Jin, Y.; Lu, L.; Tu, W.; Luo, T.; Fu, Z. Impacts of polystyrene microplastic on the gut barrier, microbiota and metabolism of mice. *Sci. Total Environ.* **2019**, *649*, 308–317. [CrossRef]
28. Jiang, P.; Yuan, G.-H.; Jiang, B.-R.; Zhang, J.-Y.; Wang, Y.-Q.; Lv, H.-J.; Zhang, Z.; Wu, J.-L.; Wu, Q.; Li, L. Effects of microplastics (MPs) and tributyltin (TBT) alone and in combination on bile acids and gut microbiota crosstalk in mice. *Ecotoxicol. Environ. Saf.* **2021**, *220*, 112345. [CrossRef]
29. Claus, S.P.; Ellero, S.L.; Berger, B.; Krause, L.; Bruttin, A.; Molina, J.; Paris, A.; Want, E.J.; de Waziers, I.; Cloarec, O.; et al. Colonization-induced host-gut microbial metabolic interaction. *MBio* **2011**, *2*, e00271-10. [CrossRef]
30. Diaz Heijtz, R.; Wang, S.; Anuar, F.; Qian, Y.; Björkholm, B.; Samuelsson, A.; Hibberd, M.L.; Forssberg, H.; Pettersson, S. Normal gut microbiota modulates brain development and behavior. *Proc. Natl. Acad. Sci. USA* **2011**, *108*, 3047–3052. [CrossRef]
31. Chen, W.; Zhu, R.; Ye, X.; Sun, Y.; Tang, Q.; Liu, Y.; Yan, F.; Yu, T.; Zheng, X.; Tu, P. Food-derived cyanidin-3-O-glucoside reverses microplastic toxicity promoting discharge and modulating the gut microbiota in mice. *Food Funct.* **2022**, *13*, 1447–1458. [CrossRef] [PubMed]
32. Chen, W.; Ye, X.; Tang, Q.; Yu, T.; Tu, P.; Zheng, X. Cyanidin-3-O-glucoside reduces nanopolystyrene-induced toxicity and accumulation: Roles of mitochondrial energy metabolism and cellular efflux. *Environ. Sci. Nano* **2022**, *7*, 2572–2586. [CrossRef]
33. Kik, K.; Bukowska, B.; Sicińska, P. Polystyrene nanoparticles: Sources, occurrence in the environment, distribution in tissues, accumulation and toxicity to various organisms. *Environ. Pollut.* **2020**, *262*, 114297. [CrossRef] [PubMed]
34. Deng, Y.; Zhang, Y.; Lemos, B.; Ren, H. Tissue accumulation of microplastics in mice and biomarker responses suggest widespread health risks of exposure. *Sci. Rep.* **2017**, *7*, 46687. [CrossRef]
35. Watts, A.J.R.; Lewis, C.; Goodhead, R.M.; Beckett, S.J.; Moger, J.; Tyler, C.R.; Galloway, T.S. Uptake and retention of microplastics by the shore crab Carcinus maenas. *Environ. Sci. Technol.* **2014**, *48*, 8823–8830. [CrossRef]
36. Hämer, J.; Gutow, L.; Köhler, A.; Saborowski, R. Fate of microplastics in the marine isopod Idotea emarginata. *Environ. Sci. Technol.* **2014**, *48*, 13451–13458. [CrossRef]
37. Eerkes-Medrano, D.; Thompson, R.C.; Aldridge, D.C. Microplastics in freshwater systems: A review of the emerging threats, identification of knowledge gaps and prioritisation of research needs. *Water Res.* **2015**, *75*, 63–82. [CrossRef]
38. Aßhauer, K.P.; Wemheuer, B.; Daniel, R.; Meinicke, P. Tax4Fun: Predicting functional profiles from metagenomic 16S rRNA data. *Bioinformatics* **2015**, *31*, 2882–2884. [CrossRef]
39. Zheng, X.; Qiu, Y.; Zhong, W.; Baxter, S.; Su, M.; Li, Q.; Xie, G.; Ore, B.M.; Qiao, S.; Spencer, M.D.; et al. A targeted metabolomic protocol for short-chain fatty acids and branched-chain amino acids. *Metabolomics* **2013**, *9*, 818–827. [CrossRef]
40. Gomaa, E.Z. Human gut microbiota/microbiome in health and diseases: A review. *Antonie Van Leeuwenhoek* **2020**, *113*, 2019–2040. [CrossRef]
41. Patterson, E.; Ryan, P.M.; Cryan, J.F.; Dinan, T.G.; Ross, R.P.; Fitzgerald, G.F.; Stanton, C. Gut microbiota, obesity and diabetes. *Postgrad. Med. J.* **2016**, *92*, 286–300. [CrossRef]
42. Scheithauer, T.P.M.; Rampanelli, E.; Nieuwdorp, M.; Vallance, B.A.; Verchere, C.B.; van Raalte, D.H.; Herrema, H. Gut Microbiota as a Trigger for Metabolic Inflammation in Obesity and Type 2 Diabetes. *Front. Immunol.* **2020**, *11*, 571731. [CrossRef] [PubMed]
43. Wu, B.; Wu, X.; Liu, S.; Wang, Z.; Chen, L. Size-dependent effects of polystyrene microplastics on cytotoxicity and efflux pump inhibition in human Caco-2 cells. *Chemosphere* **2019**, *221*, 333–341. [CrossRef] [PubMed]
44. Nguyen, A.; Bouscarel, B. Bile acids and signal transduction: Role in glucose homeostasis. *Cell. Signal.* **2008**, *20*, 2180–2197. [CrossRef] [PubMed]
45. Swann, J.R.; Want, E.J.; Geier, F.M.; Spagou, K.; Wilson, I.D.; Sidaway, J.E.; Nicholson, J.K.; Holmes, E. Systemic gut microbial modulation of bile acid metabolism in host tissue compartments. *Proc. Natl. Acad. Sci. USA* **2011**, *108* (Suppl. 1), 4523–4530. [CrossRef] [PubMed]
46. Brestoff, J.R.; Artis, D. Commensal bacteria at the interface of host metabolism and the immune system. *Nat. Immunol.* **2013**, *14*, 676–684. [CrossRef]
47. Jing, J.; Zhang, L.; Han, L.; Wang, J.; Zhang, W.; Liu, Z.; Gao, A. Polystyrene micro-/nanoplastics induced hematopoietic damages via the crosstalk of gut microbiota, metabolites, and cytokines. *Environ. Int.* **2022**, *161*, 107131. [CrossRef]
48. Lührs, H.; Gerke, T.; Müller, J.G.; Melcher, R.; Schauber, J.; Boxberge, F.; Scheppach, W.; Menzel, T. Butyrate inhibits NF-kappaB activation in lamina propria macrophages of patients with ulcerative colitis. *Scand. J. Gastroenterol.* **2002**, *37*, 458–466. [CrossRef]
49. Maeda, T.; Towatari, M.; Kosugi, H.; Saito, H. Up-regulation of costimulatory/adhesion molecules by histone deacetylase inhibitors in acute myeloid leukemia cells. *Blood* **2000**, *96*, 3847–3856. [CrossRef]

50. Arpaia, N.; Campbell, C.; Fan, X.; Dikiy, S.; van der Veeken, J.; deRoos, P.; Liu, H.; Cross, J.R.; Pfeffer, K.; Coffer, P.J.; et al. Metabolites produced by commensal bacteria promote peripheral regulatory T-cell generation. *Nature* **2013**, *504*, 451–455. [CrossRef]
51. Furusawa, Y.; Obata, Y.; Fukuda, S.; Endo, T.A.; Nakato, G.; Takahashi, D.; Nakanishi, Y.; Uetake, C.; Kato, K.; Kato, T.; et al. Commensal microbe-derived butyrate induces the differentiation of colonic regulatory T cells. *Nature* **2013**, *504*, 446–450. [CrossRef] [PubMed]

metabolites

Article

Oral Exposure to Epoxiconazole Disturbed the Gut Micro-Environment and Metabolic Profiling in Male Mice

You Weng, Ting Xu, Caihong Wang and Yuanxiang Jin *

College of Biotechnology and Bioengineering, Zhejiang University of Technology, Hangzhou 310032, China
* Correspondence: jinyx@zjut.edu.cn

Abstract: Epoxiconazole (EPX), a triazole fungicide, is widely used in agriculture to control pests and diseases. High residual and occupational exposure to EPX increases health risks, and evidence of potential harm to mammals remains to be added. In the present study, 6-week-old male mice were exposed to 10 and 50 mg/kg bw EPX for 28 days. The results showed that EPX significantly increased the liver weights. EPX also decreased the mucus secretion of the colon and altered intestinal barrier function in mice including a reduced expression of some genes (*Muc2*, *meprinβ*, *tjp1*). Moreover, EPX altered the composition and abundance of gut microbiota in the colon of mice. The alpha diversity indices (Shannon, Simpson) in the gut microbiota increased after exposure to EPX for 28 days. Interestingly, EPX increased the ratio of *Firmicutes* to *Bacteroides* and the abundance of other harmful bacteria including *Helicobacter* and *Alistipes*. Based on the untargeted metabolomic analysis, it was found that EPX altered the metabolic profiles of the liver in mice. KEGG analysis of differential metabolites revealed that EPX disrupted the pathway related to glycolipid metabolism, and the mRNA levels of related genes were also confirmed. In addition, the correlation analysis showed that the most altered harmful bacteria were associated with some significantly altered metabolites. The findings highlight that EPX exposure changed the micro-environment and lipid metabolism disturbance. These results also suggest that the potential toxicity of triazole fungicides to mammals cannot be ignored.

Keywords: mice; epoxiconazole; intestinal barrier; gut microbiota; metabolome

1. Introduction

For the past few years, large amounts of triazole fungicides have been approved for use in agriculture. With the increasing reports of fungicides, widespread safety concerns have been raised about the whereabouts of residues and the toxic effects of pesticide exposure [1,2]. Fungicides may enter the aquatic environment through spray drift, surface runoff, and rainfall [3]. Occupational exposure in agriculture increases the potential for toxic effects [4]. Epoxiconazole (1-[[3-(2-chlorophenyl)-2-(4-fluorophenyl) oxiran-2-yl] methyl]-1,2,4triazole, EPX), as a triazole fungicide, is one of the most widely used pesticides worldwide [5]. In killing fungi, it does so by directly inhibiting sterol 14α demethylase encoded by the CYP51A1 gene in humans. The blocking of sterol 14α demethylase results in a deficiency of ergosterol, which is essential for cell membranes in yeast and fungi [6–8]. It has been shown that fungicides have toxic effects on mammal organisms [9–11] and numerous studies have revealed that EPX produced brain, cardiac, liver, kidney, intestinal, and endocrine-disrupting toxicity [12–17]. There have been few studies on metabolic toxicity and intestinal barrier function impairment in mice.

Drugs entering the body by intragastric administration are common in vivo experiments of physiology, pathology, and toxicology [18]. The gastrointestinal tract is the main digestive organ, and its epithelial surface has a physical barrier that effectively absorbs nutrients from food [19]. The first probability is harmful substances in the gastrointestinal tract, which will have a toxic effect on the intestine [20]. Intestinal barrier function is influenced by endogenous and exogenous factors such as cytokines and chemicals [21]. A breakdown of the intestinal barrier can lead to tissue damage such as a disruption in mucus production [22,23]. The presence of rich flora in intestinal contents can ferment dietary fiber. The intestinal microbiome can also regulate intestinal motility and maintain the integrity of the intestinal epithelial biological barrier [24,25]. In addition, the intestinal microbiome influences gut and systemic health through its metabolites. Bacteria interact directly or indirectly with the host through secreted biologically active molecules, which may have protective or harmful effects on the intestinal barrier and various organs including the liver, kidney, and brain [26,27]. The liver interacts with the gut via the gut–liver axis.

With the development of mass spectrometry, the application of the metabolome in toxicological research has rapidly increased [28]. At environmentally relevant concentrations, exposed organisms respond more sensitively at the molecular level than at the organic level [29]. The implications of pesticides on biological metabolism should be addressed [30,31]. The metabolome based on LC-MS has been widely applied to explore changes in the liver metabolic profile [32–34]. Here, 6-week-old male C57BL/6 mice were orally administered 10 mg/kg bw, and 50 mg/kg bw EPX for 28 days. The growth phenotypes, biochemical indices, and intestinal barrier function were investigated. Gut microbiota is conducted to explore changes in the gut micro-environment. In addition, the untargeted metabolome of the liver can effectively reveal the difference in the metabolic profile of mice induced by EPX at the molecular level. This study aimed to reveal the effects of the pesticide EPX on the intestinal and hepatic toxicity in mice, mainly related to the effects of gut micro-environment and metabolic profile. The findings could provide evidence for the toxicological effects of the triazole fungicide EPX on the gut and liver in mammals, which can provide support for assessing the health risks of fungicides.

2. Materials and Methods

2.1. Chemical

EPX (CAS No.: 106325-08-0, purity $\geq$ 95%) and corn oil (CAS No.: 8001-30-7, medical-grade) were purchased from Aladdin Chemical Reagent Industry (Shanghai, China) and stored at 4 °C.

2.2. Animals and Study Design

The five-week-old male C57BL/6 mice (n = 24) were supplied by the China National Laboratory Animals Resource Center (Shanghai, China). The mice were randomly assigned to cages (a photoperiod of 12:12-h light/dark cycle, 22 $\pm$ 2 °C). Water and food were freely available throughout the feeding process. After a week of acclimatization, all mice were randomly placed into three treatment groups: Con, EPX-L (10 mg/kg bw), and the EPX-H group (50 mg/kg bw) (n = 8). The 100 µL corn oil with doses of 0, 10, 50 mg/kg bw were administered intragastrically daily, respectively, and the doses were selected based on the LD50 (>5000 mg/kg) of the rat. The duration of the gavage was 28 days, as a sub-chronic exposure test. Each mouse could drink pure water and consume commercial feed freely.

After 28 days of continuous gavage, all mice were sacrificed and dissected. The mice were fasted for 12 h, anesthetized with ether, then blood was taken from the eyeballs, and the neck was pulled to death. Serum was collected from the whole blood centrifugated (7000 rpm) at 4 °C for 10 min. The killed mice were dissected promptly, and the organs were weighed and immediately placed in liquid nitrogen. The serum and rapid-freezing organs were deposited at −80 °C for further testing. All mouse experiments were authorized by the Guiding Principles of Zhejiang University of Technology (20211117091).

2.3. Histological Analysis of the Liver and Colon

The liver and colon (n = 4) selected from the Con and EPX-H treatment group were randomly used for histological analysis. Briefly, the liver and colon were fixed in 4% paraformaldehyde overnight, embedded within paraffin wax, and cut into 5 µm-thick sections. The 5 µm-thick sections of livers were stained with hematoxylin and eosin (H&E),

and the sections of the colon were stained with Alcian blue-periodic acid Schiff (AB-PAS). ImageJ2 was used for the quantitative analysis of AB-PAS staining.

2.4. Immunohistochemical and Immunofluorescence Analysis of the Colon

The pre-processing was the same as the previous steps to obtain 5 μm-thick sections of colons. For the immunohistochemical analysis, the rabbit anti-Muc2 (GB11344, Servicebio, Wuhan, China) primary antibody was applied at 1:500, then stained with FITC-conjugated (GB25303, Sevicebio) anti-rabbit secondary antibody. For the immunofluorescence analysis, the rabbit anti-claudin-1 (GB11032, Servicebio) and -ZO-1 (GB111402, Servicebio) primary antibodies were applied at 1:500, then stained with the HRP-conjugated (GB23303, Servicebio) anti-rabbit IgG secondary antibody. All images were captured with a fluorescence microscope (Nikon, Tokyo, Japan).

2.5. Biochemical Evaluation of the Serum and Hepatic Indices

The serum aspartate aminotransferase (AST) and alanine aminotransferase (ALT) were evaluated to assess liver function. The serum and hepatic triglyceride (TG), total cholesterol (TC), glucose (Glu), pyruvate (PYR), high-density lipoprotein cholesterol (HDL-C), and low-density lipoprotein cholesterol (LDL-C) were measured to evaluate the changed biochemical indices of the mice. The serum samples were directly used to detect the liver function and biochemical indicators. Liver samples were processed by adding PBS at 1:9 for grinding and centrifuging at '1000× *g* for 10 min. The supernatant was used to assess the biochemical indicators. The above tests followed the instructions in the commercial kits purchased from Nanjing Jiancheng Institute of Biotechnology (Nanjing, China). In addition, the protein concentration in the liver was detected with a Bradford Protein Assay Kit (Beyotime, Shanghai, China) for the relative quantification of each indicator.

2.6. RNA Extraction and Quantitative Real-Time PCR

The total RNA was extracted from the liver and colon of mice with the TRIZOL reagent. All extraction steps followed the instructions, and chloroform, isopropyl alcohol, and 75% alcohol were prepared. A Nano-300 (ALLSHENG, Hangzhou, China) detected the RNA concentration and purity. Furthermore, 1000 ng RNA was used for the cDNA synthesis using a reverse transcription kit. The next step was RT-qPCR, which was performed with the SYBR green system. All reagents were purchased from Vazyme (Nanjing, China), and all operations were conducted according to the instructions. Here, the specific primer sequences of the gene are listed in Table S1, and the synthesis was performed by Sangon Biotech (Shanghai, China). The expression level of β-actin was used to normalize the expression of specific genes. The PCR protocol followed the previous study [35]. The relative quantification of genes in all treatment groups was based on an earlier study [36].

2.7. 16S rRNA (V3-V4 Region) Sequencing and Data Analysis

The V3–V4 region of 16S rRNA sequencing was used to explore the intestinal microflora changes between the Con and EPX-H treatment groups. The total genomic DNA was extracted from the colon contents using a magnetic bead extraction method in our previous study [37]. Then, the DNA was amplified with specific primers of bacteria 16S rRNA (V3–V4, 338F: 5′-ACTCCTACGGGAGGCAGGAG-3′; 806R: 5′-GGACTACHVGGGTWTCT AAT-3′). The DNA amplification program was conducted according to a previous study [35]. Finally, the PCR products were purified with a Qiagen Gel Extraction Kit (Qiagen, DUS, Germany). The next step was sequencing library construction with the TruSeq® DNA PCR-Free Sample Preparation Kit and adding index codes. The Qubit@ 2.0 Fluorometer (Thermo Scientific, Waltham, MA, USA) and Agilent Bioanalyzer 2100 system evaluated the library quality. The Illumina NovaSeq platform was used for high-throughput sequencing, and the 250 bp paired-end reads were obtained.

After data splitting and screening, sequence analysis was conducted with Uparse software (Uparse v7.0.1001) [38]. For the species annotation of 16S, the Silva Database was applied according to the Mothur algorithm. Alpha diversity analysis was calculated using the OIIME (Version 1.7.0) including PD_whole_tree, Shannon, Simpson, chao1, and ace. For the beta diversity, the OIIME (Version 1.9.1) was applied to calculate the unweighted unifrac. The linear discriminant analysis effect size (LEfSe) was used to detect the difference in taxonomies between the Con and EPX-H groups, which was convenient for finding the biomarker.

The microflora differences at the phylum level were determined by qPCR with DNA, and the primers are shown in Table S2. The protocol was performed as in the previous study [39]. The abundance of 16S was used to normalize the relative abundance of specific bacteria.

2.8. LC-MS-Based Metabolomics Analysis

The livers were used for metabolomics analysis based on LC-MS. A sample of 100 g of liver tissue was ground in tissue extract [75% (methyl alcohol: chloroform = 9:1): 25% H_2O]. After sanding (50 Hz for 60 s, twice), the homogenate was treated with ultrasound for 30 min and placed on ice for 30 min before being centrifuged at 12,000 rpm for 10 min at 4 °C to achieve the supernatant for concentration and drying. The samples were redissolved using 50% acetonitrile solution with 2-chloro-L-phenylalanine solution (internal standard), which was used for LC-MS detection. The detection included chromatography conducted by the ultra-performance liquid system (Thermo Fisher Scientific, Waltham, MA, USA) and mass spectrometry was performed by the mass spectrometer detector (Thermo Fisher Scientific, Waltham, MA, USA). The above extraction, detection, and analysis work were all supported by Suzhou PANOMIX Biomedical Tech Co., Ltd. (Suzhou, China). Details are presented in the Supplementary Materials.

2.9. Statistical Analysis

Data were executed with GraphPad Prism 7.0 software and presented as the mean ± standard error of the mean (SEM). Data analysis between the Con and treatment groups was performed with one-way ANOVA followed by Dunnett's post hoc test. Statistical significance was denoted with * ($p \leq 0.05$) and ** ($p \leq 0.01$).

3. Results

3.1. Oral Exposure to EPX Altered the Growth Phenotype of Mice

After continuous EPX administration for 28 days, the body weights of mice showed an upward trend. However, there was no statistical significance (Figure 1A). Fasted for 12 h, the body weights of mice in different treatment groups did not change significantly (Figure 1B). However, the liver weight was elevated obviously ($p < 0.05$), with the increase in concentration as well as the ratio of liver to body weight (Figure 1B). Although the kidney weight did not change significantly, the ratio of the kidney to body weight decreased significantly ($p < 0.05$) in a concentration-dependent manner. Similarly, the fat weight did not vary, and the ratio of fat to body weight decreased ($p = 0.035$) in the EPX-H group.

In addition, there was no apparent pathological damage (vacuolization) to the liver of mice after being exposed to EPX (Figure 1C). The serum AST and ALT were also not altered significantly (Figure 1D).

3.2. EPX Altered the Biochemical Indices of Mice

For the serum biochemical indices (Table 1), the TC levels decreased significantly ($p < 0.05$) with the concentration dependence, and the TG and Glu levels did not change. The levels of PYR decreased ($p = 0.03$) in the EPX-L treatment group. Furthermore, HDL ($p = 0.005$) and LDL ($p = 0.03$) were lower in the high-concentration treatment group than in the other groups.

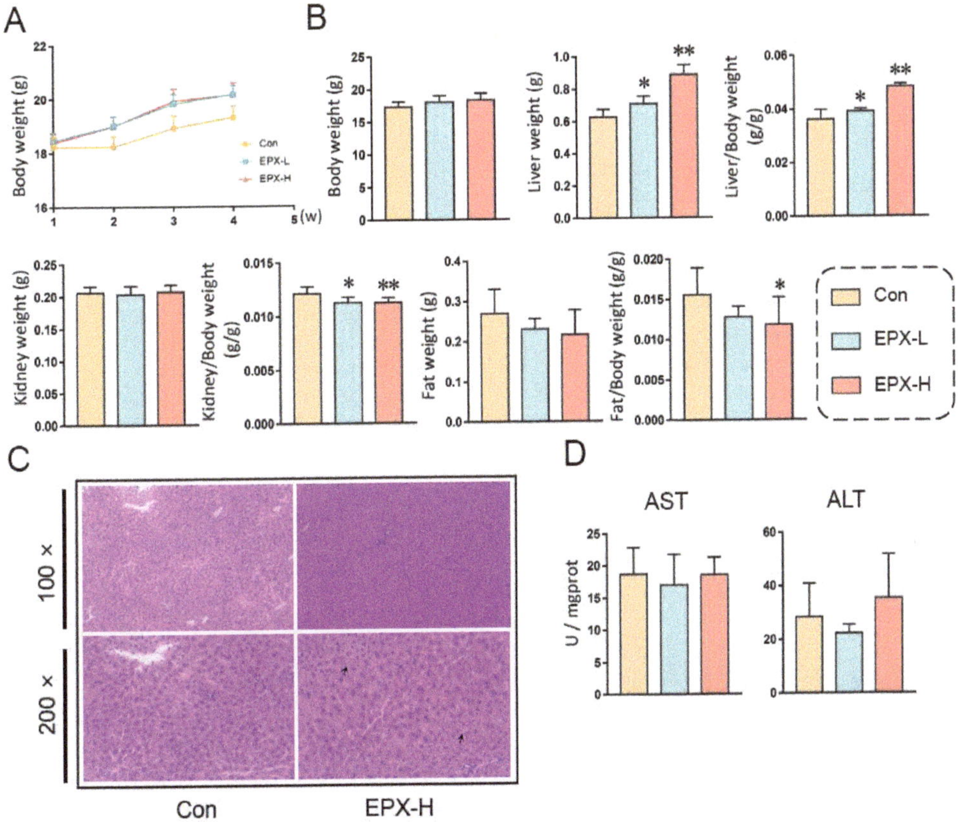

Figure 1. Effects of the oral exposure to EPX on the growth phenotype of mice. (**A**) The body weight of mice in each group weekly during exposure. (**B**) The body weight and organ weight of mice on the day of dissection. (**C**) The H&E staining of the liver. Black arrow: vacuolation. (**D**) Serum liver function indicators: AST activity, ALT activity. Values are shown as the means ± SEM (n = 8), and statistical significance: $p < 0.05$ *; $p < 0.01$ **.

Table 1. Effects of EPX exposure on the biochemical indicators of mice.

Biochemical Indicators	Con	EPX-L	EPX-H
Serum			
TC	2.267 ± 0.042	2.038 ± 0.038 *	1.914 ± 0.083 **
TG	0.133 ± 0.006	0.117 ± 0.015	0.150 ± 0.009
Glu	6.115 ± 0.357	5.782 ± 0.406	5.605 ± 0.215
PYR	0.402 ± 0.013	0.359 ± 0.007 *	0.439 ± 0.013
HDL	3.119 ± 0.152	2.651 ± 0.176	2.302 ± 0.181 **
LDL	0.774 ± 0.039	0.762 ± 0.062	0.533 ± 0.081 *
Liver			
TC	0.028 ± 0.007	0.030 ± 0.004	0.021 ± 0.005 *
TG	0.148 ± 0.051	0.151 ± 0.031	0.124 ± 0.027
Glu	0.163 ± 0.013	0.188 ± 0.033	0.137 ± 0.014
PYR	0.011 ± 0.005	0.012 ± 0.002	0.008 ± 0.002
NEFA	0.058 ± 0.016	0.055 ± 0.016	0.038 ± 0.006 *

Full name of biochemical indicators: TG, Triglyceride; TC, Total cholesterol; Glu, Glucose; PYR, Pyruvate; HDL, High-density lipoprotein cholesterol; LDL, Low-density lipoprotein cholesterol; NEFA, Non-esterified fatty acid. Values are shown as the means ± SEM (n = 8), and statistical significance: $p < 0.05$ *; $p < 0.01$ **.

In addition, the hepatic biochemical indices were also detected (Table 1). The TC level was also decreased observably ($p = 0.03$) in the EPX-H treatment group. The TG, Glu, and PYR levels did not change after exposure to EPX. The HDL and LDL were not detected in the liver (not shown). Furthermore, we found that the NEFA level also decreased ($p = 0.02$) in the liver of mice exposed to EPX-H.

3.3. EPX Affected the Mucus Secretion and Tight Junctions in the Colon of Mice

The mucus secretion of mice showed a numerical decrease in the EPX-H group compared to the Con group (Figures 2A and S1A). Immunohistochemistry revealed that the expression of Muc2, a protein associated with the mucus secretion, also showed a downward trend, although there was no significant effect (Figures 2A and S2B). We further detected the expression levels of genes involved in mucus secretion. As shown in Figure 2B, the expressions of the four genes detected were all decreased, among which, the expression level of *Muc2* in the colon of mice treated with EXP-L was significantly reduced ($p = 0.026$). Additionally, the transcriptional level of *meprinβ* was decreased in all EPX treatment groups ($p = 0.016$).

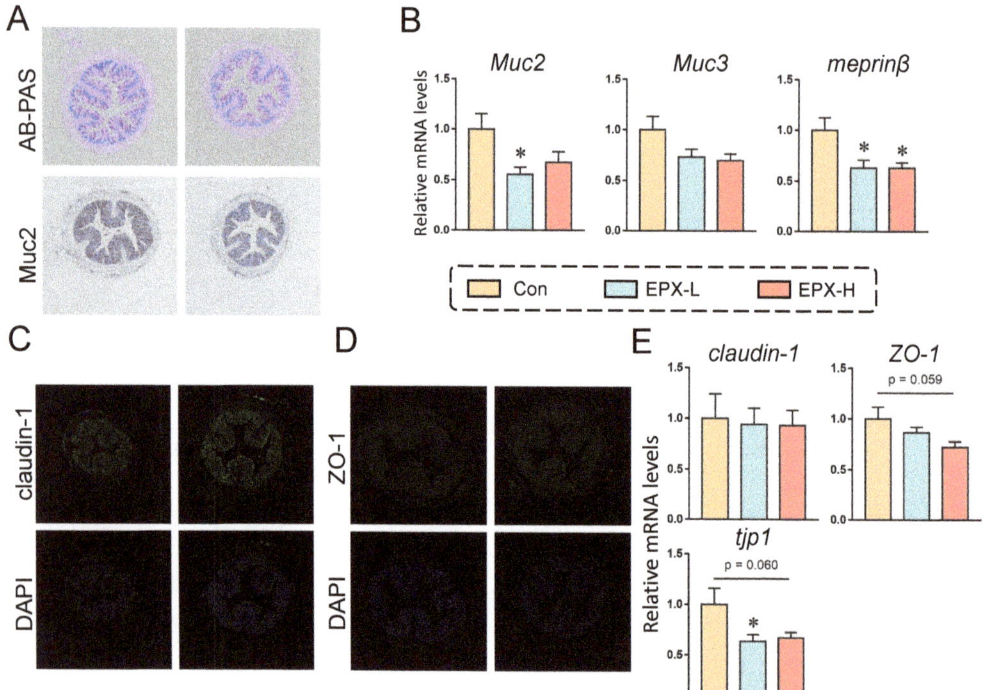

Figure 2. Effects of the oral exposure to EPX on the intestinal barrier function of mice. (**A**) AB-PAS staining and immunohistochemistry of Muc2 in the colon ($n = 3$). (**B**) The transcriptional levels of genes related to mucous secretion: *Muc2*, *Muc3*, and *meprinβ*. (**C**, **D**) Immunofluorescent staining of claudin-1 and ZO-1 ($n = 3$). (**E**) The transcriptional levels of genes related to tight junction: *claudin-1*, *ZO-1*, and *tjp1*. Values are shown as the means ± SEM ($n = 8$), and statistical significance: $p < 0.05$ *.

In addition, immunofluorescence revealed no significant changes in the claudin-1 and ZO-1 protein expression (Figure 2C,D). Similarly, the transcriptional levels of several genes associated with tight junctions were also examined. As shown in Figure 2E, the mRNA level of tjp1 was decreased in the EPX-L group ($p = 0.040$) and EPX-H ($p = 0.060$) group when compared to the Con group, and the mRNA level of ZO-1 was also decreased in the EPX-H group ($p = 0.059$).

3.4. EPX Regulated AMPs Expression and the Ionic Transport-Related Genes in the Colon of Mice

As shown in Figure S2A, the relative mRNA levels of genes related to *lyz*, *plazr4a*, and *ang4* were all increased significantly ($p < 0.05$) in the EPX-H treatment group. The other genes (*defα20*, *defα3*) did not change significantly.

Additionally, we analyzed the relative mRNA levels of genes involved in ionic transport in the colon of mice (Figure S2B). The results revealed that exposure to EPX had little effect on ion transport. Among the genes analyzed (*cftr*, *nkcc1*, *slc26a3*, *slc26a6*), only the mRNA level of *slc26a3* increased in a concentration-dependent manner ($p < 0.05$).

3.5. Oral Exposure to EPX Altered the Composition of Intestinal Microbiota in the Colon Contents

Principal coordinate analysis showed that the colonic microbiota in the EPX treatment group was different from that of the Con group (Figure 3A). After being exposed to EPX, the microbial composition changed, and the number of OTUs increased (Figure 3B). There was a difference in the affinities of species within the intestinal microbiota (PD_whole_tree: $p = 0.069$) of the two treatment groups. The Shannon and Simpson indices increased significantly ($p < 0.05$) in the EPX treatment group. Two other indices (chao1, ace) also increased, though not significantly. The above results show that EPX exposure increased the intestinal microbiota diversity in the colon content of mice. In addition, EPX disrupted the intestinal microbiota composition at the phylum level (Figure 3D). Specifically, the *Campylobacteria* and *Actinobacteria* levels increased in the EPX treatment group. Each of the other phyla had varying levels of change.

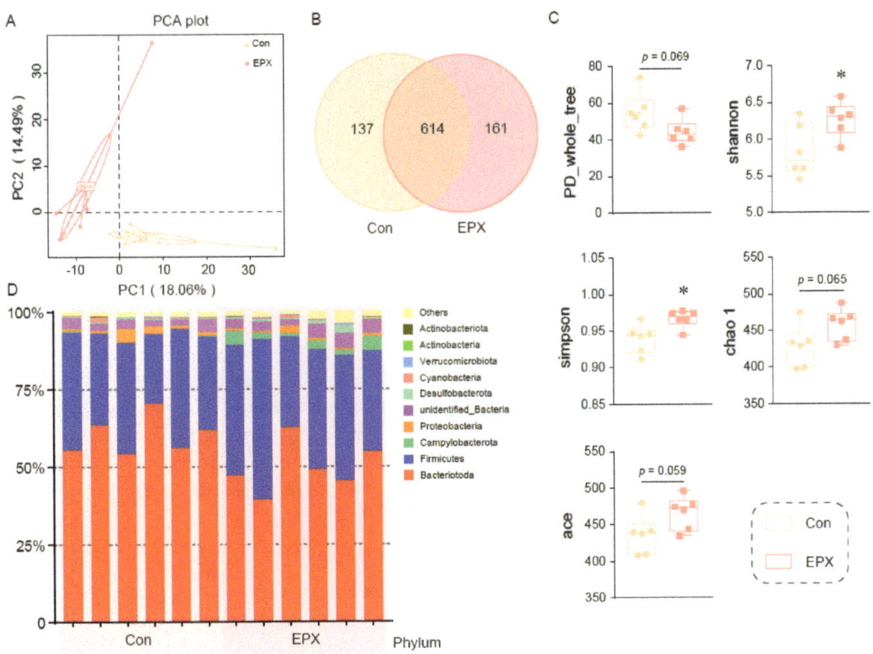

Figure 3. Effects of the oral exposure to EPX on the diversity of intestinal microbiota in the colon of mice ($n = 6$). (**A**) Principal coordinate analysis (PCoA) of OTUs. (**B**) Venn diagram of the OTUs numbers. (**C**) The relative abundance of the top—10 at the phylum level. (**D**) The indices of alpha diversity: PD_whole_tree, Shannon, Simpson, chao 1, ace. Values are shown as the means ± SEM ($n = 6$), and statistical significance: $p < 0.05$ *.

Moreover, the difference in the intestinal microbiota in the colon contents between the EPX treatment group and the Con group was evaluated at the phylum level through RT-qPCR (Figure 4A). The relative abundance of *Actinobacteria*, *Verrucomicrobia*, and *α-Proteobacteria* was elevated significantly ($p < 0.01$) in the EPX-H treatment group. Although our RT-qPCR showed that only the relative abundance of *Firmicutes* declined ($p = 0.03$) in the EPX-L treatment group, 16S rRNA sequencing showed that the relative abundance ratio of *Firmicutes* to *Bacteroides* ($p = 0.02$) was significantly increased (Figure 4B). LEfSe analysis revealed that the relative abundance of *Campylobacteria* was increased significantly (Figure 4C). The relative abundance of some harmful bacteria such as *g_Alistipes*, *g_Helicobacter* was also increased. Specifically, other bacteria genera had been significantly altered (Figure 4D) including those upregulated (*Lachnospiraceae*, *Prevotellaceae*, *Blautia*, and *Colidextribacter*) and downregulated (*Alloprevotella*, *Bacteroides*, *Odoribacter*, and *Anaerostipes*).

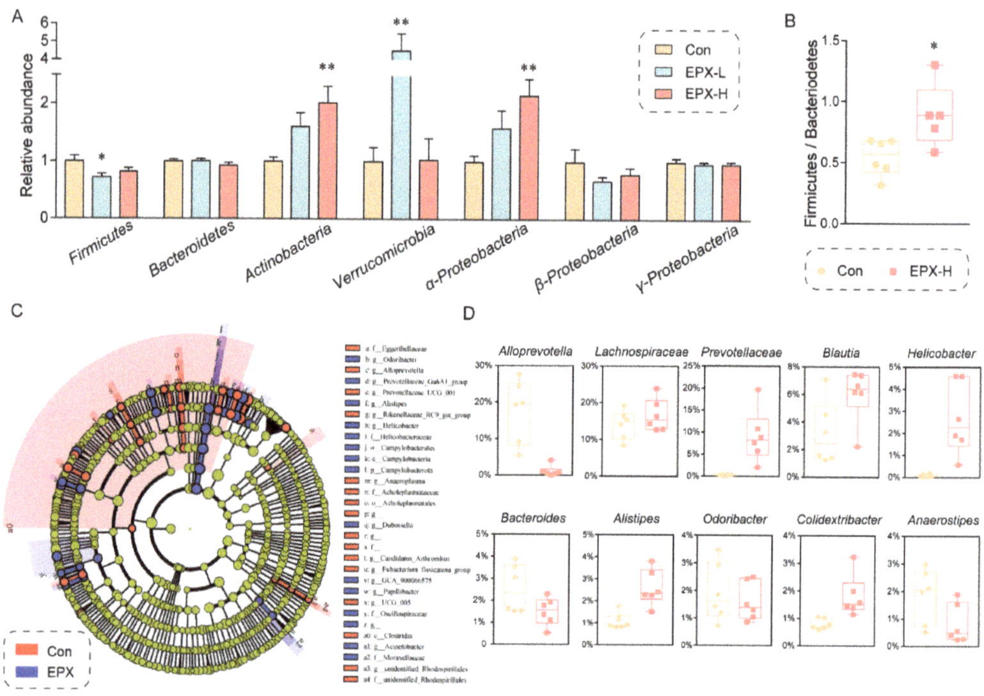

Figure 4. Effects of the oral exposure to EPX on the composition of intestinal microbiota in the colon of mice. (**A**) Relative abundance of the intestinal microbiota at the phylum level. Values are shown as the means ± SEM ($n = 8$), and statistical significance: $p < 0.05$ *; $p < 0.01$ **. (**B**) The ratio of Firmicutes to Bacteroides. (**C**) Cladogram generated from the LEfSe analysis (LDA score > 3). (**D**) The top-10 genera at the genus level.

3.6. Oral Exposure to EPX Disrupted the Metabolic Profile of the Liver

Untargeted metabolomics based on LC-MS revealed differences in the metabolic profiles between the EPX treatment group and the Con group. In the negative and positive modes, the primary differential metabolites in the EPX treatment group showed better separation than that in the Con group (Figure 5A,B), and the metabolome data were reliable (Figure S3). As shown in the heatmap, MSMS secondary analysis suggested that samples from each treatment group in biological replications were clustered together by hierarchical clustering (Figure 5C). Among the identified metabolites, there was a total of 123 differential metabolites (DEMs) including 95 upregulated DEMs and 28 downregulated DEMs (Figure 5D). Some of the various metabolites included organic acids (indoleacetic acid, deoxycholic acid, trans-ferulic acid) and saccharides (D-xylose). In addition, the enrichment analysis of the KEGG pathway based on these DEMs showed that lipid metabolism-related metabolic pathways were the most affected including the PPAR signaling pathway and sphingolipid signaling pathway (Figure 5E). Interestingly, several bacteria with significant variations at the genus level were significantly correlated with differential metabolites such as *Alloprevotella*, *Prevotellaceae*, and *Bacteroides* (Figure 5F). In addition, through the screening of specific metabolites, we found changes in some major metabolites in glycometabolism (glycolysis) and lipid metabolism pathways such as citric acid, adipic acid, and ribitol (Figure 5G).

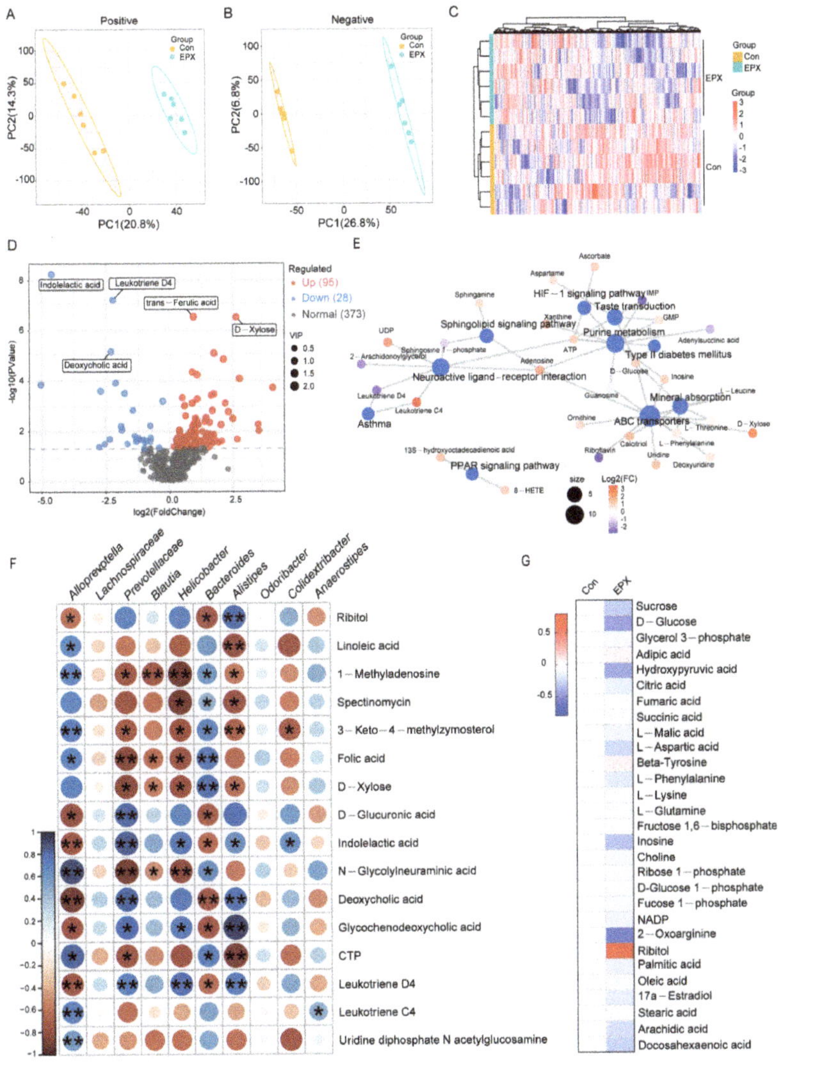

Figure 5. Effects of oral exposure to EPX on the metabolic profiles of the liver in mice. (**A**,**B**) PCA of metabolites in positive and negative ion mode. (**C**) Heatmap of DEMs. (**D**) Volcano map of DEMs. (**E**) Concept network of the KEGG pathway analysis for DEMs. (**F**) The correlation between the top genera conducted with the Spearman's rank test. (**G**) Carbohydrate and lipid metabolites. Significant correlation: $p < 0.05$ *; $p < 0.01$ **.

3.7. Oral Exposure to EPX Affected the Transcription of Genes Related to Lipid Metabolism in the Liver

Based on previous results, we further detected the transcription levels of genes involved in lipid metabolism (Figure 6). EPX downregulated the expression of *PPAR-γ*, *SREBP1c*, and *scd1*, and upregulated the expression of *FAS* and *FAT*, which are involved in fatty acid synthesis (Figure 6A). Notably, the relative expression of *coA-s*, which synthesizes and oxidizes fatty acids and pyruvate, was markedly decreased. The expression of genes involved in lipid β-oxidation was also downregulated including *PPAR-α*, *CPT1*, *acot-1*, and *MCAD* (Figure 6C). The transcription levels of *fatp1* and *fatp2* were also decreased significantly in the EPX-H treatment group, which are involved in fatty acid transport (Figure 6D). Furthermore, the transcription levels of the TG synthesis-related genes were less affected, and the *DGAH* mRNA level was decreased in the EPX-H treatment group (Figure 6E). For the glycolysis and Glu transport, the relative mRNA levels of *GK*, *PK*, and *Glut2* were increased significantly in the EPX-H treatment group (Figure 6F,G).

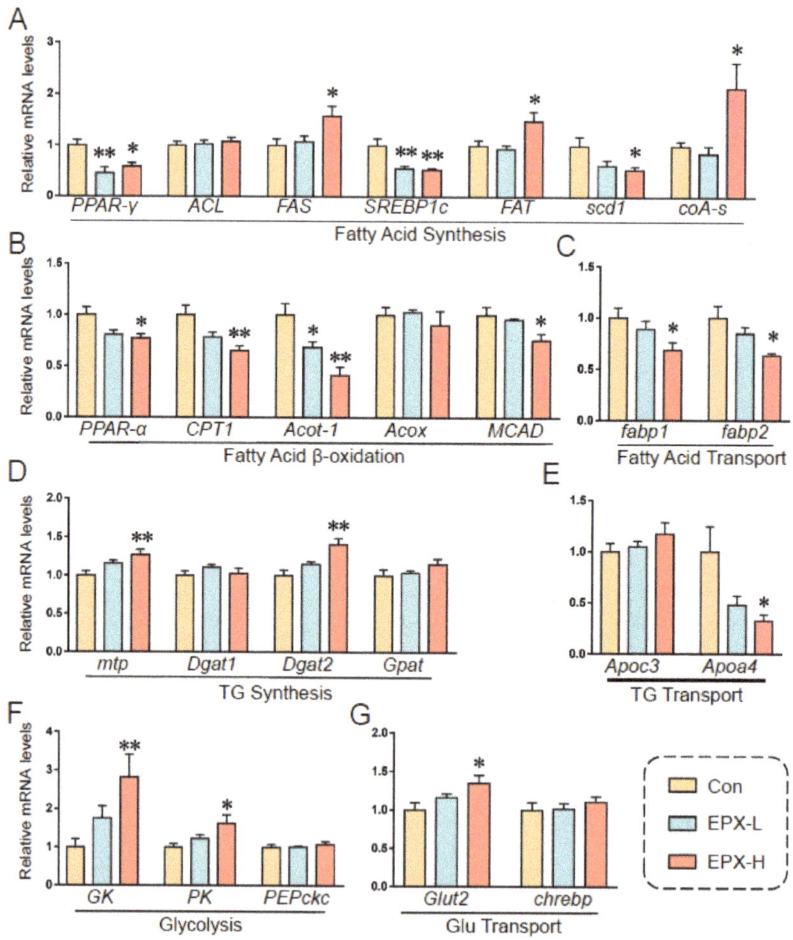

Figure 6. Effects of the oral exposure to EPX on the transcriptional levels of genes related to glycolipid metabolism. (**A**) Fatty acid synthesis. (**B**) Fatty acid β-oxidation. (**C**) Fatty acid transport. (**D**) TG synthesis. (**E**), TG transport. (**F**) Glycolysis. (**G**) Glu transport. Values are shown as the means ± SEM (n = 8), and statistical significance: $p < 0.05$ *; $p < 0.01$ **.

4. Discussion

Health risk assessments for pesticides are always in progress. As one of the most widely used agricultural fungicides, triazoles account for over a quarter of the global sales of fungicides [40]. Their structures are similar to medical triazole drugs and have raised public health concerns [2,41]. Here, we assessed the potential toxicity of the triazole fungicide EPX in mammals. We found that the oral administration of EPX altered the growth phenotype and biochemical indices of the mice. EPX also induced enterotoxicity including impairment of the intestinal barrier function in the colon and intestinal microbiota dysbiosis. More importantly, untargeted metabolomics revealed changes in the liver metabolic profiles of the mice, affecting the lipid metabolic-related metabolic pathways.

The growth phenotype can first reflect the toxic effects of exogenous substances. EPX did not change the body weight of mice throughout the exposure period (Figure 1A). However, EPX significantly increased the liver weight and liver/body weight ratios and decreased kidney/body weight ratios in a concentration-dependent manner (Figure 1B). As previously reported, the liver weights and liver/body weight ratios increased significantly in the CD-1 mice fed with EPX at 50, 200, and 500 ppm [42]. EPX also increased the absolute and relative liver weight in the mice and rats [14,43]. This also supports that EPX does induce hepatotoxic effects. Although the serum AST and ALT levels also did not change in our study (Figure 1D) as well as in the existing findings in rat [43], the hepatotoxicity of triazole fungicides has previously been established [44–47]. After 28 days of EPX exposure, the AST and ALT levels were significantly increased, and liver function was disturbed [14]. This may be due to the significant individual differences in the mice in our study, and the error of the measured results was significant, failing to obtain the changes in these two levels. In addition, the serum and hepatic biochemical indices were also detected. The serum TC levels decreased with increasing EPX exposure concentration, and the hepatic TC level decreased significantly (Table 1). The same result was confirmed by the lower serum TC levels in the mice treated with 200 and 500 ppm EPX [42]. However, the biochemical indices of the parents would have the opposite trend of change [47]. The phenomenon also had the same conclusion in our previous experiments on the passage of maternal exposure [48]. The above results indicated that EPX can change the growth phenotype and physiological indices of the mice and may cause liver toxicity.

The gut is a vital tissue that can prevent harmful substances from entering the body's circulation and acts as an intestinal barrier [49]. The intestinal barrier prevents foreign substances from entering the body through microbial recognition, mucus secretion, tight junction, AMP production, and ion transport [20]. Numerous pollutants can disrupt the intestinal barrier function and homeostasis of intestinal microbiota such as pesticides and microplastics [50,51]. The mucus layer is the physical line of defense of the gastrointestinal tract, and the Muc2 protein is the major structural component of the mucus [52,53]. We found that EPX reduced the colon mucus secretion in mice, and the expression of *Muc2* was obviously decreased in the EPX-L treatment group (Figure 2A,B). Furthermore, *meprinβ*, a gene related to mucus secretion, was significantly reduced in the EPX treatment group. An impaired tight junction is another major cause of barrier dysfunction, resulting in increased intestinal permeability. The Zo-1 anchors membrane proteins (claudin-1), the main component of tight junctions, on the actin cytoskeleton to maintain the integrity of interepithelial connections [54,55]. The mRNA levels of tight junction-related genes were decreased, and EPX had a more significant effect on the expression of *tjp1* and *zo-1* genes (Figure 2E). In addition, some genes associated with AMP expression and ion transport were also affected, mainly *lyz*, *plazr4a* and *slc26a3*. A recent study suggested that EPX alters the pathological structure of the colon of mice including the reduction in mucosal width and villus, and causes the infiltration of intestinal inflammatory cells [56]. The same triazole fungicide, difenoconazole, and prothioconazole disrupted the intestinal barrier function in mice [51,57].

The structure of the intestinal flora of the mice was similar to that of human flora. *Firmicutes* are essential in energy absorption, diabetes, and obesity development [58]. *Actinobac-*

teria can prevent systemic gastrointestinal diseases [59], while *proteobacteria* can prepare the gut for the rigorous colonization of anaerobic bacteria required for healthy intestinal function [60]. An existing study revealed that the relative abundance of *proteobacteria* and *actinobacteria* increased significantly in the intestinal flora of IgA nephropathy patients [61]. In the present study, we found that the abundance of *actinobacteria* and *α-Proteobacteria* was increased in the EPX-H treatment group, and the abundance of firmicutes was decreased in the EPX-L treatment group (Figure 4A). The ratio of *Bacteroides* to *Firmicutes* was increased (Figure 4B), which may contribute to metabolic syndromes such as obesity and diabetes [62]. In addition, *Verrucomicrobia* is a mucin-degrading bacterium that lives in the mucus layer of the large intestine and is involved in maintaining gut integrity [63]. However, abnormally proliferative *Verrucomicrobia* will survive by over-depleting mucin and may cause damage to the intestinal barrier. The abundance of *Verrumicrobia* was increased in the colon of mice exposed to EPX-H. At the genus level, the abundance of some harmful bacteria grew, and that of beneficial bacteria decreased. For example, *alist3ipes* is associated with obesity, and *helicobacter* can cause digestive tract lesions [64,65]. *Bacteroides*, as beneficial bacteria, play an important role in immune system regulation [66]. In contrast, *Bacteroides* at the phylum level were elevated in the gut of EPX-exposed rats [43]. However, the genus level of *Bacteroides* in rats exposed to EPX for 6 weeks also decreased compared to the control samples [13]. This slight difference in genus level is not sufficient to cause a change in the phylum level, which may lead to a difference in the results. This could be a result of a difference between species. In short, EPX disrupts the intestinal flora and may cause metabolic disorders.

Furthermore, we used untargeted metabolomics to explore changes in the mouse liver metabolic profiles. Metabolomics based on liquid chromatography-mass spectrometer (LC-MS) is widely used for the metabolic profiling of urine, serum samples, or animal tissue extracts [67]. The present study found that EPX resulted in apparent changes in the overall metabolite levels in mice (Figure 5A,B). The 1H-NMR-based metabolomics revealed that EPX altered the adult male zebrafish's metabolic spectrum, mainly through increased amino acid levels [68,69]. According to the results of the KEGG enrichment of DEMs, the levels of some amino acids changed significantly including ornithine, L-leucine, and L-phenylalanine (Figure 5E). The same study also found that EPX altered genes related to energy metabolism in adult zebrafish including mitochondrial respiratory chain, ATP synthesis, and fatty acid β-oxidative [68]. Consistent with our research, EPX also affected the PPAR signaling pathway in the liver of the mice offspring [47]. From the perspective of the metabolic pathway, EPX altered metabolites in the pathway associated with energy metabolism in the liver of mice such as ATP, IMP, GMP, and UDP. Based on the correlation analysis of the bacteria with the greatest variation in phylum level and the greatest differences in metabolites, we found that some bacteria were closely related to metabolites such as *Alloprevotella*, *Prevotellaceae*, *Helicobacter*, *Bacteroides*, and *Alistipes* (Figure 5F). We believe that the effects of EPX on the abundance of these harmful bacteria are strongly related to the toxic effect on the liver. In addition, numerous toxicological studies have used transcriptome analysis to effectively detect changes in biomolecular indicators of model organisms, which can explain the mechanism of toxic events [70,71]. The transcriptome is also a good method to further explore the effects of pesticides on the gene level of mice.

Some metabolites in glycolipid metabolism changed significantly (Figure 5G). Again, the transcriptional levels of genes involved in glycolipid metabolism were also affected by EPX in mice (Figure 6). Specifically, with an altered PPAR signaling pathway, we found that the gene expression of *PPAR-γ*, *scd1*, *PPAR-α*, *MCAD CPT1*, *acot-1*, and *fabp1*, decreased significantly, which classified into fatty acid synthesis, fatty acid β-oxidation, and fatty acid transport. Reduced expression levels of CPT1, AOX, and MCAD in adult zebrafish exposed to EPX were also confirmed [68]. In addition, the metabolome revealed that D-glucose decreased as did glucose in the serum and liver (no significant). It may be that the expression of genes involved in glycolysis (*GK*, *PK*) and Glu transport (*Glut2*) increased significantly. Our previous experiments on larval zebrafish have also shown that EPX reduced glucose levels and increased the expression of *GK* and *HK1* but decreased the *PK* expression levels [72]. Transcriptome analysis also suggested that EPX may alter glycolipid metabolic pathways in mice such as cholesterol metabolism as well as the PPAR signaling pathway above-mentioned [47]. It is also our finding that EPX induced glycolipid metabolism in the liver of mice.

5. Conclusions

The toxicity of EPX to the liver and gut of male C57BL/6 mice exposed for 28 days was studied with 16S sequencing and untargeted metabolomics. EPX increased the liver weight and decreased the relative kidney weight. EPX also altered the intestinal barrier function and induced gut microbiota dysbiosis. Furthermore, EPX disrupted the metabolic profile of the liver in mice and induced glycolipid metabolism disorder. These results clarify the effect of EPX on the gut, intestinal microenvironment, and liver metabolism of the mice and provide a more theoretical basis for the risk assessment of EPX.

Supplementary Materials: The following supporting information can be downloaded at: https://www.mdpi.com/article/10.3390/metabo13040522/s1, Method: LC-MS-based metabolomics analysis [73–75]; Figure S1: The ratio of the mucus secretion area of colon; Figure S2: Effects of EPX exposure on the AMPs and ionic transport of colon in mice; Figure S3: PCA score chart of quality control samples; Table S1: The primer sequence pairs used in the RT-qPCR of specific genes; Table S2: The primer sequence pairs used in the RT-qPCR of bacteria at phylum level.

Author Contributions: Conceptualization, Y.W.; methodology, Y.W. and T.X.; validation, T.X.; investigation, T.X.; data curation, C.W.; writing—original draft preparation, Y.W.; writing—review and editing, Y.J.; supervision, C.W. and Y.J.; funding acquisition, Y.J. All authors have read and agreed to the published version of the manuscript.

Funding: The research was supported by the National Natural Science Foundation of China (42277279) and the Natural Science Foundation of Zhejiang Province (LZ20B070002).

Institutional Review Board Statement: The animal study protocol was approved by the Guiding Principles of Zhejiang University of Technology (20211117091, 2021).

Informed Consent Statement: Not applicable.

Data Availability Statement: Not applicable.

Conflicts of Interest: The authors declare no conflict of interest.

Abbreviations

TG	Triglyceride
TC	Total cholesterol
Glu	Glucose
PYR	Pyruvate
HDL	High-density lipoprotein cholesterol
LDL	Low-density lipoprotein cholesterol
NEFA	Non-esterified fatty acid

References

1. Yuan, X.; Pan, Z.; Jin, C.; Ni, Y.; Fu, Z.; Jin, Y. Gut microbiota: An underestimated and unintended recipient for pesticide-induced toxicity. *Chemosphere* **2019**, *227*, 425–434. [CrossRef] [PubMed]
2. Toda, M.; Beer, K.D.; Kuivila, K.M.; Chiller, T.M.; Jackson, B.R. Trends in Agricultural Triazole Fungicide Use in the United States, 1992–2016 and Possible Implications for Antifungal-Resistant Fungi in Human Disease. *Environ. Health Perspect.* **2021**, *129*, 55001. [CrossRef] [PubMed]
3. Konwick, B.J.; Garrison, A.W.; Avants, J.K.; Fisk, A.T. Bioaccumulation and biotransformation of chiral triazole fungicides in rainbow trout (*Oncorhynchus mykiss*). *Aquat. Toxicol.* **2006**, *80*, 372–381. [CrossRef] [PubMed]
4. Hamdi, H.; Rjiba-Touati, K.; Ayed-Boussema, I.; M'nassri, A.; Chaabani, H.; Rich, S.; Abid-Essefi, S. Epoxiconazole caused oxidative stress related DNA damage and apoptosis in PC12 rat Pheochromocytoma. *Neurotoxicology* **2022**, *89*, 184–190. [CrossRef]

5. Lauschke, K.; Dalgaard, M.D.; Emneus, J.; Vinggaard, A.M. Transcriptomic changes upon epoxiconazole exposure in a human stem cell-based model of developmental toxicity. *Chemosphere* **2021**, *284*, 131225. [CrossRef]
6. Tully, D.B.; Bao, W.; Goetz, A.K.; Blystone, C.R.; Ren, H.; Schmid, J.E.; Strader, L.F.; Wood, C.R.; Best, D.S.; Narotsky, M.G.; et al. Gene expression profiling in liver and testis of rats to characterize the toxicity of triazole fungicides. *Toxicol. Appl. Pharmacol.* **2006**, *215*, 260–273. [CrossRef]
7. Price, C.L.; Parker, J.E.; Warrilow, A.G.; Kelly, D.E.; Kelly, S.L. Azole fungicides—Understanding resistance mechanisms in agricultural fungal pathogens. *Pest Manag. Sci.* **2015**, *71*, 1054–1058. [CrossRef]
8. Golianova, K.; Havadej, S.; Verebova, V.; Ulicny, J.; Holeckova, B.; Stanicova, J. Interaction of Conazole Pesticides Epoxiconazole and Prothioconazole with Human and Bovine Serum Albumin Studied Using Spectroscopic Methods and Molecular Modeling. *Int. J. Mol. Sci.* **2021**, *22*, 1925. [CrossRef]
9. Wang, X.; Weng, Y.; Geng, S.; Wang, C.; Jin, C.; Shi, L.; Jin, Y. Maternal procymidone exposure has lasting effects on murine gut-liver axis and glucolipid metabolism in offspring. *Food Chem. Toxicol.* **2023**, *174*, 113657. [CrossRef]
10. Othmène, Y.B.; Monceaux, K.; Karoui, A.; Salem, I.B.; Belhadef, A.; Abid-Essefi, S.; Lemaire, C. Tebuconazole induces ROS-dependent cardiac cell toxicity by activating DNA damage and mitochondrial apoptotic pathway. *Ecotoxicol. Environ. Saf.* **2020**, *204*, 111040. [CrossRef]
11. Albrecht, W. Highlight report: Hepatotoxicity of triazole fungicides. *Arch. Toxicol.* **2019**, *93*, 3037–3038. [CrossRef]
12. Hamdi, H.; Khlifi, A.; Hallara, E.; Houas, Z.; Najjar, M.F.; Abid-Essefi, S. Subchronic exposure to Epoxiconazole induced-heart damage in male Wistar rats. *Pestic. Biochem. Physiol.* **2022**, *182*, 105034. [CrossRef]
13. Kaziem, A.E.; He, Z.; Li, L.; Wen, Y.; Wang, Z.; Gao, Y.; Wang, M. Changes in soil and rat gut microbial diversity after long-term exposure to the chiral fungicide epoxiconazole. *Chemosphere* **2021**, *272*, 129618. [CrossRef]
14. Hamdi, H.; Othmène, Y.B.; Ammar, O.; Klifi, A.; Hallara, E.; Ghali, F.B.; Houas, Z.; Najjar, M.F.; Abid-Essefi, S. Oxidative stress, genotoxicity, biochemical and histopathological modifications induced by epoxiconazole in liver and kidney of Wistar rats. *Environ. Sci. Pollut. Res. Int.* **2019**, *26*, 7535–7547. [CrossRef]
15. Heise, T.; Schmidt, F.; Knebel, C.; Rieke, S.; Haider, W.; Pfeil, R.; Kneuer, C.; Niemann, L.; Marx-Stoelting, P. Hepatotoxic effects of (tri)azole fungicides in a broad dose range. *Arch. Toxicol.* **2015**, *89*, 2105–2117. [CrossRef]
16. Taxvig, C.; Hass, U.; Axelstad, M.; Dalgaard, M.; Boberg, J.; Andeasen, H.R.; Vinggaard, A.M. Endocrine-disrupting activities in vivo of the fungicides tebuconazole and epoxiconazole. *Toxicol. Sci.* **2007**, *100*, 464–473. [CrossRef]
17. Hamdi, H.; Graiet, I.; Abid-Essefi, S.; Eyer, J. Epoxiconazole profoundly alters rat brain and properties of neural stem cells. *Chemosphere* **2022**, *288*, 132640. [CrossRef]
18. Ishibashi, H.; Uchida, K.; Nishiyama, Y.; Yamaguchi, H.; Abe, S. Oral administration of itraconazole solution has superior efficacy in experimental oral and oesophageal candidiasis in mice than its intragastric administration. *J. Antimicrob. Chemother.* **2007**, *59*, 317–320. [CrossRef]
19. Fernandez-Tome, S.; Hernandez-Ledesma, B. Gastrointestinal Digestion of Food Proteins under the Effects of Released Bioactive Peptides on Digestive Health. *Mol. Nutr. Food Res.* **2020**, *64*, e2000401. [CrossRef]
20. Natividad, J.M.; Verdu, E.F. Modulation of intestinal barrier by intestinal microbiota: Pathological and therapeutic implications. *Pharmacol. Res.* **2013**, *69*, 42–51. [CrossRef]
21. Jacob, C.; Yang, P.C.; Darmoul, D.; Amadesi, S.; Saito, T.; Cottrell, G.S.; Coelho, A.M.; Singh, P.; Grady, E.F.; Perdue, M.; et al. Mast cell tryptase controls paracellular permeability of the intestine. Role of protease-activated receptor 2 and beta-arrestins. *J. Biol. Chem.* **2005**, *280*, 31936–31948. [CrossRef] [PubMed]
22. Eisenstein, M. Gut reaction. *Nature* **2018**, *563*, S34–S35. [CrossRef] [PubMed]
23. Van der Sluis, M.; De Koning, B.A.; De Bruijn, A.C.; Velcich, A.; Meijerink, J.P.; Van Goudoever, J.B.; Büller, H.A.; Dekker, J.; Van Seuningen, I.; Renes, I.B.; et al. Muc2-deficient mice spontaneously develop colitis, indicating that MUC2 is critical for colonic protection. *Gastroenterology* **2006**, *131*, 117–129. [CrossRef] [PubMed]
24. Preidis, G.A.; Ajami, N.J.; Wong, M.C.; Bessard, B.C.; Conner, M.E.; Petrosino, J.F. Composition and function of the undernourished neonatal mouse intestinal microbiome. *J. Nutr. Biochem.* **2015**, *26*, 1050–1057. [CrossRef]
25. Usuda, H.; Okamoto, T.; Wada, K. Leaky Gut: Effect of Dietary Fiber and Fats on Microbiome and Intestinal Barrier. *Int. J. Mol. Sci.* **2021**, *22*, 7613. [CrossRef]
26. Guarner, F.; Malagelada, J.R. Gut flora in health and disease. *Lancet* **2003**, *361*, 512–519. [CrossRef]
27. Stecher, B.; Hardt, W.D. The role of microbiota in infectious disease. *Trends Microbiol.* **2008**, *16*, 107–114. [CrossRef]
28. Yuan, M.; Breitkopf, S.B.; Yang, X.; Asara, J.M. A positive/negative ion-switching, targeted mass spectrometry-based metabolomics platform for bodily fluids, cells, and fresh and fixed tissue. *Nat. Protoc.* **2012**, *7*, 872–881. [CrossRef]
29. Pan, Y.; Chang, J.; Wan, B.; Liu, Z.; Yang, L.; Xie, Y.; Hao, W.; Li, J.; Xu, P. Integrative analysis of transcriptomics and metabolomics reveals the hepatotoxic mechanism of thiamethoxam on male Coturnix japonica. *Environ. Pollut.* **2022**, *293*, 118460. [CrossRef]
30. Yang, J.S.; Qi, W.; Farias-Pereira, R.; Choi, S.; Clark, J.M.; Kim, D.; Park, Y. Permethrin and ivermectin modulate lipid metabolism in steatosis-induced HepG2 hepatocyte. *Food Chem. Toxicol.* **2019**, *125*, 595–604. [CrossRef]
31. Ku, T.; Zhou, M.; Hou, Y.; Xie, Y.; Li, G.; Sang, N. Tebuconazole induces liver injury coupled with ROS-mediated hepatic metabolism disorder. *Ecotoxicol. Environ. Saf.* **2021**, *220*, 112309. [CrossRef]
32. Wang, C.; Zhang, Y.; Deng, M.; Wang, X.; Tu, W.; Fu, Z.; Jin, Y. Bioaccumulation in the gut and liver causes gut barrier dysfunction and hepatic metabolism disorder in mice after exposure to low doses of OBS. *Environ. Int.* **2019**, *129*, 279–290. [CrossRef]

33. Li, J.; Hu, Y.; Liu, L.; Wang, Q.; Zeng, J.; Chen, C. PM2.5 exposure perturbs lung microbiome and its metabolic profile in mice. *Sci. Total Environ.* **2020**, *721*, 137432. [CrossRef]
34. Chen, L.; Lu, W.; Wang, L.; Xing, X.; Chen, Z.; Teng, X.; Zeng, X.; Muscarella, A.D.; Shen, Y.; Cowan, A.; et al. Metabolite discovery through global annotation of untargeted metabolomics data. *Nat. Methods* **2021**, *18*, 1377–1385. [CrossRef]
35. Jin, Y.; Lu, L.; Tu, W.; Luo, T.; Fu, Z. Impacts of polystyrene microplastic on the gut barrier, microbiota and metabolism of mice. *Sci. Total Environ.* **2019**, *649*, 308–317. [CrossRef]
36. Livak, K.J.; Schmittgen, T.D. Analysis of relative gene expression data using real-time quantitative PCR and the 2(-Delta Delta C(T)) Method. *Methods* **2001**, *25*, 402–408. [CrossRef]
37. Wang, Y.; Jin, C.; Wang, D.; Zhou, J.; Yang, G.; Shao, K.; Wang, Q.; Jin, Y. Effects of chlorothalonil, prochloraz and the combination on intestinal barrier function and glucolipid metabolism in the liver of mice. *J. Hazard. Mater.* **2021**, *410*, 124639. [CrossRef]
38. Edgar, R.C. UPARSE: Highly accurate OTU sequences from microbial amplicon reads. *Nat. Methods* **2013**, *10*, 996–998. [CrossRef]
39. Wan, Z.; Wang, C.; Zhou, J.; Shen, M.; Wang, X.; Fu, Z.; Jin, Y. Effects of polystyrene microplastics on the composition of the microbiome and metabolism in larval zebrafish. *Chemosphere* **2019**, *217*, 646–658. [CrossRef]
40. ECDC (European Centre for Disease Prevention and Control). *Risk Assessment on the Impact of Environmental Usage of Triazoles on the Development and Spread of Resistance to Medical Triazoles in Aspergillus Species*; ECDC: Stockholm, Sweden, 2013.
41. Snelders, E.; Huis In't Veld, R.A.; Rijs, A.J.; Kema, G.H.; Melchers, W.J.; Verweij, P.E. Possible environmental origin of resistance of Aspergillus fumigatus to medical triazoles. *Appl. Environ. Microbiol.* **2009**, *75*, 4053–4057. [CrossRef]
42. Hester, S.; Moore, T.; Padgett, W.T.; Murphy, L.; Wood, C.E.; Nesnow, S. The Hepatocarcinogenic Conazoles: Cyproconazole, Epoxiconazole, and Propiconazole Induce a Common Set of Toxicological and Transcriptional Responses. *Toxicol. Sci.* **2012**, *127*, 54–65. [CrossRef] [PubMed]
43. Xu, C.; Liu, Q.; Huan, F.; Qu, J.; Liu, W.; Gu, A.; Wang, Y.; Jiang, Z. Changes in Gut Microbiota May Be Early Signs of Liver Toxicity Induced by Epoxiconazole in Rats. *Chemotherapy* **2014**, *60*, 135–142. [CrossRef] [PubMed]
44. Qiu, L.; Jia, K.; Huang, L.; Liao, X.; Guo, X.; Lu, H. Hepatotoxicity of tricyclazole in zebrafish (*Danio rerio*). *Chemosphere* **2019**, *232*, 171–179. [CrossRef] [PubMed]
45. Jiang, J.; Chen, L.; Wu, S.; Lv, L.; Liu, X.; Wang, Q.; Zhao, X. Effects of difenoconazole on hepatotoxicity, lipid metabolism and gut microbiota in zebrafish (*Danio rerio*). *Environ. Pollut.* **2020**, *265*, 114844. [CrossRef]
46. Marx-Stoelting, P.; Ganzenberg, K.; Knebel, C.; Schmidt, F.; Rieke, S.; Hammer, H.; Schmidt, F.; Pötz, O.; Schwarz, M.; Braeuning, A. Hepatotoxic effects of cyproconazole and prochloraz in wild-type and hCAR/hPXR mice. *Arch. Toxicol.* **2017**, *91*, 2895–2907. [CrossRef]
47. Le Corre, L.; Brulport, A.; Vaiman, D.; Chagnon, M.C. Epoxiconazole alters the histology and transcriptome of mouse liver in a transgenerational pattern. *Chem. Biol. Interact.* **2022**, *360*, 109952. [CrossRef]
48. Wang, C.; Weng, Y.; Tu, W.; Jin, C.; Jin, Y. Maternal exposure to sodium rho-perfluorous nonenoxybenzene sulfonate during pregnancy and lactation disrupts intestinal barrier and may cause obstacles to the nutrient transport and metabolism in F0 and F1 generations of mice. *Sci. Total Environ.* **2021**, *794*, 148775. [CrossRef]
49. Magalhaes, J.G.; Tattoli, I.; Girardin, S.E. The intestinal epithelial barrier: How to distinguish between the microbial flora and pathogens. *Semin. Immunol.* **2007**, *19*, 106–115. [CrossRef]
50. Luo, T.; Wang, C.; Pan, Z.; Jin, C.; Fu, Z.; Jin, Y. Maternal Polystyrene Microplastic Exposure during Gestation and Lactation Altered Metabolic Homeostasis in the Dams and Their F1 and F2 Offspring. *Environ. Sci. Technol.* **2019**, *53*, 10978–10992. [CrossRef]
51. Zhang, H.; Yang, G.; Bao, Z.; Jin, Y.; Wang, J.; Chen, J.; Qian, M. Stereoselective effects of fungicide difenoconazole and its four stereoisomers on gut barrier, microbiota, and glucolipid metabolism in male mice. *Sci. Total Environ.* **2022**, *805*, 150454. [CrossRef]
52. Johansson, M.E.; Phillipson, M.; Petersson, J.; Velcich, A.; Holm, L.; Hansson, G.C. The inner of the two Muc2 mucin-dependent mucus layers in colon is devoid of bacteria. *Proc. Natl. Acad. Sci. USA* **2008**, *105*, 15064–15069. [CrossRef]
53. Donaldson, G.P.; Lee, S.M.; Mazmanian, S.K. Gut biogeography of the bacterial microbiota. *Nat. Rev. Microbiol.* **2016**, *14*, 20–32. [CrossRef]
54. Chiba, H.; Osanai, M.; Murata, M.; Kojima, T.; Sawada, N. Transmembrane proteins of tight junctions. *Biochim. Biophys. Acta* **2008**, *1778*, 588–600. [CrossRef]
55. Suzuki, T. Regulation of intestinal epithelial permeability by tight junctions. *Cell Mol. Life Sci.* **2013**, *70*, 631–659. [CrossRef]
56. Sun, W.; Yan, S.; Meng, Z.; Tian, S.; Jia, M.; Huang, S.; Wang, Y.; Zhou, Z.; Diao, J.; Zhu, W. Combined ingestion of polystyrene microplastics and epoxiconazole increases health risk to mice: Based on their synergistic bioaccumulation in vivo. *Environ. Int.* **2022**, *166*, 107391. [CrossRef]
57. Hu, L.; Wang, X.; Bao, Z.; Xu, Q.; Qian, M.; Jin, Y. The fungicide prothioconazole and its metabolite prothioconazole-desthio disturbed the liver-gut axis in mice. *Chemosphere* **2022**, *307*, 136141. [CrossRef]
58. Ley, R.E.; Turnbaugh, P.J.; Klein, S.; Gordon, J. Human gut microbes associated with obesity. *Nature* **2006**, *444*, 1022–1023. [CrossRef]
59. Binda, C.; Lopetuso, L.R.; Rizzatti, G.; Gibiino, G.; Cennamo, V.; Gasbarrini, A. Actinobacteria: A relevant minority for the maintenance of gut homeostasis. *Dig. Liver Dis.* **2018**, *50*, 421–428. [CrossRef]
60. Shin, N.R.; Whon, T.W.; Bae, J.W. Proteobacteria: Microbial signature of dysbiosis in gut microbiota. *Trends Biotechnol.* **2015**, *33*, 496–503. [CrossRef]

61. Zhong, Z.; Tan, J.; Tan, L.; Tang, Y.; Qiu, Z.; Pei, G.; Qin, W. Modifications of gut microbiota are associated with the severity of IgA nephropathy in the Chinese population. *Int. Immunopharmacol.* **2020**, *89*, 107085. [CrossRef]
62. Grigor'eva, I.N. Gallstone Disease, Obesity and the Firmicutes/Bacteroidetes Ratio as a Possible Biomarker of Gut Dysbiosis. *J. Pers. Med.* **2020**, *11*, 13. [CrossRef] [PubMed]
63. Geerlings, S.Y.; Kostopoulos, I.; de Vos, W.M.; Belzer, C. Akkermansia muciniphila in the Human Gastrointestinal Tract: When, Where, and How? *Microorganisms* **2018**, *6*, 75. [CrossRef] [PubMed]
64. Zeng, Q.; Li, D.; He, Y.; Li, Y.; Yang, Z.; Zhao, X.; Liu, Y.; Wang, Y.; Sun, J.; Feng, X.; et al. Discrepant gut microbiota markers for the classification of obesity-related metabolic abnormalities. *Sci. Rep.* **2019**, *9*, 13424. [CrossRef] [PubMed]
65. Liu, D.; Chen, S.; Gou, Y.; Yu, W.; Zhou, H.; Zhang, R.; Wang, J.; Ye, F.; Liu, Y.; Sun, B.; et al. Gastrointestinal Microbiota Changes in Patients with Gastric Precancerous Lesions. *Front. Cell Infect. Microbiol.* **2021**, *11*, 749207. [CrossRef] [PubMed]
66. Zafar, H.; Saier, M.H., Jr. Gut Bacteroides species in health and disease. *Gut Microbes* **2021**, *13*, 1848158. [CrossRef]
67. Yang, G.; Wang, Y.; Li, J.; Wang, D.; Bao, Z.; Wang, Q.; Jin, Y. Health risks of chlorothalonil, carbendazim, prochloraz, their binary and ternary mixtures on embryonic and larval zebrafish based on metabolomics analysis. *J. Hazard. Mater.* **2021**, *404*, 124240. [CrossRef]
68. Wang, Y.; Teng, M.; Wang, D.; Yan, J.; Miao, J.; Zhou, Z.; Zhu, W. Enantioselective bioaccumulation following exposure of adult zebrafish (*Danio rerio*) to epoxiconazole and its effects on metabolomic profile as well as genes expression. *Environ. Pollut.* **2017**, *229*, 264–271. [CrossRef]
69. Jia, M.; Wang, Y.; Wang, D.; Teng, M.; Yan, J.; Yan, S.; Meng, Z.; Li, R.; Zhou, Z.; Zhu, W. The effects of hexaconazole and epoxiconazole enantiomers on metabolic profile following exposure to zebrafish (*Danio rerio*) as well as the histopathological changes. *Chemosphere* **2019**, *226*, 520–533. [CrossRef]
70. Martinez, R.; Codina, A.E.; Barata, C.; Tauler, R.; Pina, B.; Navarro-Martin, L. Transcriptomic effects of tributyltin (TBT) in zebrafish eleutheroembryos. A functional benchmark dose analysis. *J. Hazard. Mater.* **2020**, *398*, 122881. [CrossRef]
71. Fang, L.; Fang, C.; Di, S.; Yu, Y.; Wang, C.; Wang, X.; Jin, Y. Oral exposure to tire rubber-derived contaminant 6PPD and 6PPD-quinone induce hepatotoxicity in mice. *Sci. Total Environ.* **2023**, *869*, 161836. [CrossRef]
72. Weng, Y.; Huang, Z.; Wu, A.; Yu, Q.; Lu, H.; Lou, Z.; Lu, L.; Bao, Z.; Jin, Y. Embryonic toxicity of epoxiconazole exposure to the early life stage of zebrafish. *Sci. Total Environ.* **2021**, *778*, 146407. [CrossRef]
73. Zelena, E.; Dunn, W.B.; Broadhurst, D.; Francis-McIntyre, S.; Carroll, K.M.; Begley, P.; O'Hagan, S.; Knowles, J.D.; Halsall, A.; HUSERMET Consortium; et al. Development of a Robust and Repeatable UPLC-MS Method for the Long-Term Metabolomic Study of Human Serum. *Anal. Chem.* **2009**, *81*, 1357–1364. [CrossRef]
74. Gagnebin, Y.; Tonoli, D.; Lescuyer, P.; Ponte, B.; de Seigneux, S.; Martin, P.Y.; Schappler, J.; Boccard, J.; Rudaz, S. Metabolomic analysis of urine samples by UHPLC-QTOF-MS: Impact of normalization strategies. *Anal. Chima. Acta.* **2017**, *955*, 27–35. [CrossRef]
75. Kieffer, D.A.; Piccolo, B.D.; Vaziri, N.D.; Liu, S.; Lau, W.L.; Khazaeli, M.; Nazertehrani, S.; Moore, M.E.; Marco, M.L.; Martin, R.J.; et al. Resistant starch alters gut microbiome and metabolomic profiles concurrent with amelioration of chronic kidney disease in rats. *Am. J. Physiol. Renal. Physiol.* **2016**, *310*, F857–F871. [CrossRef] [PubMed]

metabolites

Review

Gut Microbiota Modulation in Osteoporosis: Probiotics, Prebiotics, and Natural Compounds

Xufeng Chu [1], Hailin Xing [1], Minghao Chao [1], Panpan Xie [1,*] and Lili Jiang [2,*]

[1] Department of Orthopedic Surgery, The Fifth Affiliated Hospital of Wenzhou Medical University, Lishui Municipal Central Hospital, 289 Kuocang Road, Lishui 323000, China; doctorchuxf@163.com (X.C.); lishuixhl@163.com (H.X.); 17772293609@163.com (M.C.)

[2] Department of Laboratory Medicine, The Fifth Affiliated Hospital of Wenzhou Medical University, Lishui Municipal Central Hospital, 289 Kuocang Road, Lishui 323000, China

* Correspondence: xpp2025qwer@163.com (P.X.); jianglili0412@163.com (L.J.)

Abstract: Osteoporosis is a multifactorial bone metabolic disorder characterized by the deterioration of bone mass and microarchitecture, leading to increased fragility and fracture risk. Recent advances have revealed the critical role of the gut microbiota in the pathogenesis of osteoporosis, primarily mediated by metabolite-driven and immune-mediated interactions along the gut–bone axis. Dysbiosis, or microbial imbalance, can influence bone health by modulating host metabolism, immune function, and endocrine responses. While growing evidence suggests that gut microbiota modulation holds therapeutic potential for osteoporosis, the underlying mechanisms remain poorly understood. This review examines the latest findings on the role of prebiotics, probiotics, and natural bioactive substances in modulating the gut microbiota to improve bone health. We discuss how these interventions may restore microbial balance, enhance gut barrier function, and reduce systemic inflammation, thereby influencing bone metabolism. A deeper understanding of the gut–bone axis will pave the way for more targeted, effective, and personalized therapeutic strategies for osteoporosis prevention and treatment.

Keywords: gut microbiota; modulation; osteoporosis; natural compounds

1. Osteoporosis and Gut Microbiota

1.1. Osteoporosis

Osteoporosis is a systemic skeletal disorder characterized by reduced bone mass and the deterioration of bone microarchitecture, which increases bone fragility and susceptibility to fractures. The pathogenesis primarily involves an imbalance in bone metabolism, where bone resorption exceeds bone formation, leading to progressive bone loss. Osteoporosis is most commonly seen in the elderly, particularly postmenopausal women and older men, as bone density decreases with age [1]. According to the World Health Organization (WHO), approximately 41.5 million people worldwide are affected by osteoporosis, making it a major global health issue [2].

Primary osteoporosis arises from age-related bone loss or estrogen deficiency (e.g., postmenopausal osteoporosis), while secondary osteoporosis results from medical conditions (e.g., hyperparathyroidism) or medications (e.g., glucocorticoids). Risk factors include genetic predisposition, age, gender, hormonal changes, poor nutrition (e.g., insufficient calcium and vitamin D), lack of physical activity, smoking, and excessive alcohol consumption [3]. Additionally, medical conditions such as hyperparathyroidism, rheumatoid arthritis, and long-term use of glucocorticoids also increase osteoporosis risk [4,5].

 https://doi.org/10.3390/metabo15050301

Osteoporosis is often asymptomatic, with many individuals unaware of their condition until they experience a fracture. Given its gradual and painless progression, osteoporosis is sometimes referred to as the "silent killer" or "iceberg disease" [6]. Globally, osteoporosis is responsible for 8.9 million fractures annually, with common fracture sites including the spine, hip, and forearm, and an increasing frequency of fragility fractures in the pelvis [7].

1.2. Gut Microbiota

The gut microbiota is a complex microbial ecosystem comprising bacteria, archaea, viruses, and fungi that reside in the human digestive tract. In adults, approximately 100 trillion microorganisms make up the gut microbiota, whose collective genetic material, the microbiome, vastly outnumbers the human genome. This microbiome is often referred to as the "second genome" [8]. In recent years, attention has focused on the gut microbiota's role in maintaining host health and regulating disease processes, including metabolic, immune, nervous, and endocrine functions.

Metabolically, the gut microbiota breaks down dietary fibers and polysaccharides that the host cannot digest, producing short-chain fatty acids (SCFAs) like acetate, propionate, and butyrate. These metabolites serve as important energy sources for intestinal epithelial cells and modulate immune responses and energy metabolism [9]. The microbiota also synthesizes essential amino acids, vitamins (such as vitamins K and B), and regulates bile acid and lipid metabolism, playing a critical role in maintaining metabolic balance [10,11]. Additionally, the gut microbiota is crucial for the development and function of the immune system, interacting with intestinal immune cells to modulate both innate and adaptive immune responses [12,13].

Homeostasis of the gut microbiota is vital for health. Dysbiosis, or an imbalance in microbial populations, is linked to various diseases, including obesity, diabetes, inflammatory bowel disease, allergic disorders, depression, neurodegenerative diseases, and certain cancers [14]. Dysbiosis is typically marked by a reduction in beneficial microbes, an overgrowth of pathogenic species, and decreased microbial diversity. Factors such as poor diet, overuse of antibiotics, infections, stress, and lack of physical activity can trigger dysbiosis [15].

In summary, the gut microbiota, often referred to as an "invisible organ" [16], plays a crucial role in both physiological and pathological processes. Further exploration of the relationship between the microbiota and host health could lead to novel therapeutic strategies and personalized health management approaches.

1.3. Gut–Bone Axis

Recent studies have emphasized the pivotal role of the gut microbiota in regulating bone metabolism through the "gut–bone axis". Dysbiosis significantly influences bone health by modulating the host's metabolic, immune, and endocrine systems, contributing to the development and progression of osteoporosis and bone loss [17,18]. The growing recognition of the gut microbiota as a potential regulator of bone health highlights its complex and multifaceted interactions with bone metabolism.

The gut microbiota can directly or indirectly affect bone mass through various mechanisms, including the modulation of host metabolism, calcium absorption, hormone levels, immune function, and the central nervous system [17,19]. However, despite progress in understanding the gut–bone axis, the underlying mechanisms remain poorly defined. The bidirectional interactions between bone and the gut microbiota, and the specific regulatory pathways involved, require further investigation. This gap in knowledge hinders our understanding of osteoporosis pathogenesis and limits the development of microbiota-based interventions.

Nevertheless, the gut microbiota presents significant potential as a therapeutic target for osteoporosis and related bone diseases. Microbiota-based approaches, such as probiotics, prebiotics, and natural bioactive compounds, could offer effective strategies for preventing and treating osteoporosis. These interventions aim to restore a healthy microbial balance early, suppress bone disease, and explore new avenues for osteoporosis prevention and treatment.

1.4. Modulation of Gut Microbiota

Gut microbiota modulation can be achieved through various interventions aimed at restoring microbial balance and optimizing host health. Key strategies include dietary changes, probiotics, prebiotics, antibiotics, FMT, and bioactive compounds [20]. These approaches influence microbiota composition and function, thereby impacting host metabolism, immune responses, and overall health.

Studies have identified significant differences in the gut microbiota of osteoporosis patients compared to healthy controls [21]. Targeting the gut microbiota in osteoporosis has thus emerged as an important area of research. Investigating gut microbiota–bone interactions can help uncover the pathophysiological mechanisms of osteoporosis and provide a scientific foundation for developing new therapeutic strategies.

Probiotics and prebiotics, as key modulators of the gut microbiota, have been shown to significantly influence bone metabolism. Probiotics enhance bone health by increasing beneficial gut bacteria, restoring microbial balance, and modulating immune responses to reduce inflammation in bone metabolism. For example, *Lactobacillus rhamnosus GG* has been shown to regulate the dysbiosis in the gut of ovariectomized mice. This probiotic promotes the secretion of anti-inflammatory cytokines, such as TGF-β and IL-10, while suppressing pro-inflammatory cytokines like TNF-α and IL-17. These effects help mitigate estrogen deficiency-induced inflammation, reduce osteoclastic activity, and improve osteoporosis [22].

Prebiotics, which serve as a "fuel" for probiotics, indirectly support bone health by promoting the growth and activity of beneficial gut bacteria. The fermentation of prebiotics in the gut produces short-chain fatty acids (SCFAs), such as butyrate, that activate gut receptors to inhibit bone resorption and stimulate bone formation [23]. Additionally, prebiotics improve intestinal barrier function, reduce gut permeability, and limit the passage of harmful bacteria and their toxins, further protecting bone health [24].

In addition, plant-derived natural compounds, particularly polyphenols and flavonoids, are increasingly recognized for their ability to regulate the gut microbiota and improve bone health. These compounds positively influence bone metabolism through various mechanisms. Oxidative stress is a key factor in osteoporosis, and polyphenolic compounds have strong antioxidant properties that help alleviate oxidative damage. Moreover, grape seed extract has been shown to modulate gut microbiota composition, increase beneficial bacterial populations, and alter their metabolic profiles, thereby protecting bone mineral density and slowing osteoporosis progression [25].

In summary, the regulation of gut microbiota plays a pivotal role in bone metabolism and may significantly influence the onset and progression of osteoporosis. While current evidence supports the potential of probiotics, prebiotics, and plant-derived compounds in bone health, further investigation is needed to understand the underlying mechanisms and optimize these strategies for clinical applications.

1.5. Potential Mechanisms of Gut Microbiota in Bone Metabolism

Although substantial evidence supports the link between gut microbiota and bone metabolism, the exact molecular mechanisms remain insufficiently understood. Key questions persist regarding the roles of microbial metabolites, signaling pathways, and immune modulators in maintaining bone health.

One significant mechanism by which gut microbiota may influence bone homeostasis is through the diffusion of intestinal metabolites into systemic circulation (Figure 1, Part ①). Among these metabolites, short-chain fatty acids (SCFAs) such as butyrate, propionate, acetate, and valerate have been central to gut–bone axis research [26]. For example, valerate supplementation in ovariectomized mice has been shown to reduce bone resorption and improve bone microstructure [27]. Valerate exerts its effects primarily by inhibiting RELA protein production and enhancing IL-10 mRNA expression, which suppresses osteoclast maturation while promoting osteoblast differentiation. Similarly, sodium butyrate has been shown to alleviate oxidative stress and inflammation in rats, promoting osteoblast differentiation and inhibiting osteoclast differentiation [28]. Additionally, sodium butyrate enhances SIRT1 expression, a NAD^+-dependent deacetylase that mitigates oxidative stress and promotes osteoblast differentiation while concurrently inhibiting histone deacetylases (HDACs) [17,28]. This dual action restores bone metabolism by balancing acetylation-dependent signaling, ultimately improving bone strength and density in osteoporotic models [17]. Taken together, these data suggest that SCFAs are effective regulators of bone cell metabolism and play a significant role in maintaining bone homeostasis [29].

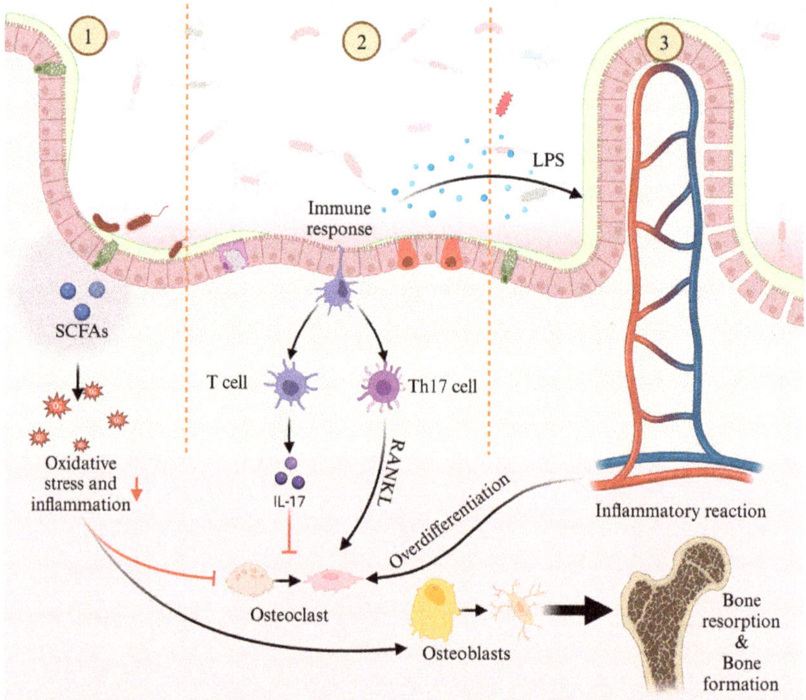

Figure 1. Potential mechanisms of gut microbiota in regulating osteoporosis: ① The metabolites of gut microbiota, short-chain fatty acids, modulate osteoblasts and osteoclasts through oxidative stress and inflammatory responses, affecting bone resorption and bone formation; ② Gut microbiota influences bone resorption by regulating the balance between T cell and Th17 cell; ③ Inflammatory factors, such as LPS, which are metabolites of gut microbiota, excessively activate osteoclasts through inflammatory responses, impacting bone resorption and bone formation.

Furthermore, gut microbiota interact with immune cells to regulate both innate and adaptive immune responses [30] (Figure 1, Part ②). CD4+ T helper (Th) cells play a pivotal role in immune function, modulating other immune cells, such as B lymphocytes, through surface receptors and cytokines. In the gut, Th cell expansion facilitates their migration to the bone marrow, where they increase the recruitment of inflammatory monocytes as osteoclast precursors [31]. Imbalances between Th17 and regulatory T (Treg) cells have been linked to bone health, as Th17 cells promote osteoclast differentiation and bone resorption via the RANKL pathway, while Treg cells suppress osteoclast differentiation [32].

Finally, increased intestinal permeability allows inflammatory mediators, such as lipopolysaccharides (LPS), to enter systemic circulation, triggering inflammation and excessive osteoclast activation (Figure 1, Part ③). LPS has been implicated in postmenopausal osteoporosis, where it promotes femoral bone loss and enhances osteoclast survival while inhibiting osteogenic differentiation of osteoclast precursors [33,34].

Understanding these mechanisms is critical for developing microbiota-targeted therapies. Therefore, the following sections explore research progress on the role of prebiotics, probiotics, and natural bioactive compounds in modulating the gut microbiota to improve osteoporosis, laying the foundation for further investigations into the specific mechanisms of these interventions, their impacts on the gut microbiota, and the development of more precise and personalized intervention strategies (Figure 2).

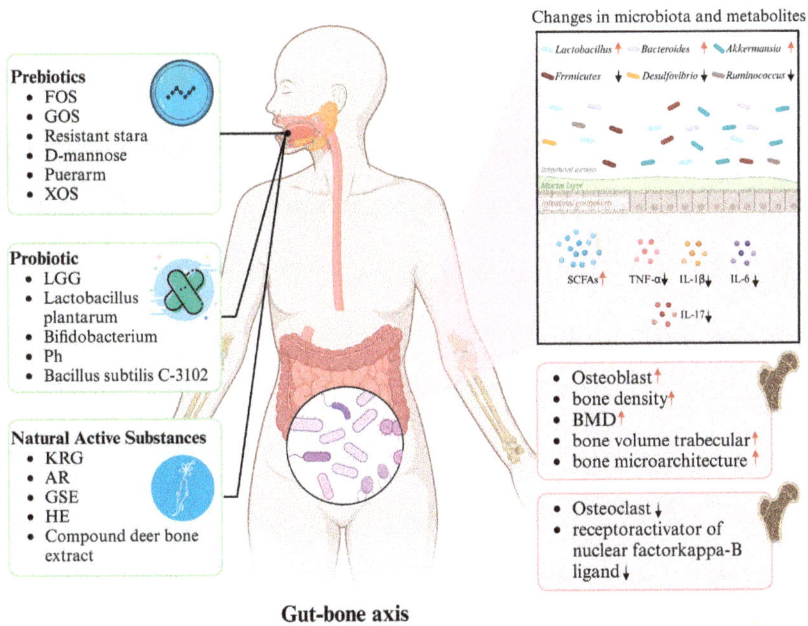

Figure 2. The pivotal role of gut microbiota in mediating the effects of prebiotics, probiotics, and natural substances on osteoporosis regulation: Upon ingestion, these bioactive agents modulate the composition and relative abundance of gut microbiota, thereby inducing alterations in metabolite profiles, which subsequently drive structural bone remodeling and contribute to the amelioration of osteoporosis.

1.6. NAFLD/MAFLD and Its Emerging Role in Bone Metabolism

Non-alcoholic fatty liver disease (NAFLD) is a multisystemic condition and has become the leading cause of chronic liver disease worldwide, affecting approximately 25% of the global population and posing a significant burden on healthcare systems [35]. To better reflect its metabolic underpinnings, an international panel of experts has proposed renaming NAFLD as metabolic dysfunction-associated fatty liver disease (MAFLD) [36].

NAFLD/MAFLD is increasingly recognized not only as a liver-centric disease but also as a condition with systemic implications. A growing body of epidemiological evidence has established a link between NAFLD/MAFLD and osteoporosis. Patients with NAFLD exhibit a significantly higher risk of developing osteoporosis [37]. In a U.S. population-based study involving individuals aged 20–59 across various races and sexes, NAFLD was independently associated with reduced bone mineral density (BMD) [38]. Similarly, Chinese cohort studies have shown a consistent inverse relationship between NAFLD and BMD, regardless of sex [39,40]. These findings suggest that NAFLD contributes to osteoporosis not only through impaired hepatic function but also via systemic metabolic alterations. Proposed mechanisms include the dysregulation of bone turnover markers, vitamin D deficiency, chronic hepatic inflammation, liver fibrosis, a disrupted lipid metabolism, and gut microbiota dysbiosis [41].

Dysbiosis of the gut microbiota is frequently observed in patients with NAFLD. Altered microbial composition compromises intestinal barrier integrity, allowing bacterial metabolites such as lipopolysaccharide (LPS), short-chain fatty acids (SCFAs), and ethanol to translocate via the portal circulation to the liver, thereby exacerbating hepatic metabolic dysfunction through the gut–liver axis [42]. In parallel, chronic low-grade inflammation and metabolic disturbances arising from gut dysbiosis can affect bone metabolism via the gut–bone axis, as demonstrated in models of osteoporosis [43].

These observations highlight a complex bidirectional interaction wherein liver health—particularly in the context of NAFLD—modulates gut microbiota composition, which, in turn, influences bone homeostasis. The gut–liver–bone axis has thus emerged as a potential physiological network of critical importance in bone metabolism regulation. Elucidating the mechanistic pathways underlying this axis may offer novel strategies for the prevention and treatment of osteoporosis in individuals with metabolic liver disease [44].

1.7. Sex Differences in Gut Microbiota and Osteoporosis Pathogenesis

Significant sex-related differences exist in gut microbiota composition, bone metabolism, and the pathogenesis of osteoporosis. In females, hormonal fluctuations related to the menstrual cycle and estrogen levels induce unique gut microbiota patterns, distinct from those observed in males [45]. Regarding bone metabolism, females generally exhibit a higher rate of bone formation and a lower rate of bone resorption, primarily due to the influence of estrogen [46]. However, with aging and the associated decline in estrogen levels, females become more susceptible to osteoporosis. In contrast, males experience relatively stable estrogen levels, with bone metabolism being predominantly influenced by androgens and other metabolic factors [47]. Although osteoporosis occurs less frequently in males, their bone mineral density significantly declines with aging, especially when testosterone levels decrease [48]. Estrogen not only affects gut microbiota composition but also modulates bone metabolism via specific immune pathways within the gut–bone axis. Therefore, the sex-dependent regulation of bone metabolism is likely linked to differences in gut microbiota composition [49].

Reproductive states such as pregnancy, lactation, and menopause exert profound effects on both gut microbiota and bone metabolism in women. During pregnancy and lactation, significant hormonal changes influence both bone metabolism and gut microbiota composition. Studies have shown that during pregnancy, hormonal levels of estrogen and progesterone rise significantly, leading to changes in gut microbiota diversity. These changes are closely associated with maternal bone metabolism [50]. Moreover, during pregnancy, increased calcium release from bones supports fetal skeletal development, yet this process also elevates the risk of osteoporosis [51]. Postmenopausal women experience a sharp decline in estrogen levels, leading to significant reductions in bone mineral density and an increased risk of osteoporosis. This decline in estrogen is closely linked to heightened inflammation and increased bone resorption [52]. Importantly, postmenopausal women often exhibit gut dysbiosis, and the decrease in estrogen levels not only directly affects bone metabolism but also indirectly influences osteoporosis by modulating gut microbiota and the immune system [53].

1.8. Literature Search Strategy

A systematic search was conducted across three major databases (PubMed, Web of Science, and Scopus) for studies published between January 2010 and March 2024, using the following Boolean search terms: ("gut microbiota" OR "gut microbiome" OR "intestinal flora") AND ("osteoporosis" OR "bone loss" OR "bone mineral density") AND ("prebiotics" OR "probiotics" OR "synbiotics" OR "natural compounds" OR "bioactive substances"). Inclusion Criteria: (1) Study types: peer-reviewed randomized controlled trials (RCTs), preclinical studies (animal models or in vitro mechanistic investigations), or observational studies with mechanistic analyses; (2) Outcomes: Focus on bone health parameters, including but not limited to bone mineral density, bone microstructure, or biochemical markers. Exclusion Criteria: (1) non-English publications; (2) studies without control groups; (3) review articles, commentaries, or meta-analyses lacking original experimental data.

2. Prebiotics and Osteoporosis

In 1995, Gibson and Roberfroid introduced the concept of prebiotics—non-digestible food ingredients that selectively stimulate the growth and/or activity of beneficial gut bacteria, thereby improving host health [54]. Clinical studies have shown that prebiotic supplements, including fructooligosaccharides (FOS) and galactooligosaccharides (GOS), can reduce body fat and alleviate obesity-related diseases [55,56]. Moreover, prebiotics have been reported to enhance calcium retention in postmenopausal women and increase bone mass in ovariectomized rats [57,58]. This has led to growing interest in the role of prebiotics in preventing bone loss and osteoporosis, as well as in understanding the underlying mechanisms (Table 1, Figure 2).

Table 1. Summary of studies related to the effects of prebiotics on osteoporosis.

Class	Study Population	Prebiotics	Function	Gut Microbiota	Reference
Animal Experiment	High-fat diet male mice (C57BL/6)	FOS, GOS	• FOS/GOS treatment significantly attenuates high-fat-induced bone loss and reverses the imbalance of osteoblast, adipocyte, and osteoclast differentiation. • FOS/GOS treatment significantly improves high-fat diet-induced downregulation of intestinal connexin (tight junction protein 1, tight junction protein 15, ZO-1, and JAM-A) expression and an increase in inflammatory factors (TNF-α, IL-6, IL-17).	• FOS/GOS treatment significantly ameliorates high-fat diet-induced ecological imbalance of the gut microbiota: increased *Firmicutes–Bacteriodetes* and decreased biodiversity.	[24]

Table 1. *Cont.*

Class	Study Population	Prebiotics	Function	Gut Microbiota	Reference
	Forty-five male senescence-accelerated mice	5% FOS, 5% GM	• Femoral calcium content was significantly higher in the FOS group than in the control group.	• The number of *Lactobacillus* and *Bacteroides* anomalies in the cecum contents of mice in the FOS group was significantly higher than that in the control group. • *Clostridium* spp. counts were significantly higher in the GM group of mice than in the control group.	[59]
	Female rats (SD)	Short-chain fructo-oligosaccharides	• Short-chain fructo-oligosaccharides treatment leads to significantly higher bone anabolic responses with no effect on catabolic parameters and enhanced peak bone mass		[23]
	4-week-old male rats (SD)	GOS	• GOS supplementation increases net calcium absorption in a dose-dependent manner. • GOS dietary supplementation increases net magnesium absorption, calcium absorption in the femur, calcium and magnesium retention, and fracture strength of the femur and tibia. • Increased total bone mineral density and area of the distal femur and bone mineral density of the proximal tibia after dietary supplementation with GOS.	• Increased relative ratio of *Bifidobacteria* to GOS. • Significant reduction in the number of gut flora bands in mice after GOS dietary supplementation.	[60]
	Female mice (ddY)	Resistant starch (RS)	• Upregulation of IL-10 mRNA expression in the colon in mice with 12% resistant starch in the diet. • Downregulation of nuclear factor kB receptor activator ligand and IL-7 receptor gene expression in bone marrow of mice with 12% resistant starch in the diet. • Dietary 12% resistant starch attenuates ovariectomy-induced bone loss.	• Significant increase in the number of *Bifidobacteria* in the feces of mice after dietary supplementation with resistant starch.	[61]
	8-week-old female mice (ddY)	Soya isoflavones (ISO), resistant starch (RS)	• Combined treatment with ISO and RS alters the expression of ovariectomy-induced inflammation-associated factors in bone marrow and prevents the loss of trabecular bone density in the distal femur. • The combined treatment of ISO and RS can weaken bone resorption in mice.	• Resistant starch treatment significantly increases the relative abundance of *Bifidobacterium* spp.	[62]

Table 1. *Cont.*

Class	Study Population	Prebiotics	Function	Gut Microbiota	Reference
	12-month-old senile female mice (C57BL/6)	D-mannose	• Significant increase in cortical bone volume and bone trabecular microarchitecture in mice following dietary addition of D-mannose. • Downregulation of cytokines related to osteoclastogenesis in the bone marrow of mice in the D-mannose group	• D-mannose reconstitutes the gut microbiota and alters metabolite composition in mice. • Higher proportions of *Verrucomicrobia*, *Akkermansia*, and *Verrucomicrobiaceae* in the feces of D-mannose-supplemented mice compared to Sham group mice. • D-mannose increases the relative abundance of *Verrucomicrobia* and *Akkermansiaceae* in feces compared to control mice.	[63]
	10-week-old female rats (SD)	Puerarin	• Increased bone mineral density in orally administered Puerarin rats.	• Oral intake of Puerarin improves intestinal mucosal integrity and reduces systemic inflammation in rats. • Oral intake of Puerarin ameliorates intestinal dysbiosis and increases the concentration of metabolites and short-chain fatty acids in rats; enrichment of amino acid metabolism, LPS biosynthesis, and butyric acid metabolism pathways.	[64]
	28-day-old male mice (ICR)	Xylo-oligosaccharides (XOS)	• Significantly higher bone mineral density and fracture strength in mice with 4% XOS in the diet. • BMD was significantly enhanced in mice with 4% XOS in the diet during late growth. • Dietary supplementation with XOS upregulates the expression of related calcium transport proteins.	• XOS significantly decreased cecum pH and increased cecum wall weight in a dose-dependent manner. • As XOS concentration increases, villus height and the ratio of villus height to crypt depth increase.	[65]
Population Studies	Healthy adolescent girls	GOS	• Both low-dose (5 g GOS/d) and high-dose (10 g GOS/d) GOS treatments significantly improved calcium absorption compared with the control group.	• Total dominant fecal flora is not affected by GOS. • *Bifidobacteria* in feces increased with treatment with GOS.	[66]

The protective effects of FOS on osteoporosis have been validated in several animal studies. Tanabe et al. fed high-fat diet (HFD) mice 5% FOS and observed significant improvements in osteoporosis and changes in gut microbiota composition [24]. Compared to the control group, FOS treatment alleviated HFD-induced bone loss, reversed the imbalance in differentiation between osteoblasts, adipocytes, and osteoclasts, and improved gut barrier function by reducing tight junction protein downregulation and inflammatory factor increases. These effects were associated with a correction of gut dysbiosis, including an increase in the *Firmicutes/Bacteroidetes* ratio and a decrease in microbial diversity. Similar findings were observed in accelerated aging male mice, where FOS supplementation led to higher bone calcium content and an increased abundance of *Lactobacillus* and *Bacteroides* compared to the control group [59]. In female SD rats, short-chain FOS supplementation increased bone metabolic response and peak bone mass compared to ovariectomized rats [23].

GOS, a naturally occurring prebiotic found in human milk, has also shown potential in improving mineral balance and bone characteristics. Weaver et al. investigated the effects of varying concentrations of GOS on bone mineral content and properties in rats [60]. GOS supplementation resulted in significantly higher mineral absorption and retention, including calcium and magnesium, compared to controls. Additionally, femur and tibia breaking strength, as well as bone mineral density (BMD) at both the distal and proximal femur ends, were significantly increased. The relative abundance of *Bifidobacteria* also increased, while microbial diversity decreased. Human studies by Whisner et al. on GOS supplementation in girls aged 10–13 years confirmed these findings, with GOS significantly improving calcium absorption and increasing *Bifidobacterium* abundance in fecal samples [66]. These results suggest that GOS-enhanced calcium absorption is mediated through *Bifidobacteria.*

In addition to FOS and GOS, other prebiotics such as resistant starch, D-mannose, puerarin, and xylooligosaccharides (XOS) have also been studied for their impact on bone resorption and loss. Tousen et al. fed female mice with resistant starch (cornstarch) to assess its effects on bone loss [61]. Compared to the control group, mice on a 12% resistant starch diet showed reduced bone loss following ovariectomy, with a downregulation of RANKL and interleukin-7 receptor gene expression in bone marrow. These effects were likely linked to increased IL-10 expression in colon tissues and an enhanced relative abundance of *Bifidobacteria*. In a follow-up study, resistant starch supplementation significantly reduced bone resorption and increased *Bifidobacterium* levels in fecal samples [62]. Similarly, Liu et al. found that D-mannose supplementation in mice significantly improved cortical bone volume and trabecular bone microstructure while downregulating osteoclastogenesis-related factors [63]. The gut microbiota of D-mannose-fed mice showed increased abundances of *Firmicutes*, *Erysipelotrichales*, *Verrucomicrobia*, and *Akkermansiaceae* compared to control and ovariectomized groups.

Studies on puerarin, a natural compound, have demonstrated its ability to increase bone density in rats, improve gut dysbiosis, and enrich gut microbial metabolism related to amino acids, lipopolysaccharide (LPS) biosynthesis, and butyrate production, while also increasing short-chain fatty acid concentrations [64]. XOS supplementation has similarly been shown to increase bone density, breaking strength, bone crystallinity, and calcium transport protein expression in mice [65]. These findings indicate that prebiotics, as dietary components, can modulate gut microbiota structure and function, influencing bone resorption and improving bone characteristics.

Through the analysis and summary of different experimental models, it was found that in the majority of ovariectomized mouse models, supplementation with prebiotics primarily increased the relative abundance of *Bifidobacteria* in the gut, as well as the concentration of short-chain fatty acids, amino acid metabolism, biosynthesis of lipopolysaccharides (LPS), and pyrimidine metabolism. This also led to a reduction in systemic inflammation and the regulation of bone metabolism. In high-fat diet-induced obese mouse models, prebiotic supplementation mainly increased the *Firmicutes–Bacteroidetes* ratio in the gut, improved intestinal barrier function, and reversed the imbalance in the differentiation of osteoblasts, adipocytes, and osteoclasts, thereby modulating bone metabolism. In growth-stage or preclinical models during the growth phase, prebiotic supplementation mainly increased the relative abundance of *Bifidobacteria* in the gut and enhanced calcium absorption in the bones. In accelerated aging mouse models, prebiotic supplementation led to an increased relative abundance of *Lactobacillus, Bacteroides,* and *Clostridium* spp. in the gut.

In summary, prebiotics such as FOS, GOS, resistant starch, D-mannose, and XOS have been shown to modulate gut microbiota composition and improve bone health through several interconnected mechanisms. These prebiotics primarily promote the growth of beneficial bacteria, including *Bifidobacteria* and *Lactobacilli,* which are associated with enhanced gut barrier function and reduced systemic inflammation. The microbiota alterations induced by prebiotic treatments lead to increased production of SCFAs, such as butyrate, which play a vital role in maintaining intestinal integrity and reducing inflammation. Additionally, some prebiotics improve calcium and mineral absorption in the gut, potentially by enhancing the microbiota-mediated breakdown of dietary fibers and boosting mineral bioavailability. In animal studies, prebiotics like FOS and GOS have demonstrated the ability to modulate bone metabolism by increasing bone mineral content, density, and osteoblast activity, potentially through the gut–bone axis. These mechanisms suggest that prebiotics may offer a synergistic approach for managing osteoporosis by simultaneously supporting gut health and regulating bone homeostasis.

3. Probiotics and Osteoporosis

Probiotics are live microorganisms that confer health benefits to the host by promoting a balanced gut microbiota and improving its characteristics [67]. Osteoporosis has been linked to changes in gut microbiota, particularly a reduction in *Lactobacillus* species and an increase in pro-inflammatory cytokines such as TNF-α, IL-6, and IL-14 in the serum [68]. The concept of bone microbiology highlights the close relationship between gut health and bone metabolism [69]; moreover, as research progresses, scientists are investigating how probiotic exposure affects bone properties and the underlying mechanisms involved (Table 2, Figure 2).

Table 2. Summary of studies related to the effects of probiotics on osteoporosis.

Class	Study Population	Probiotics	Function	Gut Microbiota	Reference
Animal Experiment	3-month-old female rats (SD)	*Lactobacillus rhamnosus* GG (*LGG*)	• Protective effects of *LGG* were found in ovariectomized rats and were more favorable in osteogenesis. • *LGG* treatment ameliorated estrogen deficiency-induced inflammation and mucosal injury and increased GLP-2R and tight junction protein expression.	• *LGG* treatment increased the abundance and community diversity of gut microbiota and improved community richness. • *LGG* treatment in rats decreased the relative abundance of *Firmicutes* and *Desulfobacterota* and increased the relative abundance of *Bacteroidetes* and decreased the relative abundance ratio of *Firmicutes/Bacteroidetes*.	[22]

Table 2. *Cont.*

Class	Study Population	Probiotics	Function	Gut Microbiota	Reference
	12-week-old female rats (SD)	*Lactobacillus plantarum*	• *Lactobacillus plantarum* treatment restored bone microstructural parameters, increased bone density, number, and thickness of bone trabeculae. • *Lactobacillus plantarum* inhibits osteoclast formation and promotes osteoblast formation.	• *Lactobacillus plantarum* treatment increased the diversity of gut microbiota, decreased the ratio of *Firmicutes/Bacteroidota*, increased the relative abundance of beneficial bacteria, and decreased the relative abundance of harmful bacteria (*Desulfovibrionaceae*). • The serum metabolites of rats in different treatment groups were significantly changed, mainly in the pentose and glucuronic acid interconversion pathway and the propionic acid metabolism pathway.	[70]
	6-week-old male mice (C57BL/6)	*Bifidobacterium lactis* BL-99	• *Bifidobacterium lactis* BL-99 treatment significantly improved bone volume percentage (BV/TV), trabecular number, and thickness in mice with ulcerative colitis.	• *Bifidobacterium lactis* BL-99 treatment significantly increased the expression of intestinal barrier-related proteins. • After *Bifidobacterium lactis* BL-99 treatment, the relative abundance of *Bacteroides* and *Firmicutes* at the family level decreased. The relative abundance of *Akkermansia* increased at the genus level.	[71]
	8-week-old female mice (C57BL/6)	*Bifidobacterium*	• *Bifidobacterium* treatment significantly improved bone mineral density, bone volume/total volume ratio (BV/TV), and trabecular number, and effectively inhibited bone loss. • *Bifidobacteria* can inhibit the expression of inflammatory cytokines in the intestinal tract, reduce intestinal inflammation, and inhibit the overproduction of osteoclasts.	• *Bifidobacterium* treatment increased the relative abundance of *Lactobacillus*, *Clostridium*, and *Bifidobacterium* and decreased the relative abundance of *Desulfovibrio* and *Ruminococcus* in the colon.	[72]
	8-week-old female mice (C57BL/6)	*Prevotella histicola* (*Ph*)	• *Ph* treatment inhibited osteoclast formation and promoted osteogenesis, reduced the release of pro-inflammatory cytokines (IL-1β and TNF-α), and reversed the expression of tight junction proteins.	• *Ph* treatment improves the composition, richness, and diversity of the gut microbiota. • *Ph* treatment can repair intestinal mucosal barrier damage and optimize intestinal permeability.	[73]

Table 2. *Cont.*

Class	Study Population	Probiotics	Function	Gut Microbiota	Reference
Population Studies	Olderwomen	*Lactobacillus reuteri* ATCC PTA 6475	• *Lactobacillus reuteri* ATCC PTA 6475 treatment can reduce bone loss in elderly women with low BMD.	• The gene richness of the gut microbiota was significantly higher after *Lactobacillus reuteri* ATCC PTA 6475 treatment.	[74]
	Patients with postmenopausal osteoporosis	*Bifidobacterium lactis* Probio-M8	• *Bifidobacterium lactis* Probio-M8 treatment can improve bone metabolism, as indicated by increased levels of vitamin D3 and decreased levels of parathyroid hormone and procalcitonin in serum.	• *Bifidobacterium lactis* Probio-M8 treatment affected the interact-related network of gut microbiota, especially bacteria that produce short-chain fatty acids. • *Bifidobacterium lactis* Probio-M8 treatment significantly increased genes encoding carbohydrate metabolic pathways and genes encoding choline phosphocytidylate transferase.	[75]
	Healthy postmenopausal women	*Bacillus subtilis* C-3102 (C-3102)	• C-3102 significantly increased BMD in the total hip. • The bone resorption marker tartrate-resistant acid phosphatase isoform 5b tended to decrease after 12 weeks of C-3102 treatment.	• After 12 weeks of C-3102 treatment, the relative abundance of the *Bifidobacterium* genus significantly increased. • The relative abundance of the *Fusobacterium* genus was significantly reduced at 12 and 24 weeks of C-3102 treatment.	[76]

Among the probiotics studied, *Lactobacillus* species have been extensively researched for their role in regulating postmenopausal osteoporosis through the gut–bone axis. Guo et al. administered *Lactobacillus rhamnosus GG (LGG)* to ovariectomized rats and observed significant changes in bone formation and gut microbiota [22]. LGG treatment reduced bone loss, improved trabecular microstructure, and increased gut microbiota diversity compared to controls. It also lowered the *Firmicutes/Bacteroidetes* ratio, improved estrogen-deficiency-induced inflammation, and enhanced the expression of tight junction proteins in the gut. Additionally, LGG treatment helped balance Th17 and Treg cells in bone tissue, counteracting the osteoporosis caused by estrogen deficiency.

Similarly, Li et al. showed that *Lactobacillus plantarum* treatment in SD rats increased bone density, trabecular number, and thickness while inhibiting osteoclast formation and promoting osteoblast activity [70]. These effects were accompanied by increased gut microbiota diversity and a reduced *Firmicutes/Bacteroidetes* ratio, as observed in previous studies. Other research has demonstrated that *Lactobacillus* treatments can prevent bone loss, accelerate fracture healing, and reduce bone density loss in various osteoporosis models [77].

In human studies, elderly women who supplemented with *Lactobacillus reuteri ATCC PTA 6475* experienced reduced bone loss, particularly those with low bone density. Gut microbiota diversity and inflammatory markers also improved [74]. In addition to *Lactobacillus*, other probiotics such as *Bacillus subtilis, Bifidobacteria,* and *Prevotella* have been shown to improve gut symptoms and alleviate osteoporosis. These probiotics help reduce gut inflammation and prevent bone loss.

For example, Lan et al. administered *Bifidobacterium lactis BL-99* to mice with ulcerative colitis, showing that BL-99 not only reduced inflammatory cytokines (TNF-α, IL-1β, IL-6, IL-17) but also improved bone volume, trabecular number, and thickness compared to controls [71]. Zhang et al. observed similar results, with *Bifidobacterium* treatment improving bone density, reducing bone loss, and positively altering gut microbiota composition [72]. In human studies, *Bifidobacterium lactis Probio-M8* improved bone metabolism in postmenopausal women by increasing vitamin D3 levels and reducing parathyroid hormone and procalcitonin levels [75].

Moreover, Zhang et al. demonstrated that *Prevotella histicola* could prevent ovariectomy-induced bone loss by promoting osteoblast formation and suppressing osteoclastogenesis in mice while improving gut microbiota composition and diversity [73]. Similarly, *Bacillus subtilis C-3102*, as shown by Takimoto et al., prevented bone loss in postmenopausal women by improving bone density and regulating gut microbiota composition [76].

In addition, probiotic metabolites also play a crucial role in the effects on bone health. Tryptophan metabolites (e.g., indole-3-propionic acid) and polyamines (e.g., spermidine) modulate osteoclastogenesis via aryl hydrocarbon receptor (AhR) signaling and mitochondrial biogenesis, respectively [78,79]. Conversely, trimethylamine N-oxide (TMAO) exacerbates bone resorption by activating the ROS-dependent NF-kappaB signaling pathway [80].

A summary and analysis of estrogen deficiency, glucocorticoid-induced, and ulcerative colitis osteoporosis models revealed that probiotic supplementation reduced the *Firmicutes/Bacteroidetes* ratio in ovariectomized mice and glucocorticoid-induced osteoporosis mouse models. Additionally, in the ovariectomized mouse model, probiotics also increased the relative abundance of *Lactobacillus*, *Clostridium*, and *Bifidobacterium* in the gut, while also decreasing the relative abundance of *Desulfovibrio* and *Ruminococcus*. This modulation improved estrogen deficiency-induced inflammatory responses, enhanced the expression of intestinal tight junction proteins, and influenced bone metabolism. In the glucocorticoid-induced osteoporosis mouse model, probiotic supplementation led to a reduction in the relative abundance of *Desulfovibrionaceae* in the gut microbiota, which, in turn, altered bone metabolism. In the ulcerative colitis mouse model, probiotic treatment resulted in a decrease in the relative abundance of *Bacteroides* and *Firmicutes* while increasing the relative abundance of *Akkermansia*. These changes ultimately led to alterations in bone volume percentage, trabecular number, and thickness in the mice.

In summary, probiotics, particularly *Lactobacillus* and *Bifidobacterium* species, play a critical role in osteoporosis management by modulating gut microbiota composition, reducing systemic inflammation, and influencing bone metabolism. Probiotics restore microbial diversity, promote the growth of beneficial gut microbes, enhance gut barrier function, and modulate immune responses. In animal models, *Lactobacillus rhamnosus GG* and *Lactobacillus plantarum* have demonstrated the ability to reduce bone loss, improve trabecular microstructure, and stimulate osteoblast activity while suppressing osteoclastogenesis. These effects are largely mediated through the reduction of pro-inflammatory cytokines like TNF-α and IL-6, which are elevated in osteoporosis and contribute to bone resorption. Additionally, probiotics can influence the gut–bone axis by modulating the *Firmicutes–Bacteroidetes* ratio, which is linked to improved bone health. The interaction between probiotics and immune cells, particularly Treg and Th17 cells, supports bone protection by regulating bone remodeling processes. Probiotics may also mitigate estrogen deficiency-induced inflammation, thus preventing bone loss in postmenopausal models. Overall, probiotics have a multifaceted impact on bone health through their ability to restore gut microbiota balance, reduce inflammation, and directly modulate bone metabolism, positioning them as promising candidates for osteoporosis treatment.

However, a limited number of randomized controlled trials have suggested that the results regarding the improvement of bone mineral density (BMD) markers with probiotic supplementation may be contradictory [81]. For instance, postmenopausal women who consumed *Lactobacillus*-rich probiotics exhibited conflicting results concerning bone density and bone turnover biomarkers [82]. Compared to the control group, the experimental group showed a reduction in bone density loss in the lumbar spine, femoral neck, and trochanter. However, no significant differences were observed in bone turnover biomarkers. One possible reason for this could be the influence of individual differences on the effectiveness of probiotic treatments. It was suggested that probiotics may only have a positive effect on bone health in populations with lower BMD, older age, or higher body mass index (BMI) [83,84]. For example, Gregori et al. recruited 292 early postmenopausal women and administered *Limosilactobacillus reuteri 6475* [83]. After two years, they measured the relative changes in tibial total volumetric BMD, as well as lumbar spine and total hip BMD. The results indicated a significant decrease in BMD for all participants, with no notable differences observed between the groups. However, a re-treatment study involving patients with high BMI found significant therapeutic effects on bone health. This discrepancy may be explained by the fact that the participants in the initial study were early postmenopausal women, whose BMD may not have experienced significant decline at the time of the study. In contrast, individuals with higher BMI may have been more responsive to probiotic treatment due to differing baseline conditions or bone metabolism rates.

4. Natural Active Substances and Osteoporosis

Several naturally active substances have shown protective effects on bone health, potentially through modulation of the gut microbiota (Table 3, Figure 2). Kang et al. treated mice with Korean Red Ginseng (KRG) extract and found that it significantly prevented antibiotic-induced bone loss, reduced gut microbiota alpha diversity, and improved gut permeability [85]. They observed shifts in the relative abundance of specific gut bacteria, including *Lactobacillus* and *Alistipes finegoldii*, which may contribute to the gut–bone axis effects of KRG. Further studies confirmed that KRG could also prevent glucocorticoid-induced bone loss in mice by restoring gut microbiota composition [86].

Table 3. Summary of studies related to the effects of naturally occurring substances on osteoporosis.

Class	Study Population	Natural-Occurring Substance	Function	Gut Microbiota	Reference
Animal Experiment	12-week-old male mice (Balb/C)	Korean Red Ginseng extract (KRG)	• KRG can prevent antibiotic-induced bone loss.	• KRG can prevent the reduction of α diversity of gut microbiota and the increase of intestinal permeability in mice. • Several genera, including *Lactobacillus*, *rc4-4*, and *Alistipesfinegoldii*, may be involved in the effect of KRG on the axis of the gut.	[85]
	7-week-old male mice (CD-1)	Korean Red Ginseng extract (KRG)	• Treatment with KRG extract prevented trabecular bone loss in the distal femur.	• It causes significant changes in gut microbiota.	[86]

Table 3. *Cont.*

Class	Study Population	Natural-Occurring Substance	Function	Gut Microbiota	Reference
	7-week-old female mice (C57BL/6)	*Agastache rugosa* ethanol extract (AR)	• AR treatment suppressed bone strength loss. • AR treatment elevated osteogenic markers in bone marrow cells and type 1 collagen $\alpha 1$ in the distal femur.	• AR treatment reversed the disturbance of gut microbiota induced by ovariectomy.	[87]
	11-week-old female mice (C57BL/6)	Grape seed extract (GSE)	• GSE treatment can inhibit the expansion of bone marrow adipose tissue, restore lipolysis of bone marrow adipose tissue, and promote bone formation, thereby improving bone loss.	• GSE treatment could reduce the number of opportunistic pathogens (*Alistipes*, *Turicibacter*, and *Romboutsia*), increase the number of beneficial bacteria *Bifidobacterium*, and regulate the imbalance of gut microbiota caused by ovariectomy. • GSE mainly affects lipid and amino acid metabolism.	[25]
	Female mice (ICR)	Compound deer bone extract	• CBDE treatment could significantly improve the microstructure of the trabecular bone in mice. The trabecular bone increased, and the reticular structure was clearer and more orderly.	• CBDE treatment makes the gut microbiota of osteoporosis model mice tend towards healthy mice in terms of type and quantity.	[88]
Population Studies	PostmenopausalWomen	Hop Extract Standardized in 8-Prenylnaringenin (HE)	• Forty-eight weeks of HE supplementation increased whole-body bone mineral density.	• In the HE group, a higher abundance of Turicibacter and Shigella genera (which are thought to be associated with whole-body bone mineral density) was observed.	[89]

Other natural substances, such as *Agastache rugosa* ethanol extract (EEAR), compound deer bone extract (CBDE), and grape seed extract (GSE), have also demonstrated potential in preventing osteoporosis. Hong et al. showed that EEAR treatment improved bone strength and reversed gut microbiota dysbiosis in ovariectomized mice, suggesting its potential for treating postmenopausal osteoporosis [87]. GSE treatment in estrogen-deficient mice inhibited bone marrow adipose tissue expansion, promoted bone formation, and modulated gut microbiota by reducing pathogenic bacteria and increasing beneficial bacteria such as *Bifidobacteria* [25]. Similarly, CBDE treatment improved trabecular microstructure and gut microbiota composition in ovariectomized mice [88].

While many studies focus on how natural substances, particularly plant-derived compounds, impact bone health through gut microbiota modulation, it is also essential to recognize that these substances may exert direct effects through molecular interactions with the body. Phytoestrogens, for example, mimic estrogen and can bind to estrogen receptors, potentially influencing bone metabolism by modulating bone resorption and formation, especially in postmenopausal women [90]. Additionally, natural compounds such as flavonoids, polyphenols, and alkaloids may act directly on bone cells or hormonal pathways to promote bone health, independent of microbiota-related mechanisms [91]. These compounds interact with various signaling pathways involved in bone remodeling, including those related to osteoblast differentiation and osteoclast activity.

In human studies, a clinical trial supplementing postmenopausal women with hops extract standardized in 8-Prenylnaringenin (HE), along with calcium and vitamin D3, demonstrated that HE supplementation increased bone density and altered gut microbiota composition, with specific bacteria showing positive correlations with bone density [89]. These findings suggest that HE extract may help prevent bone density loss in postmenopausal women, although longer-term studies are needed to confirm these results.

In conclusion, prebiotics, probiotics, and natural active substances show promise in preventing and treating osteoporosis through their effects on gut microbiota and bone metabolism. These findings highlight the need for further research into their clinical applications and potential for enhancing bone health.

5. Perspective

Modulating the gut microbiota has emerged as a promising approach for preventing and treating various diseases, including osteoporosis. This review highlights how microbiota modulation can alleviate osteoporosis and offers insights into potential therapeutic strategies. However, our current understanding remains limited.

While existing research emphasizes the critical role of the gut microbiota in osteoporosis, the precise mechanisms by which it influences bone metabolism are still unclear. Future studies should focus on how gut microbes affect bone health through their metabolites, signaling pathways, and immune modulation. Delving deeper into the molecular mechanisms of the gut–bone axis is essential for developing more targeted interventions, which will pave the way for new research directions and clinical applications in osteoporosis prevention and treatment.

Given the significant variations in gut microbiota composition among individuals, personalized intervention strategies are crucial. Future research should explore how personalized treatments, based on an individual's unique microbiota profile, can optimize therapeutic outcomes. Variations in microbiota composition and immune system function significantly influence the effectiveness of microbiota-based treatments for osteoporosis. The gut microbiota is highly personalized, shaped by factors such as genetics, diet, lifestyle, and pre-existing health conditions [92]. Additionally, immune system variations—such as differences in cytokine profiles and inflammatory responses—affect the body's response to microbiota modulation [93]. These factors underscore the need for strategies that consider both microbiota composition and immune system status to optimize osteoporosis treatment outcomes.

Much of the current research has focused on short-term effects, while the long-term impacts and safety of microbiota modulation have not been systematically evaluated. Therefore, future studies should investigate the long-term effects and potential side effects of these interventions to ensure their sustained efficacy and safety. A large portion of current research on the gut–bone axis and microbiota-based interventions relies on animal models, particularly rodents. While these models provide controlled environments to study mechanisms and therapeutic effects, they often fail to fully replicate the complexity of human microbiota and bone physiology. Differences in gut microbiota composition between species, along with variations in immune responses and bone metabolism, may limit the generalizability of findings from animal studies to human conditions. Additionally, methods of microbiota manipulation (e.g., antibiotics or fecal microbiota transplantation) used in animals may not be directly applicable to human clinical settings. Therefore, well-designed human clinical trials are crucial for assessing the long-term safety, efficacy, and potential side effects of microbiota-based interventions.

While most studies have concentrated on the immediate effects of microbiota modulation, the long-term consequences remain underexplored. Individuals with compromised immune systems, such as those with HIV or undergoing immunosuppressive treatments, may be at higher risk of opportunistic infections when using probiotics or undergoing fecal microbiota transplantation (FMT). Additionally, although FMT shows promise as a treatment, it carries the risk of transferring harmful pathogens from donor to recipient, underscoring the need for strict safety protocols and careful donor selection.

6. Conclusions

The gut microbiota plays a crucial role in the onset and progression of osteoporosis. Current evidence suggests that prebiotics, probiotics, and natural active substances have significant potential to mitigate osteoporosis. However, while early findings are promising, future research must focus on understanding the underlying mechanisms of these interventions and their long-term efficacy. By doing so, we can establish a stronger scientific foundation for microbiota-based osteoporosis treatments. Further exploration will enable the development of more precise and effective intervention strategies, offering new perspectives and methods for preventing and treating osteoporosis.

7. Lay Summary

Osteoporosis is a condition where bones become weak and fragile, leading to a higher risk of fractures. Recent studies have found that the balance of microbes in our gut may play a key role in bone health. These tiny organisms affect how our immune system works, how nutrients are absorbed, and how inflammation is controlled. This study looks at how certain dietary changes, like adding prebiotics (which feed good bacteria in the gut), probiotics (live helpful bacteria), and natural substances from plants, might help improve bone strength by supporting a healthy gut. These findings could offer new ways to protect bone health, especially for groups at higher risk, like older adults and women after menopause.

Author Contributions: Conceptualization, X.C. and L.J.; methodology, X.C. and P.X.; formal analysis, X.C. and M.C.; investigation, X.C. and H.X.; resources, H.X.; writing—original draft preparation, X.C.; writing—review and editing, L.J.; funding acquisition, H.X. All authors have read and agreed to the published version of the manuscript.

Funding: Lishui Science and Technology Plan Project (No. 2023GYX70).

Institutional Review Board Statement: Not applicable.

Informed Consent Statement: Not applicable.

Data Availability Statement: No new data were created or analyzed in this study.

Conflicts of Interest: The authors declare no conflicts of interest.

References

1. Johnston, C.B.; Dagar, M. Osteoporosis in Older Adults. *Med. Clin. N. Am.* **2020**, *104*, 873–884. [CrossRef]
2. Zhu, Z.; Yu, P.; Wu, Y.; Wu, Y.; Tan, Z.; Ling, J.; Ma, J.; Zhang, J.; Zhu, W.; Liu, X. Sex Specific Global Burden of Osteoporosis in 204 Countries and Territories, from 1990 to 2030: An Age-Period-Cohort Modeling Study. *J. Nutr. Health Aging* **2023**, *27*, 767–774. [CrossRef]
3. Pouresmaeili, F.; Kamalidehghan, B.; Kamarehei, M.; Goh, Y.M. A comprehensive overview on osteoporosis and its risk factors. *Ther. Clin. Risk Manag.* **2018**, *14*, 2029–2049. [CrossRef]

4. Williams, G.R.; Bassett, J.H.D. Thyroid diseases and bone health. *J. Endocrinol. Investig.* **2018**, *41*, 99–109. [CrossRef]
5. Xiao, P.L.; Cui, A.Y.; Hsu, C.J.; Peng, R.; Jiang, N.; Xu, X.H.; Ma, Y.G.; Liu, D.; Lu, H.D. Global, regional prevalence, and risk factors of osteoporosis according to the World Health Organization diagnostic criteria: A systematic review and meta-analysis. *Osteoporos. Int.* **2022**, *33*, 2137–2153. [CrossRef]
6. Szamatowicz, M.; Szamatowicz, J. Recent advances in prophylactics and treatment of osteoporosis. *Prz. Menopauzalny* **2022**, *21*, 133–137. [CrossRef]
7. Rommens, P.M.; Hofmann, A. Focus on fragility fractures of the pelvis. *Eur. J. Trauma Emerg. Surg.* **2021**, *47*, 1–2. [CrossRef]
8. He, F.-F.; Li, Y.-M. Role of gut microbiota in the development of insulin resistance and the mechanism underlying polycystic ovary syndrome: A review. *J. Ovarian Res.* **2020**, *13*, 73. [CrossRef]
9. Kasubuchi, M.; Hasegawa, S.; Hiramatsu, T.; Ichimura, A.; Kimura, I. Dietary gut microbial metabolites, short-chain fatty acids, and host metabolic regulation. *Nutrients* **2015**, *7*, 2839–2849. [CrossRef]
10. Visconti, A.; Le Roy, C.I.; Rosa, F.; Rossi, N.; Martin, T.C.; Mohney, R.P.; Li, W.; de Rinaldis, E.; Bell, J.T.; Venter, J.C.; et al. Interplay between the human gut microbiome and host metabolism. *Nat. Commun.* **2019**, *10*, 4505. [CrossRef]
11. Wu, J.; Wang, K.; Wang, X.; Pang, Y.; Jiang, C. The role of the gut microbiome and its metabolites in metabolic diseases. *Protein Cell* **2021**, *12*, 360–373. [CrossRef]
12. Kau, A.L.; Ahern, P.P.; Griffin, N.W.; Goodman, A.L.; Gordon, J.I. Human nutrition, the gut microbiome and the immune system. *Nature* **2011**, *474*, 327–336. [CrossRef]
13. Shi, N.; Li, N.; Duan, X.; Niu, H. Interaction between the gut microbiome and mucosal immune system. *Mil. Med. Res.* **2017**, *4*, 14. [CrossRef]
14. Shreiner, A.B.; Kao, J.Y.; Young, V.B. The gut microbiome in health and in disease. *Curr. Opin. Gastroenterol.* **2015**, *31*, 69–75. [CrossRef]
15. Ross, F.C.; Patangia, D.; Grimaud, G.; Lavelle, A.; Dempsey, E.M.; Ross, R.P.; Stanton, C. The interplay between diet and the gut microbiome: Implications for health and disease. *Nat. Rev. Microbiol.* **2024**, *22*, 671–686. [CrossRef]
16. Tu, P.; Chi, L.; Bodnar, W.; Zhang, Z.; Gao, B.; Bian, X.; Stewart, J.; Fry, R.; Lu, K. Gut Microbiome Toxicity: Connecting the Environment and Gut Microbiome-Associated Diseases. *Toxics* **2020**, *8*, 19. [CrossRef]
17. Zaiss, M.M.; Jones, R.M.; Schett, G.; Pacifici, R. The gut-bone axis: How bacterial metabolites bridge the distance. *J. Clin. Investig.* **2019**, *129*, 3018–3028. [CrossRef]
18. Tu, Y.; Yang, R.; Xu, X.; Zhou, X. The microbiota-gut-bone axis and bone health. *J. Leukoc. Biol.* **2021**, *110*, 525–537. [CrossRef]
19. Villa, C.R.; Ward, W.E.; Comelli, E.M. Gut microbiota-bone axis. *Crit. Rev. Food Sci. Nutr.* **2017**, *57*, 1664–1672. [CrossRef]
20. Schmidt, T.S.B.; Raes, J.; Bork, P. The Human Gut Microbiome: From Association to Modulation. *Cell* **2018**, *172*, 1198–1215. [CrossRef]
21. Waldbaum, J.D.H.; Xhumari, J.; Akinsuyi, O.S.; Arjmandi, B.; Anton, S.; Roesch, L.F.W. Association between Dysbiosis in the Gut Microbiota of Primary Osteoporosis Patients and Bone Loss. *Aging Dis.* **2023**, *14*, 2081–2095. [CrossRef]
22. Guo, M.; Liu, H.; Yu, Y.; Zhu, X.; Xie, H.; Wei, C.; Mei, C.; Shi, Y.; Zhou, N.; Qin, K.; et al. *Lactobacillus rhamnosus GG* ameliorates osteoporosis in ovariectomized rats by regulating the Th17/Treg balance and gut microbiota structure. *Gut Microbes* **2023**, *15*, 2190304. [CrossRef] [PubMed]
23. Porwal, K.; Pal, S.; Kulkarni, C.; Singh, P.; Sharma, S.; Singh, P.; Prajapati, G.; Gayen, J.R.; Ampapathi, R.S.; Mullick, A.; et al. A prebiotic, short-chain fructo-oligosaccharides promotes peak bone mass and maintains bone mass in ovariectomized rats by an osteogenic mechanism. *Biomed. Pharmacother.* **2020**, *129*, 110448. [CrossRef] [PubMed]
24. Zhang, Z.; Lin, T.; Meng, Y.; Hu, M.; Shu, L.; Jiang, H.; Gao, R.; Ma, J.; Wang, C.; Zhou, X. FOS/GOS attenuates high-fat diet induced bone loss via reversing microbiota dysbiosis, high intestinal permeability and systemic inflammation in mice. *Metabolism* **2021**, *119*, 154767. [CrossRef] [PubMed]
25. Lu, L.; Li, J.; Liu, L.; Wang, C.; Xie, Y.; Yu, X.; Tian, L. Grape seed extract prevents oestrogen deficiency-induced bone loss by modulating the gut microbiota and metabolites. *Microb. Biotechnol.* **2024**, *17*, e14485. [CrossRef]
26. Yang, K.L.; Mullins, B.J.; Lejeune, A.; Ivanova, E.; Shin, J.; Bajwa, S.; Possemato, R.; Cadwell, K.; Scher, J.U.; Koralov, S.B. Mitigation of Osteoclast-Mediated Arthritic Bone Remodeling By Short Chain Fatty Acids. *Arthritis Rheumatol.* **2024**, *76*, 647–659. [CrossRef]
27. Thammayon, N.; Wongdee, K.; Teerapornpuntakit, J.; Panmanee, J.; Chanpaisaeng, K.; Charoensetakul, N.; Srimongkolpithak, N.; Suntornsaratoon, P.; Charoenphandhu, N. Enhancement of intestinal calcium transport by short-chain fatty acids: Roles of Na+/H+ exchanger 3 and transient receptor potential vanilloid subfamily 6. *Am. J. Physiol. Cell Physiol.* **2024**, *326*, C317–C330. [CrossRef]
28. Tao, Z.-S.; Ma, T. Sodium butyrate protect bone mass in lipopolysaccharide-treated rats by reducing oxidative stress and inflammatory. *Redox Rep.* **2024**, *29*, 2398891. [CrossRef]

29. Wallimann, A.; Magrath, W.; Thompson, K.; Moriarty, T.; Richards, R.G.; Akdis, C.A.; O'Mahony, L.; Hernandez, C.J. Gut microbial-derived short-chain fatty acids and bone: A potential role in fracture healing. *Eur. Cell Mater.* **2021**, *41*, 454–470. [CrossRef]
30. Locantore, P.; Del Gatto, V.; Gelli, S.; Paragliola, R.M.; Pontecorvi, A. The Interplay between Immune System and Microbiota in Osteoporosis. *Mediat. Inflamm.* **2020**, *2020*, 3686749. [CrossRef]
31. Ciucci, T.; Ibáñez, L.; Boucoiran, A.; Birgy-Barelli, E.; Pène, J.; Abou-Ezzi, G.; Arab, N.; Rouleau, M.; Hébuterne, X.; Yssel, H.; et al. Bone marrow Th17 TNFα cells induce osteoclast differentiation, and link bone destruction to IBD. *Gut* **2015**, *64*, 1072–1081. [CrossRef]
32. Zhu, L.; Hua, F.; Ding, W.; Ding, K.; Zhang, Y.; Xu, C. The correlation between the Th17/Treg cell balance and bone health. *Immun. Ageing* **2020**, *17*, 30. [CrossRef]
33. Shieh, A.; Epeldegui, M.; Karlamangla, A.S.; Greendale, G.A. Gut permeability, inflammation, and bone density across the menopause transition. *JCI Insight* **2020**, *5*, e134092. [CrossRef]
34. Zhang, L.; Yu, Z.; Zhu, Y.; Zhang, C.; Su, L.; He, S.; Yin, H.; Yu, Y.; Zhu, M. Pegylation enhances the anti-osteoporosis activity of acacetin in both ovariectomized and LPS-stimulated mice. *Bioorg. Med. Chem.* **2024**, *113*, 117910. [CrossRef] [PubMed]
35. Younossi, Z.; Tacke, F.; Arrese, M.; Chander Sharma, B.; Mostafa, I.; Bugianesi, E.; Wai-Sun Wong, V.; Yilmaz, Y.; George, J.; Fan, J.; et al. Global Perspectives on Nonalcoholic Fatty Liver Disease and Nonalcoholic Steatohepatitis. *Hepatology* **2019**, *69*, 2672–2682. [CrossRef]
36. Eslam, M.; Newsome, P.N.; Sarin, S.K.; Anstee, Q.M.; Targher, G.; Romero-Gomez, M.; Zelber-Sagi, S.; Wai-Sun Wong, V.; Dufour, J.-F.; Schattenberg, J.M.; et al. A new definition for metabolic dysfunction-associated fatty liver disease: An international expert consensus statement. *J. Hepatol.* **2020**, *73*, 202–209. [CrossRef] [PubMed]
37. Cui, A.; Xiao, P.; Fan, Z.; Lei, J.; Han, S.; Zhang, D.; Wei, X.; Wang, P.; Zhuang, Y. Causal association of NAFLD with osteoporosis, fracture and falling risk: A bidirectional Mendelian randomization study. *Front. Endocrinol.* **2023**, *14*, 1215790. [CrossRef] [PubMed]
38. Xie, R.; Liu, M. Relationship Between Non-Alcoholic Fatty Liver Disease and Degree of Hepatic Steatosis and Bone Mineral Density. *Front. Endocrinol.* **2022**, *13*, 857110. [CrossRef]
39. Cui, R.; Sheng, H.; Rui, X.-F.; Cheng, X.-Y.; Sheng, C.-J.; Wang, J.-Y.; Qu, S. Low bone mineral density in chinese adults with nonalcoholic Fatty liver disease. *Int. J. Endocrinol.* **2013**, *2013*, 396545. [CrossRef]
40. Xia, M.-F.; Lin, H.-D.; Yan, H.-M.; Bian, H.; Chang, X.-X.; Zhang, L.-S.; He, W.-Y.; Gao, X. The association of liver fat content and serum alanine aminotransferase with bone mineral density in middle-aged and elderly Chinese men and postmenopausal women. *J. Transl. Med.* **2016**, *14*, 11. [CrossRef]
41. Kumar, R.; Priyadarshi, R.N.; Anand, U. Non-alcoholic Fatty Liver Disease: Growing Burden, Adverse Outcomes and Associations. *J. Clin. Transl. Hepatol.* **2020**, *8*, 76–86. [CrossRef] [PubMed]
42. Wu, M.-Y.; Fan, J.-G. Gut microbiome and nonalcoholic fatty liver disease. *Hepatobiliary Pancreat. Dis. Int.* **2023**, *22*, 444–451. [CrossRef]
43. Zhang, J.; Zhang, Q.; Liu, H.; Liu, X.; Yu, Y.; Han, D.; He, X.; Zeng, P.; Wang, J. Soy-whey dual-protein alleviates osteoporosis of ovariectomized rats via regulating bone fat metabolism through gut-liver-bone axis. *Nutrition* **2022**, *103–104*, 111723. [CrossRef] [PubMed]
44. Milosevic, I.; Vujovic, A.; Barac, A.; Djelic, M.; Korac, M.; Radovanovic Spurnic, A.; Gmizic, I.; Stevanovic, O.; Djordjevic, V.; Lekic, N.; et al. Gut-Liver Axis, Gut Microbiota, and Its Modulation in the Management of Liver Diseases: A Review of the Literature. *Int. J. Mol. Sci.* **2019**, *20*, 395. [CrossRef]
45. Zhang, X.; Zhong, H.; Li, Y.; Shi, Z.; Ren, H.; Zhang, Z.; Zhou, X.; Tang, S.; Han, X.; Lin, Y.; et al. Sex- and age-related trajectories of the adult human gut microbiota shared across populations of different ethnicities. *Nat. Aging* **2021**, *1*, 87–100. [CrossRef] [PubMed]
46. Herber, C.B.; Krause, W.C.; Wang, L.; Bayrer, J.R.; Li, A.; Schmitz, M.; Fields, A.; Ford, B.; Zhang, Z.; Reid, M.S.; et al. Estrogen signaling in arcuate Kiss1 neurons suppresses a sex-dependent female circuit promoting dense strong bones. *Nat. Commun.* **2019**, *10*, 163. [CrossRef]
47. Zhang, Y.-Y.; Xie, N.; Sun, X.-D.; Nice, E.C.; Liou, Y.-C.; Huang, C.; Zhu, H.; Shen, Z. Insights and implications of sexual dimorphism in osteoporosis. *Bone Res.* **2024**, *12*, 8. [CrossRef]
48. Bandeira, L.; Silva, B.C.; Bilezikian, J.P. Male osteoporosis. *Arch. Endocrinol. Metab.* **2022**, *66*, 739–747. [CrossRef]
49. Santos-Marcos, J.A.; Barroso, A.; Rangel-Zuñiga, O.A.; Perdices-Lopez, C.; Haro, C.; Sanchez-Garrido, M.A.; Molina-Abril, H.; Ohlsson, C.; Perez-Martinez, P.; Poutanen, M.; et al. Interplay between gonadal hormones and postnatal overfeeding in defining sex-dependent differences in gut microbiota architecture. *Aging* **2020**, *12*, 19979–20000. [CrossRef]

50. Galanis, A.; Dimopoulou, S.; Karampinas, P.; Vavourakis, M.; Papagrigorakis, E.; Sakellariou, E.; Karampitianis, S.; Zachariou, D.; Theodora, M.; Antsaklis, P.; et al. The correlation between transient osteoporosis of the hip and pregnancy: A review. *Medicine* **2023**, *102*, e35475. [CrossRef]
51. Lujano-Negrete, A.Y.; Rodríguez-Ruiz, M.C.; Skinner-Taylor, C.M.; Perez-Barbosa, L.; Cardenas de la Garza, J.A.; García-Hernández, P.A.; Espinosa-Banuelos, L.G.; Gutierrez-Leal, L.F.; Jezzini-Martínez, S.; Galarza-Delgado, D.Á. Bone metabolism and osteoporosis during pregnancy and lactation. *Arch. Osteoporos.* **2022**, *17*, 36. [CrossRef] [PubMed]
52. Walker, M.D.; Shane, E. Postmenopausal Osteoporosis. *N. Engl. J. Med.* **2023**, *389*, 1979–1991. [CrossRef] [PubMed]
53. Kverka, M.; Stepan, J.J. Associations Among Estrogens, the Gut Microbiome and Osteoporosis. *Curr. Osteoporos. Rep.* **2024**, *23*, 2. [CrossRef]
54. Gibson, G.R.; Beatty, E.R.; Wang, X.; Cummings, J.H. Selective stimulation of bifidobacteria in the human colon by oligofructose and inulin. *Gastroenterology* **1995**, *108*, 975–982. [CrossRef] [PubMed]
55. da Silva, T.F.; Casarotti, S.N.; de Oliveira, G.L.V.; Penna, A.L.B. The impact of probiotics, prebiotics, and synbiotics on the biochemical, clinical, and immunological markers, as well as on the gut microbiota of obese hosts. *Crit. Rev. Food Sci. Nutr.* **2021**, *61*, 337–355. [CrossRef] [PubMed]
56. Nicolucci, A.C.; Hume, M.P.; Martínez, I.; Mayengbam, S.; Walter, J.; Reimer, R.A. Prebiotics Reduce Body Fat and Alter Intestinal Microbiota in Children Who Are Overweight or With Obesity. *Gastroenterology* **2017**, *153*, 711–722. [CrossRef]
57. Jakeman, S.A.; Henry, C.N.; Martin, B.R.; McCabe, G.P.; McCabe, L.D.; Jackson, G.S.; Peacock, M.; Weaver, C.M. Soluble corn fiber increases bone calcium retention in postmenopausal women in a dose-dependent manner: A randomized crossover trial. *Am. J. Clin. Nutr.* **2016**, *104*, 837–843. [CrossRef]
58. Seijo, M.; Bryk, G.; Zeni Coronel, M.; Bonanno, M.; Río, M.E.; Pita Martín de Portela, M.L.; Zeni, S.N. Effect of Adding a Galacto-Oligosaccharides/Fructo-Oligosaccharides (GOS/FOS®) Mixture to a Normal and Low Calcium Diet, on Calcium Absorption and Bone Health in Ovariectomy-Induced Osteopenic Rats. *Calcif. Tissue Int.* **2019**, *104*, 301–312. [CrossRef]
59. Tanabe, K.; Nakamura, S.; Moriyama-Hashiguchi, M.; Kitajima, M.; Ejima, H.; Imori, C.; Oku, T. Dietary Fructooligosaccharide and Glucomannan Alter Gut Microbiota and Improve Bone Metabolism in Senescence-Accelerated Mouse. *J. Agric. Food Chem.* **2019**, *67*, 867–874. [CrossRef]
60. Weaver, C.M.; Martin, B.R.; Nakatsu, C.H.; Armstrong, A.P.; Clavijo, A.; McCabe, L.D.; McCabe, G.P.; Duignan, S.; Schoterman, M.H.C.; van den Heuvel, E.G.H.M. Galactooligosaccharides improve mineral absorption and bone properties in growing rats through gut fermentation. *J. Agric. Food Chem.* **2011**, *59*, 6501–6510. [CrossRef]
61. Tousen, Y.; Matsumoto, Y.; Nagahata, Y.; Kobayashi, I.; Inoue, M.; Ishimi, Y. Resistant Starch Attenuates Bone Loss in Ovariectomised Mice by Regulating the Intestinal Microbiota and Bone-Marrow Inflammation. *Nutrients* **2019**, *11*, 297. [CrossRef] [PubMed]
62. Tousen, Y.; Matsumoto, Y.; Matsumoto, C.; Nishide, Y.; Nagahata, Y.; Kobayashi, I.; Ishimi, Y. The combined effects of soya isoflavones and resistant starch on equol production and trabecular bone loss in ovariectomised mice. *Br. J. Nutr.* **2016**, *116*, 247–257. [CrossRef]
63. Liu, H.; Gu, R.; Zhu, Y.; Lian, X.; Wang, S.; Liu, X.; Ping, Z.; Liu, Y.; Zhou, Y. D-mannose attenuates bone loss in mice via Treg cell proliferation and gut microbiota-dependent anti-inflammatory effects. *Ther. Adv. Chronic Dis.* **2020**, *11*, 2040622320912661. [CrossRef]
64. Li, B.; Liu, M.; Wang, Y.; Gong, S.; Yao, W.; Li, W.; Gao, H.; Wei, M. Puerarin improves the bone micro-environment to inhibit OVX-induced osteoporosis via modulating SCFAs released by the gut microbiota and repairing intestinal mucosal integrity. *Biomed. Pharmacother.* **2020**, *132*, 110923. [CrossRef]
65. Gao, H.; Zhou, Z. Effect of Xylo-Oligosaccharides Supplementation by Drinking Water on the Bone Properties and Related Calcium Transporters in Growing Mice. *Nutrients* **2020**, *12*, 3542. [CrossRef]
66. Whisner, C.M.; Martin, B.R.; Schoterman, M.H.C.; Nakatsu, C.H.; McCabe, L.D.; McCabe, G.P.; Wastney, M.E.; van den Heuvel, E.G.H.M.; Weaver, C.M. Galacto-oligosaccharides increase calcium absorption and gut bifidobacteria in young girls: A double-blind cross-over trial. *Br. J. Nutr.* **2013**, *110*, 1292–1303. [CrossRef] [PubMed]
67. Plaza-Diaz, J.; Ruiz-Ojeda, F.J.; Gil-Campos, M.; Gil, A. Mechanisms of Action of Probiotics. *Adv. Nutr.* **2019**, *10*, S49–S66. [CrossRef]
68. Xie, H.; Hua, Z.; Guo, M.; Lin, S.; Zhou, Y.; Weng, Z.; Wu, L.; Chen, Z.; Xu, Z.; Li, W. Gut microbiota and metabonomics used to explore the mechanism of Qing'e Pills in alleviating osteoporosis. *Pharm. Biol.* **2022**, *60*, 785–800. [CrossRef] [PubMed]
69. Hsu, E.; Pacifici, R. From Osteoimmunology to Osteomicrobiology: How the Microbiota and the Immune System Regulate Bone. *Calcif. Tissue Int.* **2018**, *102*, 512–521. [CrossRef]
70. Li, S.; Han, X.; Liu, N.; Chang, J.; Liu, G.; Hu, S. *Lactobacillus plantarum* attenuates glucocorticoid-induced osteoporosis by altering the composition of rat gut microbiota and serum metabolic profile. *Front. Immunol.* **2023**, *14*, 1285442. [CrossRef]
71. Lan, H.; Liu, W.-H.; Zheng, H.; Feng, H.; Zhao, W.; Hung, W.-L.; Li, H. *Bifidobacterium lactis* BL-99 protects mice with osteoporosis caused by colitis via gut inflammation and gut microbiota regulation. *Food Funct.* **2022**, *13*, 1482–1494. [CrossRef] [PubMed]

72. Zhang, J.; Liang, X.; Tian, X.; Zhao, M.; Mu, Y.; Yi, H.; Zhang, Z.; Zhang, L. *Bifidobacterium* improves oestrogen-deficiency-induced osteoporosis in mice by modulating intestinal immunity. *Food Funct.* **2024**, *15*, 1840–1851. [CrossRef] [PubMed]
73. Zhang, Y.-W.; Cao, M.-M.; Li, Y.-J.; Sheng, R.-W.; Zhang, R.-L.; Wu, M.-T.; Chi, J.-Y.; Zhou, R.-X.; Rui, Y.-F. The Preventive Effects of Probiotic Prevotella histicola on the Bone Loss of Mice with Ovariectomy-Mediated Osteoporosis. *Microorganisms* **2023**, *11*, 950. [CrossRef]
74. Li, P.; Ji, B.; Luo, H.; Sundh, D.; Lorentzon, M.; Nielsen, J. One-year supplementation with *Lactobacillus reuteri* ATCC PTA 6475 counteracts a degradation of gut microbiota in older women with low bone mineral density. *NPJ Biofilms Microbiomes* **2022**, *8*, 84. [CrossRef]
75. Zhao, F.; Guo, Z.; Kwok, L.-Y.; Zhao, Z.; Wang, K.; Li, Y.; Sun, Z.; Zhao, J.; Zhang, H. *Bifidobacterium lactis* Probio-M8 improves bone metabolism in patients with postmenopausal osteoporosis, possibly by modulating the gut microbiota. *Eur. J. Nutr.* **2023**, *62*, 965–976. [CrossRef]
76. Takimoto, T.; Hatanaka, M.; Hoshino, T.; Takara, T.; Tanaka, K.; Shimizu, A.; Morita, H.; Nakamura, T. Effect of Bacillus subtilis C-3102 on bone mineral density in healthy postmenopausal Japanese women: A randomized, placebo-controlled, double-blind clinical trial. *Biosci. Microbiota Food Health* **2018**, *37*, 87–96. [CrossRef] [PubMed]
77. Guo, X.; Zhong, K.; Zou, L.; Xue, H.; Zheng, S.; Guo, J.; Lv, H.; Duan, K.; Huang, D.; Tan, M. Effect of *Lactobacillus casei* fermented milk on fracture healing in osteoporotic mice. *Front. Endocrinol.* **2022**, *13*, 1041647. [CrossRef]
78. Chen, C.; Cao, Z.; Lei, H.; Zhang, C.; Wu, M.; Huang, S.; Li, X.; Xie, D.; Liu, M.; Zhang, L.; et al. Microbial Tryptophan Metabolites Ameliorate Ovariectomy-Induced Bone Loss by Repairing Intestinal AhR-Mediated Gut-Bone Signaling Pathway. *Adv. Sci.* **2024**, *11*, e2404545. [CrossRef]
79. Ramsay, A.L.; Alonso-Garcia, V.; Chaboya, C.; Radut, B.; Le, B.; Florez, J.; Schumacher, C.; Fierro, F.A. Modeling Snyder-Robinson Syndrome in multipotent stromal cells reveals impaired mitochondrial function as a potential cause for deficient osteogenesis. *Sci. Rep.* **2019**, *9*, 15395. [CrossRef]
80. Wang, N.; Hao, Y.; Fu, L. Trimethylamine-N-Oxide Promotes Osteoclast Differentiation and Bone Loss via Activating ROS-Dependent NF-κB Signaling Pathway. *Nutrients* **2022**, *14*, 3955. [CrossRef]
81. Billington, E.O.; Mahajan, A.; Benham, J.L.; Raman, M. Effects of probiotics on bone mineral density and bone turnover: A systematic review. *Crit. Rev. Food Sci. Nutr.* **2023**, *63*, 4141–4152. [CrossRef] [PubMed]
82. Lambert, M.N.T.; Thybo, C.B.; Lykkeboe, S.; Rasmussen, L.M.; Frette, X.; Christensen, L.P.; Jeppesen, P.B. Combined bioavailable isoflavones and probiotics improve bone status and estrogen metabolism in postmenopausal osteopenic women: A randomized controlled trial. *Am. J. Clin. Nutr.* **2017**, *106*, 909–920. [CrossRef] [PubMed]
83. Gregori, G.; Pivodic, A.; Magnusson, P.; Johansson, L.; Hjertonsson, U.; Brättemark, E.; Lorentzon, M. Limosilactobacillus reuteri 6475 and Prevention of Early Postmenopausal Bone Loss: A Randomized Clinical Trial. *JAMA Netw. Open* **2024**, *7*, e2415455. [CrossRef] [PubMed]
84. Malmir, H.; Ejtahed, H.-S.; Soroush, A.-R.; Mortazavian, A.M.; Fahimfar, N.; Ostovar, A.; Esmaillzadeh, A.; Larijani, B.; Hasani-Ranjbar, S. Probiotics as a New Regulator for Bone Health: A Systematic Review and Meta-Analysis. *Evid. Based Complement. Altern. Med.* **2021**, *2021*, 3582989. [CrossRef]
85. Kang, H.J.; Chargo, N.; Chennupati, S.; Neugebauer, K.; Cho, J.Y.; Quinn, R.; McCabe, L.R.; Parameswaran, N. Korean Red Ginseng extract treatment prevents post-antibiotic dysbiosis-induced bone loss in mice. *J. Ginseng Res.* **2023**, *47*, 265–273. [CrossRef]
86. Chargo, N.J.; Kang, H.J.; Das, S.; Jin, Y.; Rockwell, C.; Cho, J.Y.; McCabe, L.R.; Parameswaran, N. Korean red ginseng extract prevents bone loss in an oral model of glucocorticoid induced osteoporosis in mice. *Front. Pharmacol.* **2024**, *15*, 1268134. [CrossRef]
87. Hong, S.; Cha, K.H.; Kwon, D.Y.; Son, Y.J.; Kim, S.M.; Choi, J.-H.; Yoo, G.; Nho, C.W. *Agastache rugosa* ethanol extract suppresses bone loss via induction of osteoblast differentiation with alteration of gut microbiota. *Phytomedicine* **2021**, *84*, 153517. [CrossRef]
88. Xue, C.; Pan, W.; Lu, X.; Guo, J.; Xu, G.; Sheng, Y.; Yuan, G.; Zhao, N.; Sun, J.; Guo, X.; et al. Effects of compound deer bone extract on osteoporosis model mice and intestinal microflora. *J. Food Biochem.* **2021**, *45*, e13740. [CrossRef]
89. Lecomte, M.; Tomassi, D.; Rizzoli, R.; Tenon, M.; Berton, T.; Harney, S.; Fança-Berthon, P. Effect of a Hop Extract Standardized in 8-Prenylnaringenin on Bone Health and Gut Microbiome in Postmenopausal Women with Osteopenia: A One-Year Randomized, Double-Blind, Placebo-Controlled Trial. *Nutrients* **2023**, *15*, 2688. [CrossRef]
90. Tomczyk-Warunek, A.; Winiarska-Mieczan, A.; Blicharski, T.; Blicharski, R.; Kowal, F.; Pano, I.T.; Tomaszewska, E.; Muszyński, S. Consumption of Phytoestrogens Affects Bone Health by Regulating Estrogen Metabolism. *J. Nutr.* **2024**, *154*, 2611–2627. [CrossRef]
91. Hanga-Farcaș, A.; Miere Groza, F.; Filip, G.A.; Clichici, S.; Fritea, L.; Vicaș, L.G.; Marian, E.; Pallag, A.; Jurca, T.; Filip, S.M.; et al. Phytochemical Compounds Involved in the Bone Regeneration Process and Their Innovative Administration: A Systematic Review. *Plants* **2023**, *12*, 2055. [CrossRef] [PubMed]

92. Parizadeh, M.; Arrieta, M.-C. The global human gut microbiome: Genes, lifestyles, and diet. *Trends Mol. Med.* **2023**, *29*, 789–801. [CrossRef] [PubMed]
93. Zheng, D.; Liwinski, T.; Elinav, E. Interaction between microbiota and immunity in health and disease. *Cell Res.* **2020**, *30*, 492–506. [CrossRef] [PubMed]

metabolites

MDPI

Review

Non/Low-Caloric Artificial Sweeteners and Gut Microbiome: From Perturbed Species to Mechanisms

Jiahao Feng, Jingya Peng, Yun-Chung Hsiao, Chih-Wei Liu, Yifei Yang, Haoduo Zhao, Taylor Teitelbaum, Xueying Wang and Kun Lu *

Department of Environmental Sciences and Engineering, Gillings School of Global Public Health, University of North Carolina at Chapel Hill, Chapel Hill, NC 27599, USA

* Correspondence: kunlu@unc.edu

Abstract: Background: Non/low-caloric artificial sweeteners (NAS) are recognized as chemical additives substituting sugars to avoid caloric intake and subsequent sugar-derived diseases such as diabetes and hyperglycemia. Six NAS have been claimed safe and are authorized by the US Food and Drug Administration (FDA) for public use, with acceptable daily intake information available: aspartame, acesulfame-K, saccharin, sucralose, neotame, and advantame. However, the impacts of NAS on the gut microbiome have raised potential concerns, since sporadic research revealed NAS-induced microbial changes in the gastrointestinal tracts and alterations in the microbiome–host interactive metabolism. Methods: Given the fact that the gut microbiome influences kaleidoscopic physiological functions in host health, this review aimed to decipher the impacts of NAS on the gut microbiome by implementing a comprehensive two-stage literature analysis based on each NAS. Results: This review documented disturbed microbiomes due to NAS exposure to a maximal resolution of species level using taxonomic clustering analysis, and recorded metabolism alterations involved in gut microbiome–host interactions. Conclusions: The results elucidated that specific NAS exhibited discrepant impacts on the gut microbiome, even though overlapping on the genera and species were identified. Some NAS caused glucose tolerance impairment in the host, but the key metabolites and their underlying mechanisms were different. Furthermore, this review embodied the challenges and future directions of current NAS–gut microbiome research to inspire advanced examination of the NAS exposure–gut microbiome–host metabolism axis.

Keywords: artificial sweetener; gut microbiome; microbiome–host interaction; exposure; metabolism

1. Introduction

Over the past few decades, the global population has been facing the considerable health threats of obesity and cardiovascular diseases associated with high sugar consumption [1–5]. Non/low-caloric artificial sweeteners (NAS) became a remarkable dietary sugar replacement for combating the global prevalence of obesity and hyperglycemia [6]. NAS are a category of food additives utilized in food products, which are thought to bring a sweet taste and health benefits by avoiding the substantial energy content and carbohydrate intake from table sugar [7]. The U.S. Food and Drug Administration (FDA) has authorized six NAS with provided information on their acceptable daily intake: aspartame, acesulfame-K, saccharin, sucralose, neotame, and advantame. These six authorized sweeteners are claimed to be safe for the general population under certain conditions of use by the FDA, with support from scientific research [8,9]. Contradictorily, scientific research has revealed that NAS consumption was associated with multiple disease outcomes, for instance, liver cancer and chronic liver diseases [10–12], urinary tract cancer [13], kidney injury [14], and cardiovascular diseases [14–18]. In the meantime, more and more emerging evidence has highlighted that NAS might have complicated impacts on the gut microbiota [3,19,20]. However, the specific impacts of NAS on the gut microbiota are possibly underestimated when evaluating the safety and applicability of NAS.

Humans harbor diverse and dynamic microbial communities [21,22]. The gut microbiota is a collection of endogenous microorganisms that symbiotically inhabit the digestive tract [23]. Colonization of the gut microbiota begins in the proximal gastrointestinal tract, starting in the stomach and ending with diverse microorganisms in the distal gastrointestinal tract or the colon [24]. A healthy gut microbiota is typically characterized by high taxonomic diversity, extensive microbial gene richness, and a stable core of microbial species. The gut microbiota is generally composed of six dominant phyla: Firmicutes, Bacteroidetes, Actinobacteria, Proteobacteria, Fusobacteria, and Verrucomicrobia. Of these, Firmicutes and Bacteroidetes are the most prevalent [25]. Moreover, the gut microbiota plays a significant role in numerous physiological processes, such as immune system maintenance [25–28], drug metabolism [29–31], and biotransformation [32,33]. Previous research has shown that NAS such as saccharin and aspartame traverse the human gastrointestinal tract either undigested, indicated by their existence in excrement such as feces and urine, or digested, indicated by the detection of their secondary metabolites [34–38]. Taken together, consumption of NAS can induce their exposure to intestinal microbiota.

We have found fragmented information illustrating the interaction between the gut microbiome and NAS exposure, and limited research was available to evaluate the toxicity of NAS on the gut microbiome in vivo. Notably, there is a lack of knowledge to comprehensively summarize the affected gut microbiome species and underlying host–microbe interactive metabolic mechanisms. These two components were essential for revealing the toxicology of NAS on the gut microbiome and for investigating the reciprocal influence in the NAS–gut microbiome–host axis. Consequently, the purpose of this review is to summarize the current understanding of the impacts of NAS on the gut microbiome, and to provide a guideline for future research on determining the gut microbiome–NAS interaction and gut health outcomes in the host. This review leveraged a two-stage exploratory review, primarily focused on uncovering the affected gut microbiome, followed by an illustration of the interactive metabolic mechanisms, including the metabolism pathways and signature metabolites, altered in the host as a response to NAS. Taxonomy clustering analysis was implemented to classify the biological relationships of NAS-disturbed species. This review also documented the current challenges and future directions in NAS–microbiome–host-related research. Meaningfully, this review provides fundamental information to advance the understanding of NAS interference on the gut microbiome, and contributes to decision-making on the safety of NAS in the future.

2. Materials and Methods

This review is based on a two-tier comprehensive literature analysis on the PubMed and Google Scholar databases. The purpose of the literature analysis was to investigate the associations among the use of low/no-calorie sweeteners (NAS), changes in the composition of the gastrointestinal microorganisms, and the subsequent metabolism alterations in the host. The aim of the literature search was to maximize the information input of current academic literature that examined the impact of the six officially recognized NAS on the gut microbiota, which is commonly referred to as the gut or intestinal "bacteria", "microbiome", "microbiota", "microbes", "microflora", or "microorganisms". Furthermore, the investigation encompassed the inclusion of several designations for sweetening agents, which were enumerated as follows: aspartame; sucralose, also recognized as Splenda; saccharin; acesulfame potassium, also identified as Acesulfame K or Ace-K; neotame; and advantame. A combination of phrases related to sweeteners and terms related to gut microbiota were input to the database generating the primary literature pool on 8 August 2024. The search encompassed articles of all categories, without imposing any limitations on the dates of publication, which generated 63 articles on aspartame, 104 articles on saccharin, 153 articles on sucralose, 45 articles on acesulfame potassium, 7 articles on neotame, and 4 articles on advantame. This included scholarly publications that have undergone peer review, as well as non-peer-reviewed articles such as research articles, reviews, conference abstracts, news articles, interviews, editorials and opinions, and book chapters. Neotame and advantame, as second-generation sweeteners approved more recently by the U.S. FDA—neotame in 2002 and advantame in 2014—have had a shorter timeframe for research and commercial use, potentially limiting the incentive for extensive study. Additionally, their structural similarities to aspartame, a more prevalent or contentious NAS, may have further deprioritized research on them.

This review is composed of two consecutive screening stages(Figure 1). The first screening adopted a dual exclusion strategy. Initially, the first screening excluded repetitive articles in one or more databases, which generated 133 records for this review. Published abstracts from presentations and/or conferences were matched with full articles, where applicable, and the remaining abstracts were excluded in the search. These articles were screened for the following exclusion criteria: (a) records not related to NAS exposure; (b) records that were review articles, editorial and opinion pieces, or letters responding to recent publications in the field; and (c) records focusing on multiple NAS in a single experimental setting, multiple NAS categorized as a single variable, or co-existing NAS exposed to a single independent experimental subject. Beyond this exclusion, these articles were further screened for relevance based on the following inclusion criteria: (a) in vivo studies conducted in animals and humans (all in vitro studies were excluded); (b) oral exposure to NAS, which included diet intake or drinking water exposure; and (c) reported primary measurements of the microbiome in the gut. Following application of the defined screening criteria, 22 publications were identified as relevant primary research articles investigating the administration of NAS to animals or humans and the consequent disturbance to the gut microbiota. The empirical approaches and findings of these articles, encompassing variables such as sample size, participants, control groups, perturbation related to the gut microbiota, and analytical methods, have been extracted and classified based on the type of sweetening agent. The extracted gut microbiome information was input into the National Center for Biotechnology Information (NCBI) taxonomy database to conduct taxonomic cluster analysis, which displays a hierarchical clustering of bacterial taxa (phylum to species level) that are significantly impacted by different NAS.

Based on the primary screening, we performed a secondary literature screening to collect available evidence on the association between the gut microbiota and metabolic consequences in the host. The screening aimed at documenting the altered metabolic pathways in the host and any metabolites or molecules that exhibited significant differences in the host. The exclusion criteria are listed as follows: (a) alterations in host metabolism were not reported using direct measurements derived from host biospecimen samples before or after NAS exposure; and (b) metabolites and affected metabolism pathways were not documented. This secondary screening resulted in 11 publications that served as the direct scientific evidence on integrating the triangle relationships among NAS exposure, the gut microbiome, and host metabolism. A zero match was found on advantame exposure in vivo in two-stage screening; thus, advantame was excluded in the meta-analysis. The empirical findings of these 11 articles, encompassing variables such as altered host metabolism pathways, key metabolites and their changing patterns in hosts before and after NAS exposure, and analytical methods, were extracted and classified based on the type of sweetening agent.

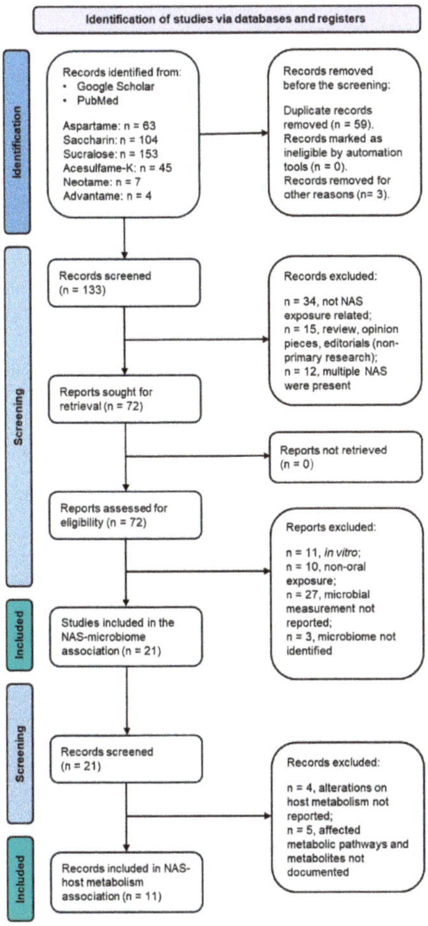

Figure 1. Flow diagram of the two-tier publication selection process for this review.

3. Gut Microbiota Species Modulated by NAS Exposure

3.1. Aspartame

The changes in the abundance of the gut microbiome affected by aspartame consumption were examined in rats after 8-week aspartame exposure in drinking water [39]. Even though the study introduced fat content in feeding as a second variable interacting with the first variable, the aspartame in water, the absolute bacterial analysis in the study concluded that *Clostridium leptum* was significantly higher ($p < 0.05$) in the low-fat feeding group fed with chow (12% kcal fat) and aspartame than in the control group fed with chow and water. *C. leptum* is known as a nitroreductase-producing bacterium, and its metabolism is associated with infective and chronic inflammatory bowel disease, including Crohn's disease and ulcerative colitis [40–42]. *C. leptum* was also identified as a prominent bacterium that elicited quantitative differences between patients with type 2 diabetes and healthy humans [43], and showed a higher abundance in overweight adolescents [44]. Aspartame exposure attenuated the increase in abundance of *Clostridium cluster XI* in both the low-fat and high-fat diet groups [39]. However, treatment with aspartame in the high-fat diet (60% kcal fat) group resulted in the highest abundance of *Clostridium cluster XI*, *Enterobacteriaceae*, and *Roseburia* spp., and total bacteria, compared to the other groups fed either the low-fat or high-fat diets, with and without aspartame [39]. An increase of *Clostridium cluster XI* was witnessed in dietary habits involving intaking high levels of carbohydrate, fat, and protein, and its increase was positively correlated with the inflammation marker pro-inflammatory mucosal IL1-β concentration [45,46]. Interestingly, *Roseburia* spp. is an anaerobic rod-shape bacterium that can produce butyrate in the colon, and it contributes to multiple diseases, including inflammatory bowel disease, type 2 diabetes mellitus, and antiphospholipid syndrome [47], which exhibits a functional overlap with the *C. leptum* mentioned above. Nevertheless, the relative bacterial abundance analysis in the study conducted by Palmnäs et al. seemed to indicate that aspartame can also induce a decrease in *Lactobacillus* spp. and no changes in *Clostridium cluster I*, regardless of the fat content in feeding [39]. The changes of *Clostridium coccoides* and *Roseburia* seemed undifferentiable in the normal fat feeding group versus the aspartame-dosed normal fat feeding group, while such changes were notable when high-fat feeding was introduced, depicted as promoting the abundance of *Roseburia* and *Clostridium coccoides*.

Acting not only in rats, aspartame consumption also altered the gut microbiome in humans, confirmed in a randomized-controlled trial encompassing 20 healthy adults administered aspartame for two weeks in a dose lower than the acceptable daily intake [48]. The results elucidated that top five glycemic responders' aspartame exposure exhibited a positive association with *Bacteroides fragilis* and *Bacteroides acidifaciens*, and an inversely negative association with *Bacteroides coprocola* [48]. However, the same studies indicated that such changes in the gut microbiome could be customized in different hosts, with an emphasis that *Akkermensia muciniphila* increased significantly in human subjects showing the lowest glycemic response to aspartame exposure [48]. Further microbiome dissimilarities were displayed with the increase of *Clostridium* sp. *CAG:7* and *Tyzzerella* sp. *Marseille-P3062* and the decrease of *Alistipes obesi* and *Eubacterium* sp. *CAG:248* in the top glycemic responders compared to the bottom glycemic responders [48]. Among all aspartame-associated gut microbiomes in the study above, enterotoxigenic *Bacteroides fragilis* was recently reviewed to be possibly associated with colorectal cancer [49], while commensal bacterium *Bacteroides acidifaciens* was proven to participate in insulin protection in serum and β-oxidation in adipose tissues, protecting the host from diabetes and obesity [50,51]. *Akkermensia muciniphila*, a next-generation probiotic, has beneficial effects on glucose and lipid metabolism and the inflammatory response in humans, as well as on endotoxemia protection [52–54]. *Clostridium* sp. *CAG:7* is a purine-degrading prebiotic [55] and *Tyzzerella* sp. *Marseille-P3062* was positively associated with Crohn's disease [56]. Altogether, aspartame exposure caused multiple changes in both probiotic and pathogenic bacteria and such bacterial consequences were customized in the host.

3.2. Saccharin

A few studies were identified reporting that saccharin exposure can modify the abundance of the gut microbiome community. It was demonstrated that saccharin exposure in humans, at a dose of 0.18 g/day for 28 days, can cause significant changes in gut microbiota in experimental subjects who showed impairment of glucose tolerance [48]. Among the five top glycemic responders, saccharin exposure was positively associated with *Prevotella copri* [48], a species contributing to glucose homeostasis through enhancing bile acid metabolism and farnesoid X receptor (FXR) signaling [57], and negatively associated with *Bacterioides xylanisolvens*, a xylanolytic anaerobe known for its dietary fiber degradation and fermentation [58]. Meanwhile, the abundance of *Alistipes onderdonkii* was significantly higher during saccharin exposure but reduced to baseline levels in follow-up measurements [48]. The shift of *Alistipes onderdonkii*, an anaerobe exhibiting known pro-inflammatory activity which modulates the inflammatory response [59], might imply that saccharin exposure caused an inflammation response in the host, and activation of *Alistipes onderdonkii* might have healed the response in a short time. Furthermore, *Firmicutes CAG:102* showed an irreversible decrease in the long term after saccharin exposure [48]. Saccharin exposure also exhibited personalized effects on the gut microbiome in different humans. Top glycemic responders had a richer abundance of *Blautia* sp. *Marselle P2398* and

Clostridium sp. *CAG:62* than bottom glycemic responders [48], in which *Blautia* sp. *Marselle P2398* was a marker reflecting major depressive disorder [60]. On the contrary, bottom glycemic responders had a higher abundance of *Bifidobacterium ruminantium, Clostridiales bacterium UBA 7739, Faecalibacterium prausnitzii,* and *Parabacteroides distasonis* than the top glycemic responders [48]. The same research team also evaluated fecal bacteria composition in mice exposed to saccharin in drinking water for 11 weeks, where they found over 40 operational taxonomic units were significantly altered in abundance using 16S RNA sequencing. At the strain level, *Bacteroides uniformis* were over-represented in the saccharin-exposed group compared with the control, while *Lactobacilluys Reuteri* were under-represented [48]. *Bacteroides uniformis* can ameliorate the metabolic and immunological dysfunction in obese mice induced by a high-fat diet [61,62], and its elevation may indicate their potential protection mechanisms via the gut microbiome to the host. Moreover, a randomized controlled trial revealed that probiotic *Lactobacilluys Reuteri* supplementation can increase the insulin sensitivity and bile acid deoxycholic acid in serum in type 2 diabetic patients [63]. Thus, decreased *Lactobacilluys Reuteri* may imply a decrease in insulin sensitivity as well as in bile acid metabolism. Results from shotgun metagenomic sequencing further exemplified the over-representation of *Bacteriodies vulgatus* and the under-representation of microbiome *Akkermansia muciniphila* in the saccharin-exposed group [7], which was coherent to the gut microbiome changes previously reported in patients with type 2 diabetes [64].

3.3. Sucralose

Sucralose, similar to saccharin, has been extensively documented for its alteration of the gut microbiome. Abou-Donia et al. reported that administration of Splenda (which contains sucralose) by oral gavage, at different concentrations of up to 1000 mg/kg for 12 weeks in rats, resulted in a significant decrease in beneficial gut bacteria [65]. Even at the lowest dose (100 mg/kg/d) the bacterial counts of *bifidobacterial, lactobacilli,* and *Bacteroides* were reduced by 36.9%, 39.1%, and 67.5%, respectively. Similarly, another study reported the alteration of 14 genera after exposing C57BL/6J to sucralose at 0.1 mg/mL in drinking water for six months, which were an increased abundance of *Turicibacteraceae turicibacter, Lachnospiraceae ruminococcus, Ruminococcaceae ruminococcus, Verrucomicrobiaceae akkermansia,* and unclassified members in the families *Clostridiaceae* and *Christensenellaceae;* and the decreased abundance of *Staphylococcaceae staphylococcus, Streptococcaceae streptococcus, Dehalobacteriaceae dehalobacterium, Lachnospiraceae anaerostipes, Lachnospiraceae roseburia,* and unassigned *Peptostreptococcaceae, Erysipelotrichaceae,* and *Order bacillales* [66]. However, these studies did not provide information on bacteria alterations at the strain level.

The impacts of sucralose on the gut microbiome exhibited solid evidence tracking back to the strain level. Human trials, by exposing healthy adults to sucralose under ADI for 28 days, revealed the alteration of three bacterial species, which were an increase of *Eubacterium CAG:352* and *Dorea longicatena,* and a decrease of *Oscillibacter ER4* [48]. However, sucralose exposure behaved differently on gut microbiome composition in different glycemic responders. *Bacteroides caccae, Bacteroides* sp. *Phil13,* and *Flavonifractor plautii* were three enriched species in the top glycemic responders, but were not shown in the bottom glycemic responders. The bottom glycemic responders accumulated more *Intestinimonas butriciproducens* than the top glycemic responders [48].

3.4. Neotame

Chi et al. reported that four-week neotame exposure in CD-1 mice facilitated the growth of two genera in the phylum *Bacteroidetes,* including *Bacteroides* and one undefined genus in *S24-7,* while significantly decreasing three genera in the family *Ruminococcaceae,* consisting of *Oscillospira, Rumniococcus,* and one undefined genus, and five genera in the family *Lachnospiraceae,* which contained *Blautia, Dorea, Ruminococcus,* and two undefined genera [67]. This is the only research focusing on neotame exposure and gut microbiome analysis.

3.5. *Acesulfame Potassium*

Acesulfame potassium can induce gut microbiome changes, but multiple variables may cause discrepancy in its gut microbiome changes. Bian et al. illustrated that acesulfame potassium induced sex-dependent alterations in gut microbiota [68]. In male mice treated with Ace-K via oral gavage at a dose of 37.5 mg/kg body weight/day, *Bacteroides* were highly increased, along with significant increases in two other genera, *Anaerostipes* and *Sutterella*. Notably, the four-week Ace-K treatment dramatically decreased the relative abundance of multiple genera in female mice, including *Lactobacillus*, *Clostridium*, an unassigned *Ruminococcaceae* genus, and an unassigned *Oxalobacteraceae* genus, and increased the abundance of *Mucispirillum*. Acesulfame potassium showed gut microbiome alteration in newborn mice when exposing their mother to acesulfame potassium. Olivier-Van Stichelen et al. illustrated that maternal exposure could induce defective *Akkermansia muciniphila* in newborns [69]. However, *Akkermansia muciniphila* showed no difference in growth when exposed to low or high acesulfame potassium in culture medium.

A comprehensive summary of the documented alterations in the gut microbiome resulting from exposure to specific NAS, as detailed in Sections 3.1–3.5, is provided in Table 1.

Table 1. Summary of documented altered gut microbiomes affected by specific NAS.

NAS	Exposure	Altered Gut Microbiome *	Analytical Methods	Reference
Aspartame ADI: 50 mg/kg Sucrose equivalence: 200× 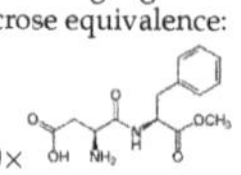	Rats, SD Normal rats (N = 10–12), 5 mg/kg/d, 8 weeks Obese rats (N = 10–12), 7 mg/kg/d, 8 weeks	Normal rats: *Clostridium leptum* ↑ Obese rats: *Clostridium cluster XI* ↓ *Enterobacteriaceae* ↑ *C. leptum* ↑ and *Roseburia* spp. ↑	Fecal DNA extraction + 16S rRNA sequencing + qRT-PCR analysis	(Palmnäs et al., 2014) [39]
	Humans, Exposed (N = 20), 0.24 g and 5.76 g glucose, 4 weeks	Top glycemic responders: *Bacteroides Fragilis* and *Bacteroides acidifaciens* ↑ *Bacteroides Coprocola* ↓ Bottom glycemic responders: *Akkenmensia muciniphila* ↑ Top compared to bottom responders: *Clostridium* sp. *CAG:7* ↑ *Tyzzerrlla* sp. *Marseille-P3062* ↑ *Alistipes obesi* and *Eubacterium* sp. *CAG:24* ↓	16S rRNA sequencing Illumina NextSeq platform	(Suez et al., 2022) [48]
	Humans, Exposed (N = 7), 1.7–33.2 mg/d, based on daily food records in four days Non-exposed (N = 24)	No significant changes	16S rRNA sequencing Length heterogeneity polymerase chain reaction (PCR) fingerprinting	(Frankenfeld et al., 2015) [70]
	Humans, Exposed (N = 15), 0.425 g/day, 14 days	No significant changes	16S rRNA sequencing Illumina MiSeq	(Ahmad et al., 2020) [71]
Saccharin ADI: 5 mg/kg Sucrose equivalence: 300× 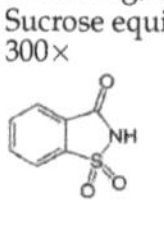	Humans, Exposed (N = 20), 0.18 g and 5.82 g glucose, 28 days	Top glycemic responders: Prevotella copri ↑ *Bacterioides xylanisolvens* ↓ *Alistipes onderdonkii* ↑ *Firmicutes CAG:102* ↓ Top compared to bottom responders: *Blautia* sp. *Marselle P2398* ↑ *Clostridium* sp. *CAG:62* ↑ *Bifidobacterium ruminantium* ↓ *Clostridiales bacterium UBA 7739* ↓ *Faecalibacterium prausnitzii* ↓ *Parabacteroides distasonis* ↓	16S rRNA sequencing Illumina NextSeq platform	(Suez et al., 2022) [48]
	Mice, C57BL/6 Exposed (N = 20), 3333 mg/kg/d, 11 weeks Control (N = 20)	*Bacteroides uniformis* ↑ *Lactobacilluys reuteri* ↓ *Bacteriodies vulgatus* ↑ *Akkermansia muciniphila* ↓	16S rRNA sequencing	(Suez et al., 2014) [7]
	Humans, (5 males and 2 females, aged 28–36) Exposed, 5 mg per kg of body weight for 5 days	*Bacteroides fragilis* ↑ *Weissella cibaria* ↑ *Candidatus arthromitus* ↓	16S rRNA sequencing	(Suez et al., 2014) [7]
	Mice, male, C57BL/6J (8 weeks old) 0.3 mg/mL in water for six months	After three-month consumption: *Anaerostipes* ↓ *Ruminococcus* ↓ *Sporosarcina* ↑ *Jeotgalicoccus* ↑ *Akkermansia* ↑ *Scillospira* and *Corynebacterium* ↑ After six-month consumption: *Ruminococcus* ↓ *Adlercreutzia* ↓ and *Dorea* ↓ *Corynebacterium* ↑, *Roseburia* ↑ and *Turicibacter* ↑.	16S rRNA gene sequencing	(Bian et al., 2017) [72]

Table 1. *Cont.*

NAS	Exposure	Altered Gut Microbiome *	Analytical Methods	Reference
	Dogs, female beagles 0.02% saccharin and eugenol, or 5% fiber blend plus 0.02% saccharin and eugenol for 10 days (N = 8)	No shifts in fecal microbial richness and diversity	16S rRNA gene sequencing	(Nogueira et al., 2019) [73]
	Mice, C57Bl\6J and Whole body T1R2-deficient mice, (eight-week-old) 250 mg/kg for 10 weeks	No alterations in microbial diversity or composition at any taxonomic level.	16S rRNA gene sequencing	(Serrano et al., 2021) [74]
	Humans, 18–45 years old (1) pulp filler/placebo (1000 mg/day 1) sodium saccharin (400 mg/day), (3) lactisole (670 mg/day), or (4) sodium saccharin (400 mg/day) + lactisole (670 mg/day) twice daily for 2 weeks	No alterations in microbial diversity or composition at any taxonomic level	16S rRNA gene sequencing	(Serrano et al., 2021) [74]
	Wistar rats, Exposed, 20 and 100 mg/kg body weight/day for 28 days	No effects on microbiome changes	16S rRNA gene sequencing	(Murali et al., 2022) [75]
	Mice, C57BL/6 J Exposed (N = 10), 9–22 mg/kg/d, 6 months Control (N = 10)	Turicibacteraceae Turicibacter ↑ *Lachnospiraceae ruminococcus* ↑ *Ruminococcaceae ruminococcus* ↑ *Verrucomicrobiaceae akkermansia* ↑ Unclassified members in Family Clostridiaceae↑*Christensenellaceae* ↑ *Staphylococcaceae Staphylococcus* ↓ *Streptococcaceae streptococcus* ↓ *Dehalobacteriaceae dehalobacterium* ↓ *Lachnospiraceae anaerostipes* ↓ *Lachnospiraceae roseburia* ↓ Unassigned *Peptostreptococcaceae* ↓ *Erysipelotrichaceae* ↓ *Bacillales* ↓	16S rRNA sequencing	(Bian, 2017) [66]
	Humans, Exposed (N = 20), 0.18 g and 5.82 g glucose, 28 days	Top glycemic responders: *Eubacterium CAG:352* ↑ *Dorea longicatena* ↑ *Oscillibacter ER4* ↓ Top compared to bottom responders: *Bacteroides caccae* ↑ *Bacteroides* sp. *Phil13* ↑ *Flavonifractor plautii* ↑ *Intestinimonas butriciproducens* ↓	16S rRNA sequencing Illumina NextSeq platform	(Suez et al., 2022) [48]
Sucralose ADI: 15 mg/kg Sucrose equivalence: 600× 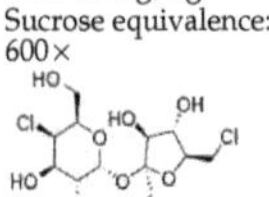	Rats, SD Exposed (N = 10/group), Splenda 1.1, 3.3, 5.5 or 11 mg/kg/d, 12 weeks Control (N = 10)	*Bifidobacterial* ↓ *Lactobacilli* ↓ *Bacteroides* ↓	Culturing plates	(Abou-Donia et al., 2008) [65]
	Mice, C57Bl/6J mice (4 weeks old) Exposed (N = 8/group), sucralose 1.4 ± 0.1 mg/kg BW/day and 14.2 ± 2.2 mg/kg BW/day Control (N = 8)	*Clostridium cluster XIVa*↓	16S rRNA sequencing	(Uebanso et al., 2017) [76]
	Mice, SAMP1/YitFc (SAMP) Exposed(N = 5–7/group), 6-week supplementation of Splenda; ingredients: sucralose/maltodextrin, 1:99, *w/w*), 1.08 mg/mL; 3.5 mg/mL; 35 mg/mL	Five classes in *Proteobacteria* phylum ↑ (*Alphaproteobacteria, Betaproteobacteria, Gammaproteobacteria, Epsilonproteobacteria, Deltaproteobacteria*) *Escherichia coli* ↑	Culturing plates + 16S rRNA sequencing	(Rodriguez-Palacios et al., 2018) [77]
	Mice, C57BL/6 (5 weeks old) Exposed (N = 8/group), 8 weeks, sucralose (2.5%, *w/v*)	In chow-only mice: *Firmicutes* ↓, *Bacteroidetes* ↓, *Bifidobacterium* ↑ In high-fat-diet mice: *Firmicutes* ↑, *Bacteroidetes* ↓	16S rDNA sequencing	(Wang et al., 2018) [78]
	Mice, Pathogen-free (SPF) C57BL/6J, male, (28 days) Exposed (N = 8/group), 0.0003 g/mL, 0.003 mg/mL, 0.03 mg/mL, 0.3 mg/mL per day for 16 weeks	In jejunum: *Tenacibaculum* ↑, *Ruegeria* ↑ In ileum: *Staphylococcus* ↑, *Corynebacterium* ↑ In cecum: *Lachnoclostridium* ↓, *Lachnospiraceae UCG-006* ↓	16S rDNA sequencing	(Zheng et al., 2022) [79]
	Humans, 18–50 years old Exposed (N = 16), 780 mg sucralose/day for 7 days Control (N = 14)	No significant changes	16S rDNA sequencing	(Thomson et al., 2019) [80]
	Humans, 18–35 years old Exposed (N = 20/group), 48 mg Splenda/day for 10 weeks	*Lactobacillus acidophilus* ↓ *Blautia coccoides* ↑	16S rRNA sequencing Quantitative polymerase chain reaction (qPCR)	(Méndez-García et al., 2022) [81]
	Mice, C57BL/6J, 8 weeks old Exposed (N = 10), 0.1 mg/mL for 6 months	Lactobacillus ↓ Ruminococcus ↓	16S rDNA sequencing	(Chi et al., 2024) [82]

Table 1. *Cont.*

NAS	Exposure	Altered Gut Microbiome *	Analytical Methods	Reference
Acesulfame potassium ADI: 15 mg/kg Sucrose equivalence: 200× 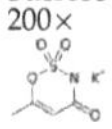	Mice, CD-1 Exposed (N = 5), 37.5 mg/kg/d, 4 weeks Control (N = 5)	Males: *Bacteroides* ↑; *Anaerostipes* ↑; *Sutterella* ↑ Females: *Mucispirillum* ↑, *Lactobacillus* ↓, *Clostridium* ↓, an unassigned *Ruminococcaceae* genus and an unassigned *Oxalobacteraceae* genus ↓	16S rRNA sequencing	(Bian, et al., 2017) [68]
	Mice, C57Bl/6J mice (4 weeks old) Exposed (N = 9/group), 15 mg/kg BW/day Control (N = 8)	No significant changes	16S rRNA sequencing	Uebanso et al., 2017) [76]
	Mice, C57BL/6J, (8 weeks old) Exposed 150 mg/kg b.w./day for 8 weeks	*Clostridiaceae* ↓ *Lachnospiraceae* ↓ *Ruminococcacea* ↓	16S rRNA sequencing	(Hanawa et al., 2021) [83]
	Humans, Exposed (N = 7), 1.7–33.2 mg/d, based on daily food records in four days Non-exposed (N = 24)	No significant changes	16S rRNA sequencing	(Frankenfeld et al., 2015) [70]
Neotame ADI: 18 mg/kg Sucrose equivalence: 7000–13,000× 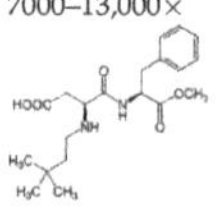	Mice, CD-1 Exposed (N = 5), 0.75 mg/kg/d, 4 weeks Control (N = 5)	*Bacteroidetes* including *Bacteroides* and one undefined genus in *S24-7* ↑ Three genera in the family *Ruminococcaceae*, consisting of *Oscillospira*, *Ruminococcus*, and one undefined genus, and five genera in family *Lachnospiraceae*, which contained *Blautia*, *Dorea*, *Ruminococcus*, and two undefined genera. ↓	16S rRNA sequencing	(Chi et al., 2018) [67]

* Note: The arrows (↑/↓) denote the modulation (increase/decrease) in the relative abundance of particular gut microbiota in response to specific NAS, as delineated in the referenced scholarly articles. These modulations are contingent upon the experimental frameworks and outcomes articulated in each study.

4. Alterations of Metabolism in NAS–Microbiome–Host Interactions

4.1. Aspartame

Aspartame and its secondary products can reach to the colon, influencing the microbiota. Previous research elucidated that aspartame can be hydrolyzed in the intestines into phenylalanine, aspartame, and methanol. Aspartame consumption was associated with fasting hyperglycemia and impaired insulin tolerance in rats in a manner independent of fat intake in the diet, indicated by no difference in an oral glucose tolerance test and an elevated area under the curve for glucose in the insulin tolerance test in rats in both low- and high-fat diets [39]. Through an approach of administrating a high physiological insulin bolus into rats, the researchers clued that aspartame was able to reduce the capacity of clearing endogenous glucose, contributed by the mechanism of inducing an impairment of insulin-mediated suppression of net hepatic glucose output, than the deduction of peripheral insulin sensitivity [39]. The interlinks between aspartame and host health were presumed to be attributed to two gut microbiota changes, which were *Enterobacteriaceae* and *Clostridium cluster XI*, revealed in a high-fat diet in rats. The increase of *Enterobacteriaceae*, a member of the potentially harmful proteobacteria, produced gases and short-chain fatty acids that have been previously reported to be associated with inflammation and insulin resistance [84–92]. The decrease of the latter, as a member of the probiotic community, caused a disadvantageous condition of the community and may consequently have increased the amount of pathogenic bacteria in the gut microbiota [39].

The metabolites produced by the gut microbiota further entailed the putative mechanisms of how aspartame affected host health via gut microbiome-related physiological changes. Gut microbiome-derived metabolites represented end products of bacteria physiological activities and were the key intermediates bridging the host and the gut microbiome [33]. Aspartame exposure is associated with changes in acetate and butyrate [39]. The decrease of butyrate in the serum of the rats can be correlated with the observed decrease in *Clostridium cluster XI*, which are known as butyrate-producing bacteria. Another signature metabolite is propionate, which exhibited large elevations under the conditions of low-fat feeding as well as high-fat feeding. This change could be attributable to the increase of *Clostridium leptum* as it produces propionate when fermenting oligosaccharides [93,94].

Human studies reported that the modulation of the gut microbiome affected by aspartame exposure induced personalized but causative impacts on the glycemic response [48]. Such impacts were further demonstrated as significant metabolomic alterations in human plasma. Kynurenine, terephthalic acid, indole-3-acetate, and benzoate were four signature metabolites altered in the most sensitive responders to aspartame in the level of the glycemic response exposed to one-week of aspartame below ADI. Among them, kynurenine, indole-3-acetate, and benzoate were increased in the plasma, while the terephthalic acid was reduced. However, the detailed mechanism correlated with the metabolomic profiles and the microbiome changes affected by aspartame were not discussed.

Previous research has mentioned that hosts who had experienced aspartame-induced gut microbiome dysbiosis developed alterations of multiple host metabolism pathways, which had the potential to be correlated with gut microbiome changes. Gluconeogenesis might be a potential mechanism through which the gut microbiome interfered with propionate production [39]. Moreover, pathways related to the urea cycle and its metabolites might be of prime consideration to understanding the interactions between gut microbiome dysbiosis and host health, as they were increased in the top aspartame responders to glycemic responses [48]. Along with these pathways, the negative association of the pathways in top glycemic responders, including phosphonate and phosphinate metabolism, flavin biosynthesis, L-histidine degradation, and L-proline degradation, may also be vital for further analysis on understanding the inhibition mechanisms of aspartame in aspartame–microbiome–host interactions [48].

4.2. Saccharin

The most direct existing evidence suggested that the mechanisms of the gut microbiome interfered with in humans who were exposed to saccharin and impaired glucose intolerance were related to 1) Uridine Monophosphate (UMP) biosynthesis and 2) glycolysis and glycan degradation [7]. The increase of UMP biosynthesis occurred simultaneously with the increase of *Prevotella copri* and shared the same pattern to an extent, which increased correspondingly along four timepoints: before exposure, Day 7 in exposure, Day 14 in exposure, and Day 28 after exposure. On the contrary, the decrease of glycolysis and glycan degradation occurred simultaneously with the decrease of *Bacteroides xylanisolvens*, and their changes were matched to each other in the exposure duration [48]. However, a study also reported that saccharin consumption also induced an increase in glycan degradation pathways and further annotation suggested that five Gram-positive and -negative bacteria species contributed to this increase, which were *Bacterioides fragilis*, *Bacteroides vulgatus*, *Parabacteroides distasonis*, *Staphylococcus aureus*, and *Providencia retteri* [7]. Such conflicts further explained that the gut microbiome was engaged in a scrimmage when exposed to saccharin. Moreover, the disturbance of metabolic pathways was displayed as a comprehensive interaction among multiple sub-pathways. The decrease of glycolysis and glycan degradation can be viewed as a weighted combined outcome of multiple sub-pathways, including homolactic fermentation, glycolysis I from glucose 6-phosphate, glycolysis II from fructose 6-phosphate, glycerol degradation to butanol, hexitol degradation, and Neu5Ac degradation [48]. Additionally, personalized differences were not neglectable between top glycemic responders and bottom glycemic responders in pathway changes. In top glycemic responders, five metabolomic pathways were significantly promoted, including (1) caprolactam degradation, (2) L-isoleucine biosynthesis II, (3) CDP-diacylglycerol biosynthesis, (4) glycerol degradation to 1.3-propanediol, and (5) mixed acid fermentation. In bottom glycemic responders, six metabolomic pathways were emphasized, which were (1) alanine, aspartame, and glutamate metabolism, (2) pentose phosphate metabolism, (3) L-serine and glycine biosynthesis I, (4) L-tryptophan biosynthesis, (5) histidine, purine, and pyrimidine biosynthesis, and (6) polyamine metabolism. Discrepancies in the affected pathways in human subjects further enlighten the complexity of deciphering the extent and magnitude to which specific bacteria strains contributed to pathway changes, thus increasing the difficulty in evaluating the interfering roles of the gut microbiome on their participation in these metabolic pathways. In the same study, serum metabolomic analysis revealed statistically significant changes in five metabolites: 4-hydroxybenzoate, benzoate, indoxyl sulfate, hexadecanedioic acid, and butyrate. Three of them, indoxyl sulfate, a metabolite related to vascular disease, 4-hydroxylbenzoate, and benzoate, increased during saccharin exposure, while hexadecanedioic acid decreased [48]. However, the study did not establish a detailed mechanism on how the gut microbiome may induce the above metabolomic changes that can result in host glucose tolerance impairment.

4.3. Sucralose

There is clear evidence to support sucralose-induced physiological changes in the host via gut microbiome alteration. Sucralose increased the abundance of bacterial genes related to pro-inflammatory mediators, which featured the increase of genes related to LPS synthesis, flagella protein synthesis, and fimbriae synthesis as well as bacterial toxin and drug resistance genes [66]. In addition, fecal metabolomic analysis confirmed that sucralose altered quorum sensing signals, amino acids and derivatives, and bile acids. Furthermore, sucralose induced elevated pro-inflammatory gene expression in the liver, including matrix metalloproteinase 2 (MMP-2) and inducible nitric-oxide synthase (iNOS), that might be due to crosstalk along the gut–liver axis [66].

Another study depicted that sucralose exposure induced a significant increase in arginine biosynthesis, a significant decrease in mixed acid fermentation, and alteration in the Tricarboxylic Acid (TCA) cycle alongside the sucralose supplementation. Three other bacterial metabolomic pathways showed synergic decreases along the sucralose exposure timeline, including urate biosynthesis/inosine 5′-phosphate degradation, adenosine deoxyribonucleotide de novo biosynthesis, and guanosine nucleotide de novo biosynthesis [48]. The same study also evaluated changes in serum metabolites, in which they found increases in isocitrate, trans-aconitate, serine, N-acetylalanine, aspartate, quinolinate, 2-C-methyl-D-erythritol 4-phosphate, galactarate, and psicose; and decreases of pseudouridine, uric acid, and sebacic acid. The researchers performed enriched pathway analysis which highlighted the alteration of host pathways, including arginine biosynthesis and glutamine metabolism, aminoacyl-tRNA biosynthesis, and the TCA cycle. In alignment with the increased abundance of the TCA cycle in the gut microbiota, the quantitative measurements of two TCA intermediates, isocitrate and trans-aconitate, increased in the serum of human subjects after experiencing sucralose exposure, enlightened the connection between the gut microbiome and host via TCA cycle changes. Furthermore, the same study discovered that three plasma metabolites exhibited significant increases in the top glycemic responders compared to the bottom glycemic responders: beta-hydroxyisobutrate, cycteate, and serine [48]. These three metabolites may play an important role in inducing glycemic responses potentially linked to gut microbiome changes.

4.4. Neotame

Neotame consumption has been proven to alter the metabolic pathways of gut microbiome. The available research revealed that streptomycin biosynthesis, amino acid metabolism, folate biosynthesis, and lipopolysaccharide biosynthesis were four pathways enriched in the neotame-treated microbiome, while seven other pathways were underrepresented, including fatty acid metabolism, sporulation, benzoate degradation, carbohydrate metabolism, lipid metabolism, bacterial chemotaxis, and ABC transporters [67]. Additionally, the functional gene analysis further illustrated that the pyruvate-derived and succinate-derived butyrate fermentation pathways were perturbed by neotame exposure by alteration of the genes in enzyme production. In the pyruvate-derived butyrate fermentation pathway, four genes were reduced, which were the genes of acetyl-CoA C-acetyltransferase, 3-hydroxybutyryl-CoA dehydrogenase, 3-hydroxybutyryl-CoA dehydratase, and butyryl-CoA dehydrogenase. In addition, the genes of phosphate butyryltransferase and butyrate kinase were increased. However, the genes of 4-hydroxybutyryl-

CoA dehydratase, butyryl-CoA dehydrogenase, and acetate CoA-transferase were decreased in succinate-derived butyrate fermentation pathways [67].

4.5. Acesulfame Potassium

Acesulfame potassium induced carbohydrate metabolism changes in the gut microbiota, but exhibited sex-dependent features. The direct evidence, reported by Bian et al., claimed sex-specific alterations of functional genes in the gut microbiome exposed to acesulfame potassium [68]. In female mice, many of the key genes related to energy metabolism were decreased, consistent with the decrease of multiple microbiome bacteria in the female mice. These genes were involved in carbohydrate absorption and transportation, fermentation and degradation, polysaccharide hydrolysis, glycolysis, and the TCA cycle. There were no exceptions in that all these genes tended to be inhibited in the female mice, thus restricting the expression of multiple proteins such as glucose uptake proteins, lactose permease, sugar and D-allose transporters, different phosphotransferases, L-xylulokinase, α-amylase, and D-xylonolactonase. In contrast, the carbohydrate absorption and metabolism pathways were activated in male mice, in alignment with their increase in gut microbiome *Bacteroides*. The genes involved in carbohydrate metabolism and fermentation pathways, sugar and xylose transportation, glycolysis, and the TCA cycle were all increased. Altogether, acesulfame potassium might interfere with the gut microbiome–host interaction by inducing sex-specific gene alterations [68].

Furthermore, acesulfame potassium induced increasing gene abundance correlated to lipopolysaccharide synthesis (LPS synthesis) that might further increase the risk of inflammation in the host. Bian et al. demonstrated that genes involved in LPS synthesis also underwent sex-specific alterations after acesulfame potassium exposure. In female mice, some LPS synthesis-related genes as well as LPS-export genes were decreased, including UDP-glucose:(Heptosyl) LPS alpha-1,3-glucosyltransferase, ADP-L-glycero-D-manno-heptose 6-epimerase, amino-4-deoxy-L-arabinose transferase, UDP-D-GlcNAcA oxidase and UDP-GlcNAc3NAcA epimerase. Meanwhile, genes encoding flagella components were also increased. However, in male mice, only two LPS synthesis genes, which were glycosyltransferase and UDP-perosamine 4-acetyltransferase, and one bacterial toxin synthesis gene, which was thiol-activated cytolysin, were enriched. In addition, Bian et al. further demonstrated the changes in key fecal metabolites [68]. In female mice, three metabolites, which were D-lactic acid, succinic acid, and 2-oleoylglycerol, were significantly decreased. Conversely, two metabolites, pyruvic acid and cholic acid, were increased and one metabolite, deoxycholic acid, was decreased in the male mice after acesulfame potassium exposure. Even though the study did not correlate the LPS synthesis alteration and key fecal metabolites to the changes in the gut microbiome on the species level, changes in the bacterial genes supported by the fecal metabolites provided solid support for interlinking gut microbiome–host interaction mechanisms with LPS synthesis.

A comprehensive overview of the documented metabolic alterations associated with specific NAS, as discussed in Sections 4.1–4.5, is provided in Table 2.

Table 2. Summary of documented altered metabolisms affected by specific NAS.

NAS	Altered Host Metabolism Pathways *	Key Metabolites Changes in Host **	Analytical Methods	Source
Aspartame ADI: 50 mg/kg Sucrose equivalence: 200× 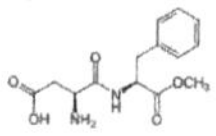	Fasting hyperglycemia and impaired insulin tolerance Pathways: Gluconeogenesis (?)	Acetate (?), butyrate ↓, propionate ↑	Serum metabolomic proton nuclear magnetic resonance spectroscopy (1H NMR)	(Palmnäs et al., 2014) [39]
	Altered glycemic response Pathways: Fatty acid degradation ↓ L-methionine biosynthesis ↓ Peptidoglycan biosynthesis ↓ Top compared to bottom responders: Pyrimidine nucleobases salvage ↑ L-omithine biosynthesis ↑ Heme biosynthesis ↑ Urea Cycle ↑ Phosphonate and phosphinate metabolism ↓ Flavin biosynthesis ↓ Pyridoxal 5′-phosphate biosynthesis and salvage ↓ L-histidine degradation ↓ L-proline biosynthesis ↓	Kynurenine ↑, terephthalic acid ↓, indole-3-acetate ↑, benzoate ↑	Serum metabolomic UPLC + Q-ToF mass spectrometry	(Suez et al., 2022) [48]
Saccharin ADI: Sucrose equivalence: 400× 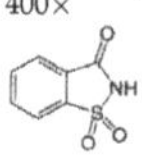	Altered glycemic response UMP biosynthesis ↑ glycolysis and glycan degradation ↓ homolactic fermentation ↓ glycolysis I from glucose 6-phosphate ↓ glycolysis II from fructose 6-phosphate ↓ glycerol degradation to butanol ↓ hexitol degradation ↓ Neu5Ac degradation ↓	4-hydroxybenzoate ↑, benzoate ↑, indoxyl sulfate ↑, hexadecanedioic acid ↓	Serum metabolomic UPLC + Q-ToF mass spectrometry	(Suez et al., 2022) [48]
	Induced glucose intolerance glycolysis and glycan degradation ↑	Propionate ↑ acetate ↑	Fecal metabolomic HPLC	(Suez et al., 2014) [7]
	Increased inflammation in the host by increasing the abundance of bacterial genes involved in elevating the pro-inflammatory mediators LPS synthesis ↑ Bacterial toxins ↑ Flagellar assembly protein ↑ Fimbrial protein ↑ Drug resistance ↑	Daidzein ↑ dihydrodaidzein ↑ O-desmethylangolensin ↑ Equol ↓ linoleoyl ethanolamide ↓ palmitoleoyl ethanolamide ↓ N,N-Dimethylsphingosine ↓ quinolinic acid ↑	Fecal metabolite analysis HPLC-Q-ToF	(Bian et al., 2017) [72]
	-	Acetate ↑, propionate ↑, butyrate ↑	Fecal metabolomic HPLC	(Serrano et al., 2021) [74]
	Altered amino acids, lipids, energy metabolism and bile acids in the plasma	-	Targeted MS-based metabolome profiling	(Murali et al., 2022) [75]
Sucralose ADI: 15 mg/kg Sucrose equivalence: 600× 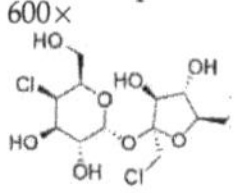	LPS synthesis ↑ Flagella protein synthesis ↑ Fimbriae synthesis ↑ Bacterial toxins and drug resistance genes ↑ Quorum sensing signals ↓ Amino acids and derivatives ↓ Bile acids (?)	N-butanoyl-l-homoserine lactone ↓, N-(3-oxo-hexanoyl)-homoserine lactone ↓, N-tetradecanoyl-L-homoserine lactone ↓, and N-pentadecanoyl-L-homoserine lactone ↓ L-tryptophan (Trp) ↑, quinolinic acid ↑, kynurenic acid ↓, and 2-aminomuconic acid ↑ L-tyrosine ↑, p-hydroxyphenylacetic acid ↓, and cinnamic acid ↓ 3-Oxo-4,6-choladienoic acid ↑, 3β,7α-dihydroxy-5-cholestenoate ↓, 3α,7β,12α-trihydroxyoxocholanyl-glycine ↓, and lithocholic acid/isoallolithocholic acid/allolithocholic acid/isolithocholic acid ↓	Fecal metabolomic HPLC-Q-ToF	(Bian, et al., 2017) [66]
	Arginine biosynthesis ↑ Mixed acid fermentation ↓ TCA cycle ↓ Urate biosynthesis/inosine 5′-phosphate degradation ↓ Adenosine deoxyribonucleotide de novo biosynthesis ↓ Guanosine nucleotide de novo biosynthesis ↓	Isocitrate ↑, trans-aconitate↑, serine ↑, N-acetylalanine ↑, aspartate ↑, quinolinate ↑, 2-C-methyl-D-erythritol 4-phosphate ↑, galactarate ↑, psicose ↑, pseudouridine ↓, uric acid ↓, and sebacic acid ↓	Serum metabolomic UPLC + Q-ToF mass spectrometry	(Suez et al., 2022) [48]
	Cholesterol–bile acid metabolism	Hepatic cholesterol ↑ cholic acid ↑, ratio of secondary bile acids (dehydrocholic acids (DCA) and lithocholic acid (LCA)) to primary bile acids (CA and CDCA) ↑	Liver/cecal metabolomic LC-MS	(Uebanso et al., 2017) [76]

Table 2. *Cont.*

NAS	Altered Host Metabolism Pathways *	Key Metabolites Changes in Host **	Analytical Methods	Source
	1. Richness of bile salt hydrolase gene (choloylglycine hydrolase) ↓, secondary bile acid synthesis pathway ↓ 2. Bile acid compositions and Farnesoid X Receptor (FXR) activation: (1)Ratios of various free bile acids and taurine-conjugated bile acids, including αMCA/TαMCA, ωMCA/TωMCA, CDCA/TCDCA and DCA/TDCA ↓, moderately for βMCA/TβMCA ($p < 0.07$) and CA/TCA ($p < 0.06$) ↓ (2)Expression of genes of Farnesoid X Receptor (FXR) signaling in livers, including *Shp*, *Cyp7a1*, *Cyp27a1*, and *Ntcp* ↓ 3. Altered hepatic cholesterol homeostasis Expressions of genes encoding three major cholesterol efflux transporters, including *Abca1*, *Abcg5*, and *Abcg8* ↑ Expression of genes associated with reverse cholesterol transport (RCT), including Ldlr and Scarb1 ↓ 4. Disrupted hepatic lipid homeostasis expression of two nuclear receptors, Srebp1c and Chrebp ↑ Acc1 gene and Cd36 gene ↑	Hepatic lipid ↑, ceramide ↑, hosphatidylethanolamines ↑, phosphatidylserines (PS) ↑, phosphatidylcholines (PC) ↑↓	Metabolomics and hepatic lipidomic UHPLC-ESI-TSQ Quantis triple quadrupole mass spectrometer	(Chi et al., 2024) [82]
Acesulfame potassium ADI: 15 mg/kg Sucrose equivalence: 200× 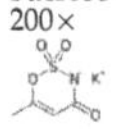	Female: Carbohydrate metabolism ↓ Lipopolysaccharide synthesis ↑ Male: Carbohydrate adsorption and metabolism ↑ Lipopolysaccharide synthesis ↑	Female: lactic acid ↓, succinic acid ↓, 2-Oleoylglycerol ↓ Male: pyruvic acid ↑, cholic acid ↑, deoxycholic acid ↓	Fecal metabolomic LC-MS	(Bian et al., 2017) [68]
Neotame ADI: 18 mg/kg Sucrose equivalence: 7000–13,000× 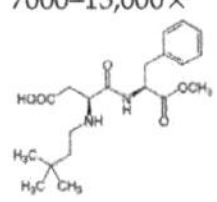	Streptomycin biosynthesis ↑ Amino acid metabolism ↑ Folate biosynthesis ↑ Lipopolysaccharide biosynthesis ↑ Fatty acid metabolism ↓ Sporulation ↓ Benzoate degradation ↓ Carbohydrate metabolism ↓ Lipid metabolism ↓ Bacterial chemotaxis ↓ ABC transporters ↓ Butyrate fermentation pathways	Malic acid ↓, mannose-6-phosphate ↓, 5-aminovaleric acid ↓and glyceric acid ↓, 1,3-dipalmitate ↑, 1-monopalmitin ↑, linoleic acid↑, stearic acid ↑, cholesterol ↑, campesterol ↑, stigmastanol ↑	Fecal metabolomic GC-MS	(Chi et al., 2018) [67]

* Note: The arrows (↑/↓) signify the upregulation or downregulation of metabolic pathways influenced by specific NAS, as detailed in the referenced studies. ** Note: The arrows (↑/↓) represent alterations (increases/decreases) in the levels of key metabolites detected in the host following exposure to specific NAS, as reported in the referenced studies and subject to the specific experimental conditions.

5. Challenges in Deciphering Underlying NAS–Gut Microbiome Mechanisms

It remains debatable if NAS modulations on the gut microbiome were due to direct interactions between NAS and the gut microbiome when NAS was orally administered. Scientific evidence has emerged revealing that each NAS exhibits unique absorption, distribution, metabolism, and excretion (ADME) patterns in organisms, and thus their kaleidoscopic alterations on the gut microbiome described in Section 3 can be the consequences of the host response in relation to their ADME differences. Clued by its unique chemical conformation, sucralose is a stable NAS that cannot be digested into monosaccharides or metabolized for energy. The substitution of hydroxyl groups in sucralose to chlorine in sucralose resisted sucralose from being cleaved by glycosidic enzymes that are capable of hydrolyzing sucrose and other carbohydrates. Combined with other evidence reported in toxicological studies targeting sucralose ADME among several species, not only in human [34], but also in mouse [36], rat [37], dog [35], and rabbit [36] studies, the biological fate of sucralose has been evaluated to be similar, depicted as low absorption and little to no metabolism in the host, regardless of species differences. Quantitatively, orally administered sucralose was proved to be excreted for 68.4% to 99% of the total dose in the feces, and only 2% to 26.5% was excreted in the urine [34–38], which illustrates that sucralose is unlikely to enter bodily circulation. Even though the high sucralose excretion percentage in the feces indicated that sucralose was barely absorbed or metabolized in the host, it can also serve as evidence that the majority of sucralose reached the large intestine and further implied that the gut microbiome was forced to be exposed directly to unmetabolized sucralose. Furthermore, a small portion of sucralose, varying from 2% to 35%, was absorbed [34–38], and its fate remains mysterious, even though sucralose-associated metabolites have been detected in urine, feces, and tissues [95]. Unlike sucralose, approximately 85% to 95% of administered saccharin was absorbed to the host plasma and was eventually eliminated unchanged in urine, as the principal method of plasma clearance [96]. Saccharin was absorbed by the gut lumen, where saccharin binds reversibly to plasma proteins and is distributed via the blood to the host body organs and eventually excreted in the urine [97–101]. However, in humans, the plasma concentration of saccharin and its time profile after oral dosage was shown to be complex, as its initial elimination was rapid during the first 10 h and then slowed down, and the slow phase was determined by prolonged absorption from the gastrointestinal tract, known as flip-flop kinetics [98]. This provided a novel insight that saccharin exposure is likely to induce long-term chronic toxicity via direct exposure to the gut microbiome, corresponding to its plasma clearance patterns, especially when active tubular transport, a primary mechanism of the renal elimination of saccharin, is saturated and causes excessive accumulation of saccharin [102]. Furthermore, research has consolidated that saccharin was excreted without undergoing metabolism in animals and humans [98], which increased the likelihood that saccharin impacted the gut microbiome via direct interference. Acesulfame potassium displayed similar absorption and excretion to saccharin, as the majority of acesulfame potassium via oral administration was absorbed rapidly and completely into the systematic circulation, and at least 82% of the absorbed acesulfame potassium was excreted in the urine within 24 h after consumption [34–37,96]. Aspartame has unique pharmacokinetics because it is quickly digested and hydrolyzed into methanol, phenylalanine, and aspartate in the gastrointestinal tract [103,104]. Aspartame is broken down in both the gastrointestinal lumen and inside intestinal mucosal cells by esterase and peptidases [103–106]. Thus, direct exposure of aspartame on the gut microbiome is unlikely to be the dominant mechanism [107]. Two second generation amino-acid based sweeteners, neotame and advantame, as two analogs of aspartame [108], may exhibit similar metabolism patterns as aspartame does in the gut microbiome–host interaction, and the impacts of their subsequent metabolites in the stomach and intestines may outweigh their direct exposure on the gut microbiome. However, for these two NAS, current knowledge is not adequate to depict whether their direct effects on the gut microbiome exist.

The systematic effects of NAS exposure on the gut microbiome remain largely inscrutable. However, causality has been established through the utilization of germ-free rodent models, which unequivocally demonstrate that the gut microbiome is the principal driving force behind NAS-induced metabolomic perturbations and glucose intolerance in the host [48,109]. A comprehensive evaluation of gut microbiome succession in conjunction with NAS exposure is urgently warranted to unravel potential bacteria–bacteria interactions within the host gut microbiome in response to NAS. As inferred from extant literature, the alterations in gut microbiome composition are predominantly depicted through comparative analyses of microbiome profiles before and after NAS exposure. However, this approach fails to elucidate critical insights regarding the mechanisms by which the microbiome engages in NAS exposure. For instance, saccharin exposure has been observed to modulate the abundance of *Bacteroides uniformis* and *Lactobacillus reuteri* [48]; nevertheless, it remains equivocal whether these variations are the result of an abrupt shift due to NAS toxicity or a progressive transition driven by microbial metabolism in response to NAS assimilation. Therefore, understanding the dynamic transformations within the bacterial community is of paramount importance for accurately interpreting the long-term ramifications of NAS toxicity on the gut microbiome. Changes in specific species within the gut microbiome community may not provide an accurate assessment of NAS modulation, given the intrinsic nature of interspecies interactions. Among these, mutualistic effects and bacterial antagonisms represent two significant modes of bacterial interaction [110].

Mutualism describes scenarios wherein multiple bacterial species synergistically benefit from collaborative development and degradation of growth substrates, while antagonism underscores the competitive dynamics among bacterial species. Consequently, when interpreting a detectable alteration in a specific bacterial species following NAS exposure, multiple causative factors must be considered, including the following: (1) direct exposure to NAS (particularly if NAS is not metabolized in the stomach and intestines, such as with saccharin); (2) indirect interferences stemming from metabolomic alterations related to NAS absorption, distribution, metabolism, and excretion (if NAS is metabolized); (3) mutualistic effects involving dominant and other non-dominant bacterial species; and (4) competition with antagonistic species within the microbial community. The subset mechanisms underlying bacterial competition within the gut microbiome may further entail competition for beneficial substrates and competition for the optimal physiological environment. Validating which hypothesis predominantly contributes to NAS modulation of the gut microbiome is exceedingly challenging, as it necessitates an intricate analysis of community-level changes without compromising the resolution required to quantify the abundance of specific strains. The taxonomic clustering and mapping of bacterial species modulated by NAS exposure, as explored in this review (Figure 2), may provide the initial clues necessary to elucidate potential bacteria–bacteria cross-feeding mechanisms within the gut microbiome in response to NAS exposure.

Elucidating the endpoints of metabolic alterations induced by NAS remains a significant challenge, particularly in understanding how these sweeteners interfere with downstream gut microbiota-mediated physiological and biochemical processes in the host. Advances in metabolomics, especially through the use of high-resolution mass spectrometry (HRMS), have been pivotal in uncovering the complex metabolic chain reactions triggered by NAS exposure. HRMS-based metabolomics provides extraordinary precision in the quantification and identification of metabolites within the gut microbiome, offering detailed insights into how NAS can profoundly alter the microbial metabolic landscape. This high-resolution approach enables the detection of subtle yet critical changes in metabolic pathways that might otherwise be overlooked, such as the reprogramming of carbohydrate metabolism within the microbial community in response to NAS. The gut microbiota, when exposed to NAS, is forced to adapt by shifting from traditional sugar metabolism, like that of sucrose, to alternative pathways. This metabolic shift leads to significant changes in the production of short-chain fatty acids (SCFAs) and other vital metabolites [7,39]. HRMS has been instrumental in detecting these nuanced shifts, providing comprehensive metabolite profiles that help elucidate how the altered metabolic pathways influence key endpoints in host metabolism, including glycemic control, lipid metabolism, and insulin sensitivity—factors that are critical for maintaining overall metabolic health [111]. The ability of HRMS to reveal disruptions in these pathways is crucial, as it helps identify potential triggers of low-grade systemic inflammation, which may arise from metabolic byproducts that provoke immune responses, thereby contributing to conditions such as obesity, type 2 diabetes, and other metabolic disorders [112,113]. Despite the insights gained, the direct links between NAS-induced microbial changes and specific disease phenotypes in the host remain elusive, underscoring the need for further research. Integrating HRMS-based metabolomic pathway analysis with microbial gene expression and host metabolic data could provide a more comprehensive understanding of the intricate interactions between NAS, the gut microbiome, and host metabolism. This approach holds the potential to identify biomarkers of NAS-induced metabolic dysregulation and clarify the long-term health implications of these widely consumed sweeteners.

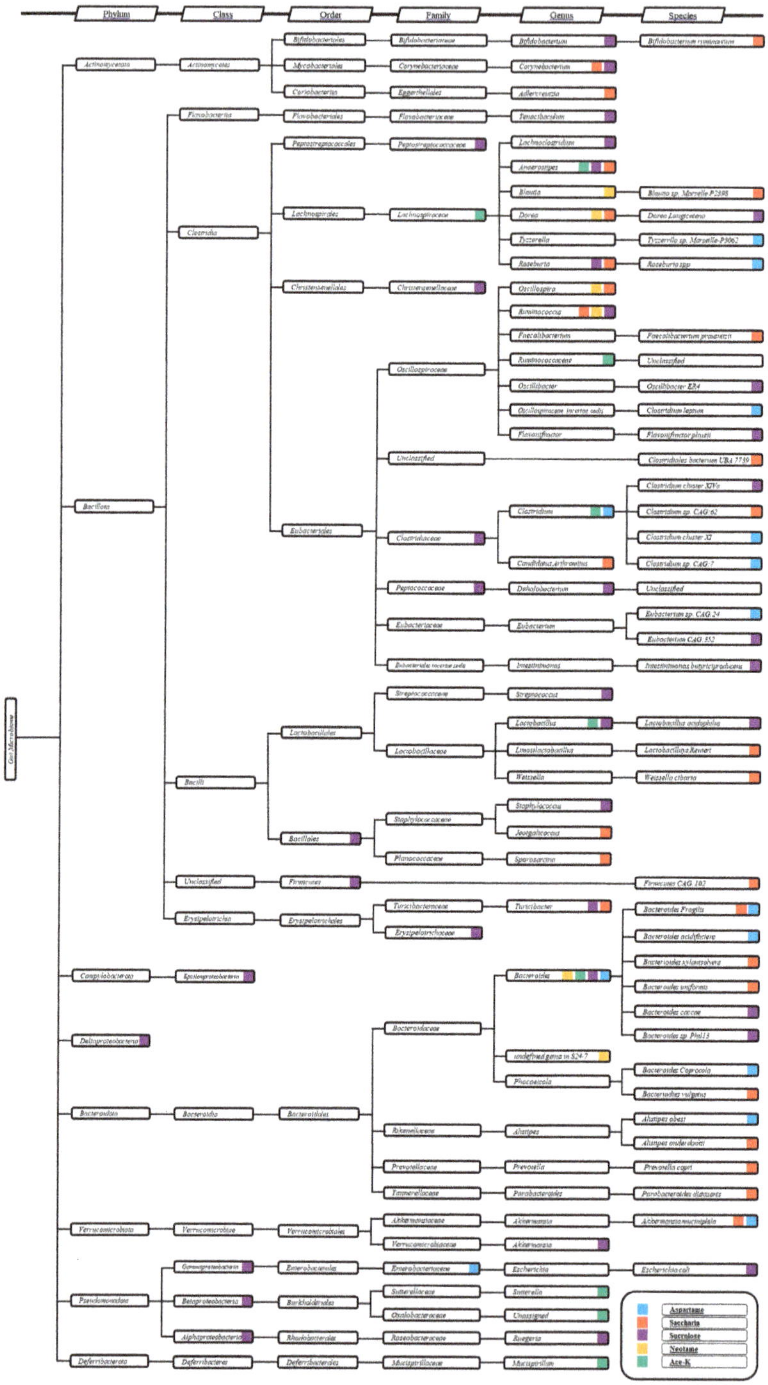

Figure 2. Taxonomic cluster analysis of the altered gut microbiome by specific NAS. The vertical axis represents the taxonomic hierarchy, with columns indicating the taxonomic rank from phylum to species. Color codes indicate the NAS associated with alterations in specific bacterial taxa.

The physiological functions of SCFAs modulated by the gut microbiome to NAS exposure remain largely elusive, posing a critical challenge in understanding their broader impact on host health. SCFAs, primarily acetate, propionate, and butyrate, are key metabolic end products of gut microbiome activity and serve as essential mediators in host–microbiome interactions [33]. These fatty acids, produced through the fermentation of dietary fibers by gut bacteria, play vital roles in maintaining gut health, regulating immune responses, and influencing metabolic processes such as lipid metabolism and glucose homeostasis [114,115]. Emerging research suggests that NAS can alter gut microbiota composition, consequently affecting SCFA production, as documented in Table 2. Some studies indicate that NAS affect SCFA-producing bacteria, leading to alterations of SCFA in serum or in the excretes [7,39]. These alterations in SCFAs could have significant implications for host metabolism, as SCFAs are involved in promoting insulin sensitivity, regulating lipid metabolism, and modulating inflammation [112,116]. SCFAs also enter systemic circulation, where they influence physiological processes such as appetite regulation via the gut–brain axis, modulation of insulin sensitivity, and adipose tissue function [111,113]. The potential of NAS to disrupt these processes through altered SCFA production from the gut microbiome could profoundly impact metabolic health, contributing to conditions like obesity, type 2 diabetes, and cardiovascular disease [112]. Moreover, SCFAs play a crucial role in maintaining gut barrier integrity by enhancing tight junction protein expression and providing energy to colonocytes, preventing gut permeability and systemic inflammation [117,118]. The modulation of SCFAs by NAS through the gut microbiome has the potential to disrupt gut homeostasis, leading to further dysbiosis, increased gut permeability, and a heightened risk of inflammatory conditions such as inflammatory bowel disease (IBD) and irritable bowel syndrome (IBS) [119]. Additionally, NAS may influence SCFA production by altering fermentation processes among the gut microbiome, potentially favoring non-SCFA metabolites that exacerbate metabolic imbalances [39]. This shift in microbial metabolism could impair the host from utilizing SCFAs efficiently, undermining their beneficial effects on energy metabolism and immune modulation [115]. Thus, understanding the relationship between NAS consumption, gut microbiome alteration, SCFA production, and host metabolic outcomes is imperative for elucidating the long-term health implications of NAS and the mechanisms through which they may impact health and disease.

Heterogeneity in the gut microbiome of organisms exposed to NAS can result in varied microbial alterations and distinct changes in metabolite profiles, leading to differential impacts on host physiology. Research by Suez et al. (2014) highlighted that NAS exposure leads to individualized effects, particularly in glycemic response impairment, which varies significantly across subjects [7]. This variability in response may be attributed to a complex interplay of extrinsic and intrinsic factors. Extrinsic factors such as diet composition [120], stress levels [121], and environmental exposures [122] play a crucial role in shaping the gut microbiome during NAS exposure, potentially leading to the promotion or suppression of specific microbial taxa. These alterations can influence metabolic pathways differently, thus contributing to the observed discrepancies in glycemic control. On the other hand, intrinsic factors, including individual differences in ADME [123], further complicate the host response to NAS. These intrinsic factors might result in the distinct bioavailability of NAS and its metabolites, thereby affecting the gut microbiome and metabolic outcomes in a highly personalized manner. Taken together, the heterogeneity between gut microbiome composition, metabolic processes, and host response underscores the need for innovative approaches to assess the impacts of NAS on human health in the future.

6. Limitation

Co-exposure to multiple NAS was not the primary focus of this review. It is undeniable that co-exposure research on multiple NAS in a strict experimental setting may also be beneficial to reveal the systematic alterations of the gut microbiome and host metabolism mechanism, which could reveal potential synergistic or antagonistic effects on gut microbiota and their subsequent impacts on metabolic health. The outcomes of these studies would be crucial for understanding the broader health implications of NAS consumption, especially in the context of metabolic syndrome and other related disorders [124]. However, without specified and detailed understanding of single NAS exposure and its consequences, the host metabolism alteration from simultaneous multiple NAS exposures in a laboratory setting cannot be attributed to the consequences of a single NAS, and thus it is not appropriate to categorize their outcomes as a single NAS in this review. Moreover, even though exposure to a NAS mixture is likely the dominant condition in human populations, NAS intake is always accompanied with other nutrients or ingredients, which undermines the credibility of whether the host metabolism alteration is induced specifically by NAS. As a consequence, the authors suspected that an investigation between the gut microbiome and multiple NAS exposure might not correspond to the collection of single NAS exposures to the gut microbiome. In the modern food industry, various NAS are commonly incorporated into foods and beverages, either individually or in combination, resulting in humans frequently consuming multiple NAS simultaneously [96]. The processing of different NAS during food production can lead to varied exposure levels and potentially synergistic or antagonistic effects within the gut microbiome, which complicates the understanding of the health impacts associated with NAS. Furthermore, when multiple NAS are consumed together, their combined effects on the gut microbiome and host metabolism become far more intricate, involving complex interactions between different microbial species and metabolic pathways [125]. These interactions due to co-exposure of multiple NAS can result in unpredictable changes in microbial composition and function, which subsequently influence the understanding of the gut microbiome–metabolic syndrome axis.

Author Contributions: Conceptualization, J.F. and K.L.; literature search, J.F. and J.P.; writing—original draft preparation, J.F. and J.P.; writing—review and editing, K.L., J.F., J.P., Y.-C.H., C.-W.L., Y.Y., H.Z., T.T. and X.W., visualization, J.F. and J.P.; supervision, K.L.; project administration, K.L.; funding acquisition, K.L. All authors have read and agreed to the published version of the manuscript.

Funding: This research was partially supported by the UNC Superfund Research program (P42ES031007) and University of North Carolina Center for Environmental Health and Susceptibility grant (P30ES010126).

Institutional Review Board Statement: Not applicable.

Data Availability Statement: Data sharing is not applicable.

Conflicts of Interest: The authors declare no conflicts of interest.

References

1. Abarca-Gómez, L.; Abdeen, Z.A.; Hamid, Z.A.; Abu-Rmeileh, N.M.; Acosta-Cazares, B.; Acuin, C.; Adams, R.J.; Aekplakorn, W.; Afsana, K.; Aguilar-Salinas, C.A.; et al. Worldwide Trends in Body-Mass Index, Underweight, Overweight, and Obesity from 1975 to 2016: A Pooled Analysis of 2416 Population-Based Measurement Studies in 128·9 Million Children, Adolescents, and Adults. *Lancet* **2017**, *390*, 2627–2642. [CrossRef] [PubMed]
2. Malik, V.S.; Hu, F.B. The Role of Sugar-Sweetened Beverages in the Global Epidemics of Obesity and Chronic Diseases. *Nat. Rev. Endocrinol.* **2022**, *18*, 205–218. [CrossRef]
3. Yu, L.; Zhou, H.; Zheng, F.; Song, J.; Lu, Y.; Yu, X.; Zhao, C. Sugar Is the Key Cause of Overweight/Obesity in Sugar-Sweetened Beverages (SSB). *Front. Nutr.* **2022**, *9*, 885704. [CrossRef]
4. AlFaris, N.A.; Alshwaiyat, N.M.; Alkhalidy, H.; AlTamimi, J.Z.; Alagal, R.I.; Alsaikan, R.A.; Alsemari, M.A.; BinMowyna, M.N.; AlKehayez, N.M. Sugar-Sweetened Beverages Consumption in a Multi-Ethnic Population of Middle-Aged Men and Association with Sociodemographic Variables and Obesity. *Front. Nutr.* **2022**, *9*, 987048. [CrossRef]
5. Lin, W.-T.; Kao, Y.-H.; Li, M.S.; Luo, T.; Lin, H.-Y.; Lee, C.-H.; Seal, D.W.; Hu, C.; Chen, L.-S.; Tseng, T.-S. Sugar-Sweetened Beverages Intake, Abdominal Obesity, and Inflammation among US Adults without and with Prediabetes—An NHANES Study. *Int. J. Environ. Res. Public Health* **2022**, *20*, 681. [CrossRef] [PubMed]
6. Arshad, S.; Rehman, T.; Saif, S.; Rajoka, M.S.R.; Ranjha, M.M.; Hassoun, A.; Cropotova, J.; Trif, M.; Younas, A.; Aadil, R.M. Replacement of Refined Sugar by Natural Sweeteners: Focus on Potential Health Benefits. *SSRN J.* **2022**. [CrossRef]
7. Suez, J.; Korem, T.; Zeevi, D.; Zilberman-Schapira, G.; Thaiss, C.A.; Maza, O.; Israeli, D.; Zmora, N.; Gilad, S.; Weinberger, A.; et al. Artificial Sweeteners Induce Glucose Intolerance by Altering the Gut Microbiota. *Nature* **2014**, *514*, 181–186. [CrossRef]

8. Pino-Seguel, P.; Moya, O.; Borquez, J.C.; Pino-de La Fuente, F.; Díaz-Castro, F.; Donoso-Barraza, C.; Llanos, M.; Troncoso, R.; Bravo-Sagua, R. Sucralose Consumption Ameliorates High-Fat Diet-Induced Glucose Intolerance and Liver Weight Gain in Mice. *Front. Nutr.* **2022**, *9*, 979624. [CrossRef]
9. Guru, S.K.; Li, Y.; Savinova, O.V.; Zhang, Y. Long-Term Consumption of Artificial Sweeteners Does Not Affect Cardiovascular Health and Survival in Rats. *PeerJ* **2022**, *10*, e13071. [CrossRef]
10. Ayada, I.; Li, J.; Pan, Q. Sugar-Sweetened Beverages and Risk of Liver Disease. *JAMA* **2023**, *330*, 2217. [CrossRef]
11. Jones, G.S.; Graubard, B.I.; Ramirez, Y.; Liao, L.M.; Huang, W.-Y.; Alvarez, C.S.; Yang, W.; Zhang, X.; Petrick, J.L.; McGlynn, K.A. Sweetened Beverage Consumption and Risk of Liver Cancer by Diabetes Status: A Pooled Analysis. *Cancer Epidemiol.* **2022**, *79*, 102201. [CrossRef] [PubMed]
12. Rosenberg, K. Sugar-Sweetened Beverage Intake Linked to Early-Onset Colorectal Cancer Risk. *AJN Am. J. Nurs.* **2021**, *121*, 52. [CrossRef] [PubMed]
13. Ringel, N.E.; Hovey, K.M.; Andrews, C.A.; Mossavar-Rahmani, Y.; Shadyab, A.H.; Snetselaar, L.G.; Howard, B.V.; Iglesia, C.B. Association of Artificially Sweetened Beverage Consumption and Urinary Tract Cancers in the Women's Health Initiative Observational Study. *Eur. Urol. Open Sci.* **2023**, *47*, 80–86. [CrossRef] [PubMed]
14. Liu, M.; Yang, S.; Ye, Z.; Zhang, Y.; Zhang, Y.; He, P.; Zhou, C.; Hou, F.F.; Qin, X. Tea Consumption and New-Onset Acute Kidney Injury: The Effects of Milk or Sweeteners Addition and Caffeine/Coffee. *Nutrients* **2023**, *15*, 2201. [CrossRef] [PubMed]
15. Debras, C.; Chazelas, E.; Sellem, L.; Julia, C.; Kesse-Guyot, E.; Allès, B.; Deschasaux-Tanguy, M.; Huybrechts, I.; Srour, B.; Touvier, M. Artificial Sweeteners and Risk of Cardiovascular Diseases in the Prospective NutriNet-Santé Cohort. *Eur. J. Public Health* **2022**, *32*, ckac129.013. [CrossRef]
16. Gomez-Delgado, F.; Torres-Peña, J.D.; Gutierrez-Lara, G.; Romero-Cabrera, J.L.; Perez-Martinez, P. Artificial Sweeteners and Cardiovascular Risk. *Curr. Opin. Cardiol.* **2023**, *38*, 344–351. [CrossRef]
17. Singh, S.; Kohli, A.; Trivedi, S.; Kanagala, S.G.; Anamika, F.N.U.; Garg, N.; Patel, M.A.; Munjal, R.S.; Jain, R. The Contentious Relationship between Artificial Sweeteners and Cardiovascular Health. *Egypt. J. Intern. Med.* **2023**, *35*, 43. [CrossRef]
18. Yang, B.; Glenn, A.J.; Liu, Q.; Madsen, T.; Allison, M.A.; Shikany, J.M.; Manson, J.E.; Chan, K.H.K.; Wu, W.-C.; Li, J.; et al. Added Sugar, Sugar-Sweetened Beverages, and Artificially Sweetened Beverages and Risk of Cardiovascular Disease: Findings from the Women's Health Initiative and a Network Meta-Analysis of Prospective Studies. *Nutrients* **2022**, *14*, 4226. [CrossRef]
19. Hosseini, A.; Barlow, G.M.; Leite, G.; Rashid, M.; Parodi, G.; Wang, J.; Morales, W.; Weitsman, S.; Rezaie, A.; Pimentel, M.; et al. Consuming Artificial Sweeteners May Alter the Structure and Function of Duodenal Microbial Communities. *iScience* **2023**, *26*, 108530. [CrossRef]
20. AL-Ishaq, R.K.; Kubatka, P.; Büsselberg, D. Sweeteners and the Gut Microbiome: Effects on Gastrointestinal Cancers. *Nutrients* **2023**, *15*, 3675. [CrossRef]
21. Panthee, B.; Gyawali, S.; Panthee, P.; Techato, K. Environmental and Human Microbiome for Health. *Life* **2022**, *12*, 456. [CrossRef]
22. Aggarwal, N.; Kitano, S.; Puah, G.R.Y.; Kittelmann, S.; Hwang, I.Y.; Chang, M.W. Microbiome and Human Health: Current Understanding, Engineering, and Enabling Technologies. *Chem. Rev.* **2023**, *123*, 31–72. [CrossRef]
23. Shi, D.; Turroni, S.; Gong, L.; Wu, W.; Yim, H.C.H. Editorial: Manipulation of Gut Microbiota as a Key Target to Intervene on the Onset and Progression of Digestive System Diseases. *Front. Med.* **2022**, *9*, 999005. [CrossRef]
24. McCallum, G.; Tropini, C. The Gut Microbiota and Its Biogeography. *Nat. Rev. Microbiol.* **2024**, *22*, 105–118. [CrossRef]
25. Li, Q.; Li, N.; Cai, W.; Xiao, M.; Liu, B.; Zeng, F. Fermented Natural Product Targeting Gut Microbiota Regulate Immunity and Anti-Inflammatory Activity: A Possible Way to Prevent COVID-19 in Daily Diet. *J. Funct. Foods* **2022**, *97*, 105229. [CrossRef] [PubMed]
26. Luo, Y.; Liu, Y.; Li, H.; Zhao, Y.; Wright, A.-D.G.; Cai, J.; Tian, G.; Mao, X. Differential Effect of Dietary Fibers in Intestinal Health of Growing Pigs: Outcomes in the Gut Microbiota and Immune-Related Indexes. *Front. Microbiol.* **2022**, *13*, 843045. [CrossRef]
27. Campbell, C.; Kandalgaonkar, M.R.; Golonka, R.M.; Yeoh, B.S.; Vijay-Kumar, M.; Saha, P. Crosstalk between Gut Microbiota and Host Immunity: Impact on Inflammation and Immunotherapy. *Biomedicines* **2023**, *11*, 294. [CrossRef]
28. Lu, Y.; Yuan, X.; Wang, M.; He, Z.; Li, H.; Wang, J.; Li, Q. Gut Microbiota Influence Immunotherapy Responses: Mechanisms and Therapeutic Strategies. *J. Hematol. Oncol.* **2022**, *15*, 47. [CrossRef]
29. Kim, D.-H. Gut Microbiota-Mediated Drug-Drug Interactions. *Drug Metab. Pharmacokinet.* **2017**, *32*, S18–S19. [CrossRef]
30. Dhurjad, P.; Dhavaliker, C.; Gupta, K.; Sonti, R. Exploring Drug Metabolism by the Gut Microbiota: Modes of Metabolism and Experimental Approaches. *Drug. Metab. Dispos.* **2022**, *50*, 224–234. [CrossRef]
31. Mishima, E.; Abe, T. Role of the Microbiota in Hypertension and Antihypertensive Drug Metabolism. *Hypertens. Res.* **2022**, *45*, 246–253. [CrossRef] [PubMed]
32. Farhat, E.K.; Sher, E.K.; Džidić-Krivić, A.; Banjari, I.; Sher, F. Functional Biotransformation of Phytoestrogens by Gut Microbiota with Impact on Cancer Treatment. *J. Nutr. Biochem.* **2023**, *118*, 109368. [CrossRef] [PubMed]
33. Rooks, M.G.; Garrett, W.S. Gut Microbiota, Metabolites and Host Immunity. *Nat. Rev. Immunol.* **2016**, *16*, 341–352. [CrossRef] [PubMed]
34. Roberts, A.; Renwick, A.G.; Sims, J.; Snodin, D.J. Sucralose Metabolism and Pharmacokinetics in Man. *Food Chem. Toxicol.* **2000**, *38*, 31–41. [CrossRef] [PubMed]
35. Wood, S.G.; John, B.A.; Hawkins, D.R. The Pharmacokinetics and Metabolism of Sucralose in the Dog. *Food Chem. Toxicol.* **2000**, *38*, 99–106. [CrossRef]

36. John, B.A.; Wood, S.G.; Hawkins, D.R. The Pharmacokinetics and Metabolism of Sucralose in the Mouse. *Food Chem. Toxicol.* **2000**, *38*, 107–110. [CrossRef]
37. Sims, J.; Roberts, A.; Daniel, J.W.; Renwick, A.G. The Metabolic Fate of Sucralose in Rats. *Food Chem. Toxicol.* **2000**, *38*, 115–121. [CrossRef]
38. Sylvetsky, A.C.; Bauman, V.; Blau, J.E.; Garraffo, H.M.; Walter, P.J.; Rother, K.I. Plasma Concentrations of Sucralose in Children and Adults. *Toxicol. Environ. Chem.* **2017**, *99*, 535–542. [CrossRef]
39. Palmnäs, M.S.A.; Cowan, T.E.; Bomhof, M.R.; Su, J.; Reimer, R.A.; Vogel, H.J.; Hittel, D.S.; Shearer, J. Low-Dose Aspartame Consumption Differentially Affects Gut Microbiota-Host Metabolic Interactions in the Diet-Induced Obese Rat. *PLoS ONE* **2014**, *9*, e109841. [CrossRef]
40. Zhang, M.; Liu, B.; Zhang, Y.; Wei, H.; Lei, Y.; Zhao, L. Structural Shifts of Mucosa-Associated Lactobacilli and *Clostridium leptum* Subgroup in Patients with Ulcerative Colitis. *J. Clin. Microbiol.* **2007**, *45*, 496–500. [CrossRef]
41. Kabeerdoss, J.; Sankaran, V.; Pugazhendhi, S.; Ramakrishna, B.S. Clostridium Leptum Group Bacteria Abundance and Diversity in the Fecal Microbiota of Patients with Inflammatory Bowel Disease: A Case–Control Study in India. *BMC Gastroenterol.* **2013**, *13*, 20. [CrossRef] [PubMed]
42. Kedia, S.; Rampal, R.; Paul, J.; Ahuja, V. Gut Microbiome Diversity in Acute Infective and Chronic Inflammatory Gastrointestinal Diseases in North India. *J. Gastroenterol.* **2016**, *51*, 660–671. [CrossRef] [PubMed]
43. Chen, P.-C.; Chien, Y.-W.; Yang, S.-C. The Alteration of Gut Microbiota in Newly Diagnosed Type 2 Diabetic Patients. *Nutrition* **2019**, *63–64*, 51–56. [CrossRef] [PubMed]
44. Santacruz, A.; Marcos, A.; Wärnberg, J.; Martí, A.; Martin-Matillas, M.; Campoy, C.; Moreno, L.A.; Veiga, O.; Redondo-Figuero, C.; Garagorri, J.M.; et al. Interplay Between Weight Loss and Gut Microbiota Composition in Overweight Adolescents. *Obesity* **2009**, *17*, 1906–1915. [CrossRef] [PubMed]
45. Koleva, P.T.; Valcheva, R.S.; Sun, X.; Gänzle, M.G.; Dieleman, L.A. Inulin and Fructo-Oligosaccharides Have Divergent Effects on Colitis and Commensal Microbiota in HLA-B27 Transgenic Rats. *Br. J. Nutr.* **2012**, *108*, 1633–1643. [CrossRef]
46. Yamaguchi, Y.; Adachi, K.; Sugiyama, T.; Shimozato, A.; Ebi, M.; Ogasawara, N.; Funaki, Y.; Goto, C.; Sasaki, M.; Kasugai, K. Association of Intestinal Microbiota with Metabolic Markers and Dietary Habits in Patients with Type 2 Diabetes. *Digestion* **2016**, *94*, 66–72. [CrossRef]
47. Nie, K.; Ma, K.; Luo, W.; Shen, Z.; Yang, Z.; Xiao, M.; Tong, T.; Yang, Y.; Wang, X. Roseburia Intestinalis: A Beneficial Gut Organism from the Discoveries in Genus and Species. *Front. Cell. Infect. Microbiol.* **2021**, *11*, 757718. [CrossRef]
48. Suez, J.; Cohen, Y.; Valdés-Mas, R.; Mor, U.; Dori-Bachash, M.; Federici, S.; Zmora, N.; Leshem, A.; Heinemann, M.; Linevsky, R.; et al. Personalized Microbiome-Driven Effects of Non-Nutritive Sweeteners on Human Glucose Tolerance. *Cell* **2022**, *185*, 3307–3328.e19. [CrossRef]
49. Scott, N.; Whittle, E.; Jeraldo, P.; Chia, N. A Systemic Review of the Role of Enterotoxic Bacteroides Fragilis in Colorectal Cancer. *Neoplasia* **2022**, *29*, 100797. [CrossRef]
50. Yang, J.-Y.; Lee, Y.-S.; Kim, Y.; Lee, S.-H.; Ryu, S.; Fukuda, S.; Hase, K.; Yang, C.-S.; Lim, H.S.; Kim, M.-S.; et al. Gut Commensal Bacteroides Acidifaciens Prevents Obesity and Improves Insulin Sensitivity in Mice. *Mucosal Immunol.* **2017**, *10*, 104–116. [CrossRef]
51. Chen, K.; Nakasone, Y.; Yi, S.; Ibrahim, H.R.; Sakao, K.; Hossain, M.A.; Hou, D.-X. Natural Garlic Organosulfur Compounds Prevent Metabolic Disorder of Lipid and Glucose by Increasing Gut Commensal *Bacteroides acidifaciens*. *J. Agric. Food Chem.* **2022**, *70*, 5829–5837. [CrossRef] [PubMed]
52. Li, J.; Yang, G.; Zhang, Q.; Liu, Z.; Jiang, X.; Xin, Y. Function of Akkermansia Muciniphila in Type 2 Diabetes and Related Diseases. *Front. Microbiol.* **2023**, *14*, 1172400. [CrossRef] [PubMed]
53. Depommier, C.; Everard, A.; Druart, C.; Plovier, H.; Van Hul, M.; Vieira-Silva, S.; Falony, G.; Raes, J.; Maiter, D.; Delzenne, N.M.; et al. Supplementation with Akkermansia Muciniphila in Overweight and Obese Human Volunteers: A Proof-of-Concept Exploratory Study. *Nat. Med.* **2019**, *25*, 1096–1103. [CrossRef] [PubMed]
54. Perraudeau, F.; McMurdie, P.; Bullard, J.; Cheng, A.; Cutcliffe, C.; Deo, A.; Eid, J.; Gines, J.; Iyer, M.; Justice, N.; et al. Improvements to Postprandial Glucose Control in Subjects with Type 2 Diabetes: A Multicenter, Double Blind, Randomized Placebo-Controlled Trial of a Novel Probiotic Formulation. *BMJ Open Diab. Res. Care* **2020**, *8*, e001319. [CrossRef] [PubMed]
55. He, S.; Xiong, Q.; Tian, C.; Li, L.; Zhao, J.; Lin, X.; Guo, X.; He, Y.; Liang, W.; Zuo, X.; et al. Inulin-Type Prebiotics Reduce Serum Uric Acid Levels via Gut Microbiota Modulation: A Randomized, Controlled Crossover Trial in Peritoneal Dialysis Patients. *Eur. J. Nutr.* **2022**, *61*, 665–677. [CrossRef]
56. Olaisen, M.; Flatberg, A.; Granlund, A.V.B.; Røyset, E.S.; Martinsen, T.C.; Sandvik, A.K.; Fossmark, R. Bacterial Mucosa-Associated Microbiome in Inflamed and Proximal Noninflamed Ileum of Patients with Crohn's Disease. *Inflamm. Bowel Dis.* **2021**, *27*, 12–24. [CrossRef]
57. Péan, N.; Le Lay, A.; Brial, F.; Wasserscheid, J.; Rouch, C.; Vincent, M.; Myridakis, A.; Hedjazi, L.; Dumas, M.-E.; Grundberg, E.; et al. Dominant Gut Prevotella Copri in Gastrectomised Non-Obese Diabetic Goto–Kakizaki Rats Improves Glucose Homeostasis through Enhanced FXR Signalling. *Diabetologia* **2020**, *63*, 1223–1235. [CrossRef]
58. Mirande, C.; Kadlecikova, E.; Matulova, M.; Capek, P.; Bernalier-Donadille, A.; Forano, E.; Béra-Maillet, C. Dietary Fibre Degradation and Fermentation by Two Xylanolytic Bacteria *Bacteroides xylanisolvens* XB1A ^{T} and *Roseburia intestinalis* XB6B4 from the Human Intestine. *J. Appl. Microbiol.* **2010**, *109*, 451–460. [CrossRef]

59. Hajjar, R.; Gonzalez, E.; Fragoso, G.; Oliero, M.; Alaoui, A.A.; Calvé, A.; Vennin Rendos, H.; Djediai, S.; Cuisiniere, T.; Laplante, P.; et al. Gut Microbiota Influence Anastomotic Healing in Colorectal Cancer Surgery through Modulation of Mucosal Proinflammatory Cytokines. *Gut* **2023**, *72*, 1143–1154. [CrossRef]
60. Yang, J.; Zheng, P.; Li, Y.; Wu, J.; Tan, X.; Zhou, J.; Sun, Z.; Chen, X.; Zhang, G.; Zhang, H.; et al. Landscapes of Bacterial and Metabolic Signatures and Their Interaction in Major Depressive Disorders. *Sci. Adv.* **2020**, *6*, eaba8555. [CrossRef]
61. Gauffin Cano, P.; Santacruz, A.; Moya, Á.; Sanz, Y. Bacteroides Uniformis CECT 7771 Ameliorates Metabolic and Immunological Dysfunction in Mice with High-Fat-Diet Induced Obesity. *PLoS ONE* **2012**, *7*, e41079. [CrossRef] [PubMed]
62. López-Almela, I.; Romaní-Pérez, M.; Bullich-Vilarrubias, C.; Benítez-Páez, A.; Gómez Del Pulgar, E.M.; Francés, R.; Liebisch, G.; Sanz, Y. *Bacteroides uniformis* Combined with Fiber Amplifies Metabolic and Immune Benefits in Obese Mice. *Gut Microbes* **2021**, *13*, 1–20. [CrossRef] [PubMed]
63. Mobini, R.; Tremaroli, V.; Ståhlman, M.; Karlsson, F.; Levin, M.; Ljungberg, M.; Sohlin, M.; Bertéus Forslund, H.; Perkins, R.; Bäckhed, F.; et al. Metabolic Effects of *Lactobacillus reuteri* DSM 17938 in People with Type 2 Diabetes: A Randomized Controlled Trial. *Diabetes Obes. Metab.* **2017**, *19*, 579–589. [CrossRef]
64. Remely, M.; Hippe, B.; Zanner, J.; Aumueller, E.; Brath, H.; Haslberger, A.G. Gut Microbiota of Obese, Type 2 Diabetic Individuals Is Enriched in Faecalibacterium Prausnitzii, Akkermansia Muciniphila and Peptostreptococcus Anaerobius after Weight Loss. *Endocr. Metab. Immune Disord.-Drug Targets* **2016**, *16*, 99–106. [CrossRef] [PubMed]
65. Abou-Donia, M.B.; El-Masry, E.M.; Abdel-Rahman, A.A.; McLendon, R.E.; Schiffman, S.S. Splenda Alters Gut Microflora and Increases Intestinal P-Glycoprotein and Cytochrome P-450 in Male Rats. *J. Toxicol. Environ. Health Part A* **2008**, *71*, 1415–1429. [CrossRef]
66. Bian, X.; Chi, L.; Gao, B.; Tu, P.; Ru, H.; Lu, K. Gut Microbiome Response to Sucralose and Its Potential Role in Inducing Liver Inflammation in Mice. *Front. Physiol.* **2017**, *8*, 487. [CrossRef]
67. Chi, L.; Bian, X.; Gao, B.; Tu, P.; Lai, Y.; Ru, H.; Lu, K. Effects of the Artificial Sweetener Neotame on the Gut Microbiome and Fecal Metabolites in Mice. *Molecules* **2018**, *23*, 367. [CrossRef]
68. Bian, X.; Chi, L.; Gao, B.; Tu, P.; Ru, H.; Lu, K. The Artificial Sweetener Acesulfame Potassium Affects the Gut Microbiome and Body Weight Gain in CD-1 Mice. *PLoS ONE* **2017**, *12*, e0178426. [CrossRef]
69. Olivier-Van Stichelen, S.; Rother, K.I.; Hanover, J.A. Maternal Exposure to Non-Nutritive Sweeteners Impacts Progeny's Metabolism and Microbiome. *Front. Microbiol.* **2019**, *10*, 1360. [CrossRef]
70. Frankenfeld, C.L.; Sikaroodi, M.; Lamb, E.; Shoemaker, S.; Gillevet, P.M. High-Intensity Sweetener Consumption and Gut Microbiome Content and Predicted Gene Function in a Cross-Sectional Study of Adults in the United States. *Ann. Epidemiol.* **2015**, *25*, 736–742.e4. [CrossRef]
71. Ahmad, S.Y.; Friel, J.K.; MacKay, D.S. The Effect of the Artificial Sweeteners on Glucose Metabolism in Healthy Adults: A Randomized, Double-Blinded, Crossover Clinical Trial. *Appl. Physiol. Nutr. Metab.* **2020**, *45*, 606–612. [CrossRef]
72. Bian, X.; Tu, P.; Chi, L.; Gao, B.; Ru, H.; Lu, K. Saccharin Induced Liver Inflammation in Mice by Altering the Gut Microbiota and Its Metabolic Functions. *Food Chem. Toxicol.* **2017**, *107*, 530–539. [CrossRef] [PubMed]
73. Nogueira, J.P.D.S.; He, F.; Mangian, H.F.; Oba, P.M.; De Godoy, M.R.C. Dietary Supplementation of a Fiber-Prebiotic and Saccharin-Eugenol Blend in Extruded Diets Fed to Dogs. *J. Anim. Sci.* **2019**, *97*, 4519–4531. [CrossRef] [PubMed]
74. Serrano, J.; Smith, K.R.; Crouch, A.L.; Sharma, V.; Yi, F.; Vargova, V.; LaMoia, T.E.; Dupont, L.M.; Serna, V.; Tang, F.; et al. High-Dose Saccharin Supplementation Does Not Induce Gut Microbiota Changes or Glucose Intolerance in Healthy Humans and Mice. *Microbiome* **2021**, *9*, 11. [CrossRef] [PubMed]
75. Murali, A.; Giri, V.; Cameron, H.J.; Sperber, S.; Zickgraf, F.M.; Haake, V.; Driemert, P.; Walk, T.; Kamp, H.; Rietjens, I.M.; et al. Investigating the Gut Microbiome and Metabolome Following Treatment with Artificial Sweeteners Acesulfame Potassium and Saccharin in Young Adult Wistar Rats. *Food Chem. Toxicol.* **2022**, *165*, 113123. [CrossRef] [PubMed]
76. Uebanso, T.; Ohnishi, A.; Kitayama, R.; Yoshimoto, A.; Nakahashi, M.; Shimohata, T.; Mawatari, K.; Takahashi, A. Effects of Low-Dose Non-Caloric Sweetener Consumption on Gut Microbiota in Mice. *Nutrients* **2017**, *9*, 560. [CrossRef]
77. Rodriguez-Palacios, A.; Harding, A.; Menghini, P.; Himmelman, C.; Retuerto, M.; Nickerson, K.P.; Lam, M.; Croniger, C.M.; McLean, M.H.; Durum, S.K.; et al. The Artificial Sweetener Splenda Promotes Gut Proteobacteria, Dysbiosis, and Myeloperoxidase Reactivity in Crohn's Disease–Like Ileitis. *Inflamm. Bowel Dis.* **2018**, *24*, 1005–1020. [CrossRef]
78. Wang, Q.-P.; Browman, D.; Herzog, H.; Neely, G.G. Non-Nutritive Sweeteners Possess a Bacteriostatic Effect and Alter Gut Microbiota in Mice. *PLoS ONE* **2018**, *13*, e0199080. [CrossRef]
79. Zheng, Z.; Xiao, Y.; Ma, L.; Lyu, W.; Peng, H.; Wang, X.; Ren, Y.; Li, J. Low Dose of Sucralose Alter Gut Microbiome in Mice. *Front. Nutr.* **2022**, *9*, 848392. [CrossRef]
80. Thomson, P.; Santibañez, R.; Aguirre, C.; Galgani, J.E.; Garrido, D. Short-Term Impact of Sucralose Consumption on the Metabolic Response and Gut Microbiome of Healthy Adults. *Br. J. Nutr.* **2019**, *122*, 856–862. [CrossRef]
81. Méndez-García, L.A.; Bueno-Hernández, N.; Cid-Soto, M.A.; De León, K.L.; Mendoza-Martínez, V.M.; Espinosa-Flores, A.J.; Carrero-Aguirre, M.; Esquivel-Velázquez, M.; León-Hernández, M.; Viurcos-Sanabria, R.; et al. Ten-Week Sucralose Consumption Induces Gut Dysbiosis and Altered Glucose and Insulin Levels in Healthy Young Adults. *Microorganisms* **2022**, *10*, 434. [CrossRef] [PubMed]

82. Chi, L.; Yang, Y.; Bian, X.; Gao, B.; Tu, P.; Ru, H.; Lu, K. Chronic Sucralose Consumption Inhibits Farnesoid X Receptor Signaling and Perturbs Lipid and Cholesterol Homeostasis in the Mouse Livers, Potentially by Altering Gut Microbiota Functions. *Sci. Total Environ.* **2024**, *919*, 169603. [CrossRef] [PubMed]
83. Hanawa, Y.; Higashiyama, M.; Kurihara, C.; Tanemoto, R.; Ito, S.; Mizoguchi, A.; Nishii, S.; Wada, A.; Inaba, K.; Sugihara, N.; et al. Acesulfame Potassium Induces Dysbiosis and Intestinal Injury with Enhanced Lymphocyte Migration to Intestinal Mucosa. *J. Gastroenterol. Hepatol.* **2021**, *36*, 3140–3148. [CrossRef] [PubMed]
84. Baldelli, V.; Scaldaferri, F.; Putignani, L.; Del Chierico, F. The Role of Enterobacteriaceae in Gut Microbiota Dysbiosis in Inflammatory Bowel Diseases. *Microorganisms* **2021**, *9*, 697. [CrossRef] [PubMed]
85. Morgan, X.C.; Tickle, T.L.; Sokol, H.; Gevers, D.; Devaney, K.L.; Ward, D.V.; Reyes, J.A.; Shah, S.A.; LeLeiko, N.; Snapper, S.B.; et al. Dysfunction of the Intestinal Microbiome in Inflammatory Bowel Disease and Treatment. *Genome Biol.* **2012**, *13*, R79. [CrossRef] [PubMed]
86. Lupp, C.; Robertson, M.L.; Wickham, M.E.; Sekirov, I.; Champion, O.L.; Gaynor, E.C.; Finlay, B.B. Host-Mediated Inflammation Disrupts the Intestinal Microbiota and Promotes the Overgrowth of Enterobacteriaceae. *Cell Host Microbe* **2007**, *2*, 204. [CrossRef]
87. Gophna, U.; Sommerfeld, K.; Gophna, S.; Doolittle, W.F.; Veldhuyzen Van Zanten, S.J.O. Differences between Tissue-Associated Intestinal Microfloras of Patients with Crohn's Disease and Ulcerative Colitis. *J. Clin. Microbiol.* **2006**, *44*, 4136–4141. [CrossRef]
88. Martinez-Medina, M.; Aldeguer, X.; Gonzalez-Huix, F.; Acero, D.; Garcia-Gil, J.L. Abnormal Microbiota Composition in the Ileocolonic Mucosa of Crohn's Disease Patients as Revealed by Polymerase Chain Reaction-Denaturing Gradient Gel Electrophoresis: *Inflamm. Bowel Dis.* **2006**, *12*, 1136–1145. [CrossRef]
89. Wang, M.; Molin, G.; Ahrné, S.; Adawi, D.; Jeppsson, B. High Proportions of Proinflammatory Bacteria on the Colonic Mucosa in a Young Patient with Ulcerative Colitis as Revealed by Cloning and Sequencing of 16S rRNA Genes. *Dig. Dis. Sci.* **2007**, *52*, 620–627. [CrossRef]
90. Carvalho, F.A.; Barnich, N.; Sauvanet, P.; Darcha, C.; Gelot, A.; Darfeuille-Michaud, A. Crohn's Disease-Associated Escherichia Coli LF82 Aggravates Colitis in Injured Mouse Colon via Signaling by Flagellin: *Inflamm. Bowel Dis.* **2008**, *14*, 1051–1060. [CrossRef]
91. Mukhopadhya, I.; Hansen, R.; El-Omar, E.M.; Hold, G.L. IBD—What Role Do Proteobacteria Play? *Nat. Rev. Gastroenterol. Hepatol.* **2012**, *9*, 219–230. [CrossRef]
92. Lavelle, A.; Lennon, G.; O'Sullivan, O.; Docherty, N.; Balfe, A.; Maguire, A.; Mulcahy, H.E.; Doherty, G.; O'Donoghue, D.; Hyland, J.; et al. Spatial Variation of the Colonic Microbiota in Patients with Ulcerative Colitis and Control Volunteers. *Gut* **2015**, *64*, 1553–1561. [CrossRef] [PubMed]
93. Xu, J.; Liu, W.; Wu, J.; Wang, W.; Wang, Z.; Yu, X.; Zhang, H.; Zhu, L.; Zhan, X. Metabolic Profiles of Oligosaccharides Derived from Four Microbial Polysaccharides by Faecal Inocula from Type 2 Diabetes Patients. *Int. J. Food Sci. Nutr.* **2021**, *72*, 1083–1094. [CrossRef]
94. Sarbini, S.R.; Kolida, S.; Gibson, G.R.; Rastall, R.A. In Vitro Fermentation of Commercial α-Gluco-Oligosaccharide by Faecal Microbiota from Lean and Obese Human Subjects. *Br. J. Nutr.* **2012**, *109*, 1980–1989. [CrossRef] [PubMed]
95. Bornemann, V.; Werness, S.C.; Buslinger, L.; Schiffman, S.S. Intestinal Metabolism and Bioaccumulation of Sucralose in Adipose Tissue In The Rat. *J. Toxicol. Environ. Health Part A* **2018**, *81*, 913–923. [CrossRef]
96. Magnuson, B.A.; Carakostas, M.C.; Moore, N.H.; Poulos, S.P.; Renwick, A.G. Biological Fate of Low-Calorie Sweeteners. *Nutr. Rev.* **2016**, *74*, 670–689. [CrossRef]
97. Matthews, H.B.; Fields, M.; Fishbein, L. Saccharin. Distribution and Excretion of a Limited Dose in the Rat. *J. Agric. Food Chem.* **1973**, *21*, 916–919. [CrossRef]
98. Renwick, A.G. The Metabolism of Intense Sweeteners. *Xenobiotica* **1986**, *16*, 1057–1071. [CrossRef]
99. Byard, J.L.; McChesney, E.W.; Golberg, L.; Coulston, F. Excretion and Metabolism of Saccharin in Man. II. Studies with 14C-Labelled and Unlabelled Saccharin. *Food Cosmet. Toxicol.* **1974**, *12*, 175–184. [CrossRef]
100. McChesney, E.W.; Golberg, L. The Excretion and Metabolism of Saccharin in Man. I. Methods of Investigation and Preliminary Results. *Food Cosmet. Toxicol.* **1973**, *11*, 403–414. [CrossRef]
101. Lethco, E.J.; Wallace, W.C. The Metabolism of Saccharin in Animals. *Toxicology* **1975**, *3*, 287–300. [CrossRef] [PubMed]
102. Sweatman, T.W.; Renwick, A.G. The Tissue Distribution and Pharmacokinetics of Saccharin in the Rat. *Toxicol. Appl. Pharmacol.* **1980**, *55*, 18–31. [CrossRef] [PubMed]
103. Stegink, L.D.; Filer, L.J. *Aspartame: Physiology and Biochemistry*, 1st ed.; CRC Press: Boca Raton, FL, USA, 2020; ISBN 978-1-00-014652-3.
104. Stegink, L.D. The Aspartame Story: A Model for the Clinical Testing of a Food Additive. *Am. J. Clin. Nutr.* **1987**, *46*, 204–215. [CrossRef] [PubMed]
105. Butchko, H.H.; Stargel, W.W.; Comer, C.P.; Mayhew, D.A.; Benninger, C.; Blackburn, G.L.; De Sonneville, L.M.J.; Geha, R.S.; Hertelendy, Z.; Koestner, A.; et al. Aspartame: Review of Safety. *Regul. Toxicol. Pharmacol.* **2002**, *35*, S1–S93. [CrossRef]
106. Hooper, N.M.; Hesp, R.J.; Tieku, S. Metabolism of Aspartame by Human and Pig Intestinal Microvillar Peptidases. *Biochem. J.* **1994**, *298*, 635–639. [CrossRef]
107. Ranney, R.E.; Oppermann, J.A.; Muldoon, E.; McMahon, F.G. Comparative Metabolism of Aspartame in Experimental Animals and Humans. *J. Toxicol. Environ. Health* **1976**, *2*, 441–451. [CrossRef]
108. O'donnell, K. Aspartame, Neotame and Advantame. In *Sweeteners and Sugar Alternatives in Food Technology*; O'Donnell, K., Kearsley, M.W., Eds.; Wiley: Hoboken, NJ, USA, 2012; pp. 117–136. ISBN 978-0-470-65968-7.

109. Nettleton, J.E.; Cho, N.A.; Klancic, T.; Nicolucci, A.C.; Shearer, J.; Borgland, S.L.; Johnston, L.A.; Ramay, H.R.; Noye Tuplin, E.; Chleilat, F.; et al. Maternal Low-Dose Aspartame and Stevia Consumption with an Obesogenic Diet Alters Metabolism, Gut Microbiota and Mesolimbic Reward System in Rat Dams and Their Offspring. *Gut* **2020**, *69*, 1807–1817. [CrossRef]
110. Fons, F.; Gomez, A.; Karjalainen, T. Mechanisms of Colonisation and Colonisation Resistance of the Digestive Tract Part 2: Bacteria/Bacteria Interactions. *Microb. Ecol. Health Dis.* **2000**, *12*, 240–246. [CrossRef]
111. Zhao, L.; Zhang, F.; Ding, X.; Wu, G.; Lam, Y.Y.; Wang, X.; Fu, H.; Xue, X.; Lu, C.; Ma, J.; et al. Gut Bacteria Selectively Promoted by Dietary Fibers Alleviate Type 2 Diabetes. *Science* **2018**, *359*, 1151–1156. [CrossRef]
112. Canfora, E.E.; Jocken, J.W.; Blaak, E.E. Short-Chain Fatty Acids in Control of Body Weight and Insulin Sensitivity. *Nat. Rev. Endocrinol.* **2015**, *11*, 577–591. [CrossRef]
113. Chambers, E.S.; Preston, T.; Frost, G.; Morrison, D.J. Role of Gut Microbiota-Generated Short-Chain Fatty Acids in Metabolic and Cardiovascular Health. *Curr. Nutr. Rep.* **2018**, *7*, 198–206. [CrossRef] [PubMed]
114. Tan, J.; McKenzie, C.; Potamitis, M.; Thorburn, A.N.; Mackay, C.R.; Macia, L. The Role of Short-Chain Fatty Acids in Health and Disease. In *Advances in Immunology*; Elsevier: Amsterdam, The Netherlands, 2014; Volume 121, pp. 91–119. ISBN 978-0-12-800100-4.
115. Koh, A.; De Vadder, F.; Kovatcheva-Datchary, P.; Bäckhed, F. From Dietary Fiber to Host Physiology: Short-Chain Fatty Acids as Key Bacterial Metabolites. *Cell* **2016**, *165*, 1332–1345. [CrossRef] [PubMed]
116. Byrne, C.S.; Chambers, E.S.; Morrison, D.J.; Frost, G. The Role of Short Chain Fatty Acids in Appetite Regulation and Energy Homeostasis. *Int. J. Obes.* **2015**, *39*, 1331–1338. [CrossRef] [PubMed]
117. Bach Knudsen, K.E. Microbial Degradation of Whole-Grain Complex Carbohydrates and Impact on Short-Chain Fatty Acids and Health. *Adv. Nutr.* **2015**, *6*, 206–213. [CrossRef] [PubMed]
118. Parada Venegas, D.; De La Fuente, M.K.; Landskron, G.; González, M.J.; Quera, R.; Dijkstra, G.; Harmsen, H.J.M.; Faber, K.N.; Hermoso, M.A. Short Chain Fatty Acids (SCFAs)-Mediated Gut Epithelial and Immune Regulation and Its Relevance for Inflammatory Bowel Diseases. *Front. Immunol.* **2019**, *10*, 277. [CrossRef]
119. Hamer, H.M.; Jonkers, D.; Venema, K.; Vanhoutvin, S.; Troost, F.J.; Brummer, R.-J. Review Article: The Role of Butyrate on Colonic Function. *Aliment. Pharmacol. Ther.* **2008**, *27*, 104–119. [CrossRef] [PubMed]
120. Zhang, C.; Yin, A.; Li, H.; Wang, R.; Wu, G.; Shen, J.; Zhang, M.; Wang, L.; Hou, Y.; Ouyang, H.; et al. Dietary Modulation of Gut Microbiota Contributes to Alleviation of Both Genetic and Simple Obesity in Children. *EBioMedicine* **2015**, *2*, 968–984. [CrossRef]
121. Knowles, S.R.; Nelson, E.A.; Palombo, E.A. Investigating the Role of Perceived Stress on Bacterial Flora Activity and Salivary Cortisol Secretion: A Possible Mechanism Underlying Susceptibility to Illness. *Biol. Psychol.* **2008**, *77*, 132–137. [CrossRef]
122. Rothschild, D.; Weissbrod, O.; Barkan, E.; Kurilshikov, A.; Korem, T.; Zeevi, D.; Costea, P.I.; Godneva, A.; Kalka, I.N.; Bar, N.; et al. Environment Dominates over Host Genetics in Shaping Human Gut Microbiota. *Nature* **2018**, *555*, 210–215. [CrossRef]
123. Milani, C.; Duranti, S.; Bottacini, F.; Casey, E.; Turroni, F.; Mahony, J.; Belzer, C.; Delgado Palacio, S.; Arboleya Montes, S.; Mancabelli, L.; et al. The First Microbial Colonizers of the Human Gut: Composition, Activities, and Health Implications of the Infant Gut Microbiota. *Microbiol. Mol. Biol. Rev.* **2017**, *81*, e00036-17. [CrossRef]
124. Ruiz-Ojeda, F.J.; Plaza-Díaz, J.; Sáez-Lara, M.J.; Gil, A. Effects of Sweeteners on the Gut Microbiota: A Review of Experimental Studies and Clinical Trials. *Adv. Nutr.* **2019**, *10*, S31–S48. [CrossRef] [PubMed]
125. Moya-Pérez, A.; Neef, A.; Sanz, Y. Bifidobacterium Pseudocatenulatum CECT 7765 Reduces Obesity-Associated Inflammation by Restoring the Lymphocyte-Macrophage Balance and Gut Microbiota Structure in High-Fat Diet-Fed Mice. *PLoS ONE* **2015**, *10*, e0126976. [CrossRef] [PubMed]

metabolites

Review

Hunting Metabolic Biomarkers for Exposure to Per- and Polyfluoroalkyl Substances: A Review

Xue Ma, Delei Cai, Qing Chen, Zhoujing Zhu, Shixin Zhang, Ziyu Wang, Zhengyan Hu, Haitao Shen and Zhen Meng *

Zhejiang Provincial Center for Disease Control and Prevention, 3399 Binsheng Road, Hangzhou 310051, China
* Correspondence: zhmeng@cdc.zj.cn

Abstract: Per- and polyfluoroalkyl substances (PFAS) represent a class of persistent synthetic chemicals extensively utilized across industrial and consumer sectors, raising substantial environmental and human health concerns. Epidemiological investigations have robustly linked PFAS exposure to a spectrum of adverse health outcomes. Altered metabolites stand as promising biomarkers, offering insights into the identification of specific environmental pollutants and their deleterious impacts on human health. However, elucidating metabolic alterations attributable to PFAS exposure and their ensuing health effects has remained challenging. In light of this, this review aims to elucidate potential biomarkers of PFAS exposure by presenting a comprehensive overview of recent metabolomics-based studies exploring PFAS toxicity. Details of PFAS types, sources, and human exposure patterns are provided. Furthermore, insights into PFAS-induced liver toxicity, reproductive and developmental toxicity, cardiovascular toxicity, glucose homeostasis disruption, kidney toxicity, and carcinogenesis are synthesized. Additionally, a thorough examination of studies utilizing metabolomics to delineate PFAS exposure and toxicity biomarkers across blood, liver, and urine specimens is presented. This review endeavors to advance our understanding of PFAS biomarkers regarding exposure and associated toxicological effects.

Keywords: per- and polyfluoroalkyl substances (PFAS); biomarker; metabolomics; toxicity

1. Introduction

Per- and polyfluoroalkyl substances (PFASs) encompass both perfluorinated and partially fluorinated alkyl compounds, characterized by the replacement or partial replacement of hydrogen atoms by fluorine atoms on all carbon atoms. These synthetic chemicals consist of a carbon backbone ranging from 4 to 18 carbon atoms in length, terminating with functional groups [1]. Depending on the terminal functional groups, PFASs can be categorized into carboxylates, perfluorosulfonates, and phosphonates, among others. Furthermore, they can be classified based on carbon chain length, including perfluorooctanoic acid (PFOA), perfluorononanoic acid (PFNA), and perfluorodecanoic acid (PFDA). Owing to their exceptional industrial properties, such as hydrophobicity, oleophobicity, heat resistance, high stability, and low surface tension, PFASs have been extensively utilized since the 1950s in various industrial and consumer applications. These applications include the production of polytetrafluoroethylene, leather goods, waterproof textiles, firefighting foams, shampoos, cosmetics, and food packaging materials [2,3].

Unfortunately, the widespread use of PFASs and their desirable properties in commercial products have also led to their pervasive detection in the environment [4–6]. Currently, various concentrations of PFASs have been detected in environmental samples, animal serum, tissue samples, and human bodies worldwide, positioning PFASs as a new class of persistent organic pollutants, alongside polychlorinated biphenyls, organochlorine pesticides, and dioxins. Epidemiological and toxicological studies have shown that PFAS exposure results in toxic effects on the liver, nervous system, immune system, reproductive and developmental processes, genetics, and endocrine function, and may even induce tumors, although the metabolic toxicity remains unclear [7–13].

In recent years, representative PFASs such as perfluorooctane sulfonate (PFOS) have been listed under the Stockholm Convention [14]. Although these regulatory interventions have significantly reduced the emissions of PFOS and PFOA, human exposure to these compounds persists due to their persistence and ongoing production in some developing countries [15]. Moreover, the high-energy C-F covalent bond renders PFASs extremely resistant to degradation, leading to their persistence in the environment. They can contaminate water and food through direct discharge, migration, bioaccumulation, and biomagnification within marine ecosystems [16]. Evidently, even at low exposure levels, the potential health hazards of PFASs cannot be overlooked. As the adverse effects of PFASs on the environment and human health become more widely recognized, and as they migrate and biomagnify through the food chain, PFASs have emerged as a significant environmental and health concern.

Metabolomics, the large-scale study of metabolites within organisms, tissues, and cells, is considered a promising tool for elucidating the associations between environmental pollutants and health, as well as the etiology of certain diseases [17]. As an emerging technology, metabolomics characterizes small molecule metabolites present in cells and their roles in various biological processes, offering a novel and powerful approach to understanding the complex molecular responses induced by PFAS exposure. Despite extensive research on PFAS detection and toxicity, studies on the metabolic toxicity and biomarkers of PFASs remain limited. As metabolomics rapidly advances, reviewing the current state of research on PFAS metabolic toxicity and its biomarkers will provide timely guidance for future research in this crucial field.

This review briefly introduces the types of PFASs, sources of human exposure, and exposure characteristics. It summarizes the latest advancements in studying PFAS toxicity using metabolomics, reviews metabolic disturbances, potential exposure biomarkers, and effect biomarkers, and emphasizes metabolic toxicity and related metabolic biomarkers.

2. Introduction to PFAS

PFASs gained widespread application in consumer and industrial products during the 1990s due to their surfactant properties. Around the year 2000, the recognition of their potential toxicity led to the gradual phasing-out of legacy PFASs, primarily long-chain perfluoroalkyl carboxylic acids or perfluoroalkyl sulfonic acids. Currently, industrial applications of PFASs are increasingly shifting towards short-chain PFASs or those with chlorinated carbon or ether bond structures, giving rise to emerging PFASs.

2.1. Classification

PFASs are classified based on their functional groups and molecular structures into carboxylic acids and sulfonic acids. The carbon chain length for carboxylic acids generally ranges from 6 to 18, while for sulfonic acids it typically ranges from 5 to 10. Due to restrictions on the use of traditional long-chain PFAS, several emerging substitutes for conventional PFASs have gradually appeared. These substitutes include short-chain compounds and those incorporating fluorinated chlorinated carbons or ether bonds. Consequently, long-chain PFCA and PFSA are referred to as "legacy PFAS", while the substitutes are termed "emerging PFAS". Common types of PFASs are listed in Table 1.

Table 1. Common PFAS compounds.

Legacy PFAS		
Name	**Structural Formula**	**CASRN**
Long-chain PFCA	$C_nF2_{n+1}COOH$, $n \geq 7$	-
Long-chain PFSA	$C_nF2_{n+1}SO_3H$, $n \geq 6$	-

Table 1. *Cont.*

Emerging PFAS Name	Structural Formula	CASRN
Short-chain PFCA	$C_nF_{2n+1}COOH$, n = 2–5	-
Short-chain PFSA	$C_nF_{2n+1}SO_3H$, n = 2–5	-
HFPO-DA or "GenX"	$C_3F_7O(CF_3)CFCOOH$	13252-13-6
NBP2	$CF_3CHFOC_3F_6OC_2F_4SO_3H$	749836-20-2
6:2 FTS	$C_8H_4F_{13}SO_3H$	27619-97-2
6:2 Cl-PFESA	$Cl(CF_2)_6O(CF_2)_2SO_3^-$	73606-19-6
8:2 Cl-PFESA	$Cl(CF_2)_8O(CF_2)_2SO_3^-$	83329-89-9
PFOSA	$C_8H_{17}SO_2NH_2$	754-91-6
EtFOSAA	$C_8F_{17}SO_2N(C_3H_7)COOH$	2991-50-6
MeFOSAA	$C_8F_{17}SO_2N(C_2H_5)COOH$	2355-31-9
6:2 FTEOs	$C_6F_{13}(C_2H_4O)_nH$, n = 4–12	1640092-35-8 [a]

PFCA, perfluoroalkyl carboxylic acid; PFSA, perfluoroalkane sulfonic acid; HFPO-DA or "GenX", hexafluoropropylene oxide dimer acid; NBP2, nafion byproduct 2; 6:2 FTS, 6:2 fluorotelomer sulfonic acid; 6:2 Cl-PFESA, 6:2 chlorinated polyfluorinated ether sulfonate; 8:2 Cl-PFESA, 8:2 chlorinated polyfluorinated ether sulfonate; PFOSA, perfluorooctane sulfonamide; EtFOSAA, 2-(N-ethyl-perfluorooctane sulfonamido) acetic acid; MeFOSAA, 2-(N-methyl-perfluorooctane sulfonamido) acetic acid; 6:2 FTEOs, 6:2 fluorotelomer ethoxylates. [a] Capstone FS-30, commercial mixture.

2.2. *Sources*

The widespread presence of PFASs in industrial emissions, untreated domestic wastewater, sewage treatment plants, and aqueous film-forming foams has led to their detection in the environment, food, and human bodies [18–21]. The most significant pathways for human exposure to PFASs include food, drinking water, skin contact, indoor dust, and outdoor air, with food being the primary route [22,23]. Consuming contaminated food and drinking water, such as vegetables, crops, fish, meat, or processed food affected by PFAS, as well as drinking PFAS-contaminated water, can lead to exposure [19,24–27]. Additionally, contact with food packaging materials containing PFASs (e.g., food packaging paper and non-stick cookware) also poses a risk [28,29]. PFASs in soil may indirectly harm human health through bioaccumulation in crops [26].

PFASs can transfer into the human body via the food chain, accumulating and magnifying at each trophic level [30]. In plants, PFASs can be absorbed by roots from contaminated soil or water and then translocated to flowers, leaves, and fruits [31,32]. Long-chain PFASs tend to remain in roots, whereas short-chain PFASs are more readily transported to fruits [32,33]. Aquatic organisms ingest PFAS-contaminated sediments and seawater, with studies showing that short-chain PFASs easily accumulate in these organisms [34,35]. In livestock, animals ingest PFASs through water and feed, which then accumulate in muscle and milk [36,37], subsequently entering the human food chain and increasing human exposure risk.

2.3. *Human Exposure Pathway and Characteristics*

Multiple PFAS compounds have been detected in human tissues such as blood, kidneys, brain, cerebrospinal fluid, liver, lungs, and placenta [38–42]. This necessitates urgent research into PFAS exposure and its effects on these human samples.

2.3.1. Population Characteristics

Occupational groups such as firefighters using aqueous film-forming foam have higher PFAS exposure risk than the general population, where diet is a critical factor influencing PFAS exposure [43]. High seafood consumption correlates with increased human PFAS levels [19]. PFAS serum levels exhibit gender differences, generally higher in males than females [44,45]. Menstruation and breastfeeding help reduce PFAS levels in females [46]. PFASs can also be transmitted to infants through umbilical cord blood, placenta, and breast milk [47,48]. Children are at higher risk of PFAS exposure than adults due to the presence of PFASs in formula milk and dairy products [27,49].

2.3.2. Distribution Trend

With the gradual elimination of legacy PFASs, their levels in human serum are declining [50]. However, emerging PFASs can still be detected in serum, with some linked to adverse pregnancy outcomes and other toxic effects [51,52].

2.3.3. Absorption and Distribution

PFAS absorption and distribution in organisms are influenced by carbon chain length, half-life, species, and gender. Animal studies show that perfluorohexanesulfonic acid (PFHxS), perfluorobutanesulfonic acid (PFBS), PFOA, and PFOS are almost completely absorbed after oral administration [53]. PFASs primarily enter the bloodstream through the intestinal barrier, binding to blood albumin and low-density lipoprotein, and then distribute to extra-intestinal organs [54]. Due to the high affinity for liver fatty acid-binding proteins, PFAS accumulation in the liver is well-documented [55]. PFASs can also undergo enterohepatic circulation, re-entering the intestines via bile and returning to the liver, leading to high liver concentrations [56]. Generally, longer carbon chains correspond to longer half-lives, ranging from hours to decades [57,58]. The half-life also varies by gender and species [59,60]. Current research mainly focuses on PFCA and PFSA, with the absorption, distribution, toxicokinetics, and metabolic toxicity of emerging PFASs requiring further investigation.

3. Toxic Effects of PFAS

Existing research has demonstrated that PFASs exhibit a range of toxic effects, including thyroid toxicity, hepatotoxicity, immunotoxicity, endocrine disruption, neurotoxicity, reproductive and developmental toxicity, and even carcinogenic and cancer-promoting properties. Additionally, elevated cholesterol levels, obesity, and endocrine disorders are associated with PFAS exposure. Although the connection between PFASs and toxicity in specific tissues and organs is recognized, the mechanisms through which PFASs influence metabolism and subsequently affect organ toxicity remain to be fully elucidated. This review summarizes the epidemiological and rodent model evidence linking PFAS-induced liver injury, reproductive and developmental toxicity, cardiovascular diseases, renal toxicity, and cancer, with a focus on metabolomics insights to better understand the metabolic mechanisms underlying PFAS exposure and toxic effects.

3.1. Liver Injury

Epidemiological studies have frequently associated PFASs, such as PFOA and PFOS, with elevated serum markers of liver injury (e.g., alanine aminotransferase (ALT), aspartate aminotransferase (AST), and alkaline phosphatase (ALP)), non-alcoholic fatty liver disease (NAFLD) grading, staging, and histopathological severity [61–64]. Rodent studies indicate that PFAS-induced hepatotoxicity manifests as liver enlargement, inflammation, steatosis, oxidative stress, apoptosis, and dysregulation of glucose and lipid metabolism [65–68].

Metabolomic studies have revealed that metabolic disturbances are critical in PFAS-induced liver damage. PFASs disrupt hepatic metabolism, particularly bile acid, amino acid, and lipid metabolism, leading to liver injury or influencing the development and susceptibility to NAFLD and liver damage [62,63,69,70]. Sen et al., through a correlation analysis of PFAS (PFOS, PFNA, PFOA) exposure levels with metabolic perturbations in the liver in 105 individuals with NAFLD, found that PFASs perturbed key metabolic pathways of bile acid metabolism and lipid metabolism in NAFLD, and that there were gender differences [63]. Similarly, it has been noted that high PFAS exposure levels were associated with more severe histology of liver damage in children with NAFLD, which may be related to elevated plasma levels of phosphoethanolamine, tyrosine, phenylalanine, aspartic acid, and creatine, and altered metabolites with reduced betaine levels [62]. Additionally, it has been found that prenatal PFAS exposure increases susceptibility to liver damage in childhood, associated with branched-chain amino acids, aromatic amino acids, and glycerophospholipid metabolism [69]. In addition, PFAS mixtures were found to be positively correlated with polipoprotein B (APOB) and γ-glutamyltransferase (GGT), and negatively correlated with direct bilirubin (DBIL) and total bilirubin (TBIL) [70].

Researchers analyzing metabolite changes in the liver or blood after PFAS exposure in rodents also show that PFASs affect bile acids, sterols, fatty acids, purine, and amino acid metabolism, contributing to liver injury [66,68,71–73]. For example, exposure to PFAS mixtures leads to liver injury in mice, such as increased liver weight, inflammation, and elevated plasma ALT levels, with significant changes in bile acid and sterol metabolism [71]. Lipidomic studies suggest that PFOS disrupts ceramide and lysophosphatidylcholine (LPC) metabolism, causing apoptosis and triglyceride depletion in liver cells, leading to morphological liver damage [72]. Functional characterization of lipid species showed that the fatty acid metabolic pattern of A/J mice was significantly altered by PFAS mixture exposure, especially the biosynthetic pathways of glycerophosphocholines (PC), glycerophosphoethanolamines (PE), and glycerophosphoserine (PS) in the liver were significantly affected [73]. In addition to bile acids, sterol, fatty acid, amino acid, and purine metabolism were also affected [66,68]. Jiang et al. examined liver nontargeted metabolomics after PFHxA exposure in mice and found that PFHxA substantially downregulated xanthine and uric acid, increased GSH, and decreased GSSH, which presumably caused oxidative stress by interfering with purine metabolism and glutathione metabolism, thereby causing liver damage [66]. Additionally, PFASs may influence liver damage by affecting gut microbiota metabolism, as evidenced by altered fecal microbiota and arginine metabolism in PFOS-exposed mice [68].

3.2. *Reproductive and Developmental Toxicity*

Studies indicate that PFASs adversely affect male reproductive function, including semen quality and male reproductive hormones [74]. Similarly, PFASs impact female reproductive functions such as menstrual cycle regulation, hormone levels, and fertility [75]. Furthermore, PFASs can cross the placental barrier, thereby influencing fetal development and contributing to adverse pregnancy outcomes like preterm birth [76], reduced birth weight [77,78], postnatal growth [79], and neurodevelopmental disorders [80,81]. Additionally, PFAS exposure is linked to increased disease risks in offspring, such as liver damage susceptibility and type 1 diabetes [69,82]. Research has identified a correlation between maternal PFAS co-exposure and decreased sperm concentration, with varying contributions from different PFASs, with perfluoroheptanoic acid as a primary factor. However, no clear association was found between maternal PFAS exposure and testicular volume or reproductive hormones [74]. Contrarily, another study showed a significant negative correlation between PFAS mixtures and reproductive hormones, particularly estradiol (E2) and the E2/total testosterone (TT) ratio, with perfluoro-n-undecanoic acid (PFUdA) being a major contributor [83]. Higher serum concentrations of PFOA and PFHxS in pregnant women are associated with increased odds of preterm birth [76]. PFASs also adversely affect prenatal and postnatal neurodevelopment. Additionally, prenatal PFAS exposure is linked to neurodevelopmental disorders in children, such as reduced IQ performance and impaired executive function [80].

Population studies using metabolomics found that the effects of PFASs on preterm birth, fetal growth, childhood obesity trajectory, susceptibility to liver injury, and other reproductive developmental processes were related to amino acid, glycerophospholipid lipid and fatty acid, bile acid, uric acid, and carbohydrate metabolism [69,76,79,84,85]. Two studies found that PFAS exposure was associated with preterm delivery or fetal growth in pregnant women and that the effects were associated with metabolite alterations, e.g., one study found that the impact of prenatal PFAS exposure on the shortening of the gestation period was associated with eight metabolomic pathways and 52 metabolites in the neonatal blood spot [76,84]. Another study also found that higher concentrations of PFOA and PFNA in maternal serum were associated with reduced fetal growth and were strongly associated with disturbances in amino acid, lipid and fatty acid, bile acid, and androgen metabolism [84]. Two other studies found that prenatal exposure to PFASs was associated with increased susceptibility to liver damage and an obesity trajectory in children, as well as with metabolites such as amino acids, glycerophospholipids, sphingolipids, and octanoylcarnitine [69,79]. Consistent with human studies, laboratory research on rodents has demonstrated that the effects of PFASs on reproductive development were associated with metabolite disruption. For instance, pregnant Sprague–Dawley rats exposed to nafion byproduct 2 (NBP2), a type of perfluoroalkyl ether sulfonate, experienced neonatal mortality, reduced pup weight, and decreased liver glycogen in pups, as well as significant alterations in lipid and carbohydrate metabolism in dams and offspring [85].

3.3. Cardiovascular Toxicity

Cardiac metabolism, including conditions such as obesity, hypertension, and hyperlipidemia, has been extensively studied in relation to PFAS exposure. This review focuses on lipid and cholesterol abnormalities. Research indicates that twelve PFAS-related characteristic metabolites are associated with cardiovascular diseases [86]. A study examining the relationship between Cl-PFESA, a typical PFAS substitute in China, and the lipid profiles of 1336 residents in Guangzhou revealed that increased Cl-PFESA levels were positively associated with total cholesterol (TC), triglycerides (TG), low-density lipoprotein cholesterol (LDL-C), and high-density lipoprotein cholesterol (HDL-C) levels. The significant positive correlation between Cl-PFESA and lipid abnormalities suggests a detrimental effect of Cl-PFESA on lipid profiles, potentially exhibiting a nonlinear association [87].

Metabolomic studies have shown that PFASs primarily cause lipid and cholesterol abnormalities by disrupting bile acid and sterol metabolism [71,88–90]. Two population-based studies using metabolomics have found that PFASs were associated with plasma triglycerides and that this association was related to metabolite alterations [88,89]. Another study revealed that PFAS exposure was linked to a higher risk of dyslipidemia, and associations between PFASs and TC and LDL were found to be primarily related to glycerophospholipid metabolism, primary bile acid biosynthesis, and linoleic acid metabolism [90]. Animal studies have also demonstrated that PFASs interfere with metabolism, leading to increased circulating cholesterol and cholesterol metabolism abnormalities. For instance, C57BL/6J mice fed a high-fat diet and exposed to a PFAS mixture exhibited elevated fasting plasma cholesterol levels [71]. The levels of several sterol metabolites, including 4-cholesten-3-one, were upregulated, while most primary bile acids, such as chenodeoxycholic acid and β-muricholate, were significantly reduced in the liver metabolome. These findings suggest that PFASs may disrupt cholesterol metabolism and transport, contributing to a PFAS-induced cholesterol imbalance.

3.4. Glucose Homeostasis Disruption

PFASs are closely associated with endocrine disorders such as diabetes, impaired glucose tolerance, overweightness, or obesity [91]. Epidemiologic studies have found that PFOA and PFOS exposure were associated with higher 30 min blood glucose levels and area under the glucose curves (AUCs) during oral glucose tolerance tests (OGTTs), as well as higher levels of HbA1c [89,92]. Additionally, research indicates that PFOA and PFHxS were associated with increased 2 h glucose levels and increased area under the glucose curve, as well as altered lipid (e.g., sphingolipids, linoleic acid, and de novo lipogenesis) and amino acid (e.g., aspartic acid and aspartate, tyrosine, arginine, and proline) metabolism [93]. Laboratory evidence further suggests that gastric gavage of pregnant rats with low doses of PFOSs disrupts critical metabolic products, such as glycerol-3-phosphate and lactosylceramide, affecting fasting blood glucose (FBG) in mothers and thus impacting glucose homeostasis [94].

3.5. Other Toxicological Impact

Studies have demonstrated that PFASs can also interfere with metabolism and impact the development of kidney diseases and cancer [95–97]. Research indicates a causal relationship between PFASs and declining kidney function and chronic kidney disease [95]. A study identified eight potential biomarkers associated with PFAS exposure, all significantly correlated with kidney function markers, suggesting adverse effects of high PFAS exposure on kidney function [96]. Furthermore, PFASs have been implicated in liver cancer risk, possibly through metabolic interference. For instance, plasma PFAS concentrations are positively correlated with hepatocellular carcinoma (HCC) risk, with PFOSs showing the strongest correlation, associated with increased HCC incidence [97]. Metabolomic analysis revealed 433 metabolites in plasma associated with PFOSs and 499 metabolites associated with HCC.

4. Potential Biomarkers

The metabolome encompasses the complete set of small molecules that participate in physiological processes within the body or in cells and tissues. The application of high-throughput analytical techniques has emerged as a novel tool for the comprehensive examination of endogenous metabolites in vivo. Research has demonstrated that pollutants can disrupt metabolic pathways, leading to alterations in metabolite levels within the human body. These changes in metabolites may indicate defects in specific pathways or the activation of certain signals, and the dynamic alterations of these metabolites can serve as biomarkers of pollutant-induced damage to the human body. Metabolomics can reveal toxicological changes and related mechanisms at an earlier stage and can be utilized as a tool for discovering biomarkers of pollutant exposure and effects. Metabolomics primarily focuses on the quantitative analysis of metabolites under specific physiological conditions following organismal exposure, understanding metabolic changes under varying conditions, and elucidating the associations between these changes and the organism's health status and disease. Therefore, comprehending the relationship between PFAS exposure and metabolites within the body can provide additional evidence for elucidating the health risks posed by pollutant exposure.

4.1. Exposure Biomarkers

Metabolomics methodologies emphasize the analysis of primary and secondary metabolites, which are crucial for obtaining the phenotypic fingerprints of organisms in their environments. These metabolic gatekeepers can also elucidate molecular connections between exposure biomarkers and health outcomes, which might otherwise be obscured by the complex interactions present in direct measurements [98]. By focusing on the quantitative analysis of metabolites in organisms exposed to PFASs under specific physiological conditions, metabolomics can identify metabolic changes induced by PFASs and their associations with health status. Consequently, we concentrate on the quantitative analysis of metabolites in blood, liver, and urine under specific physiological states of PFAS exposure, summarizing the disturbances in metabolites in both human populations and rodent models post-PFAS exposure to identify potential PFAS-related exposure biomarkers.

Current research on PFAS exposure and metabolic alterations primarily involves human and rodent subjects. The studies cover various PFAS compounds, with the most extensive research on PFOA, PFOS, PFHxS, PFNA, and PFDA. There are also some metabolomics studies on emerging PFASs, such as 6:2 FST, 6:2 Cl-PFESA, GenX, NBP2, and FTEOs, though most research focuses on the effects of individual PFAS compounds, with some addressing the impact of mixtures. In the study of PFAS-related metabolites, alterations in amino acids (with branched-chain amino acids being more prominently affected), lipids, bile acid metabolism, and the urea cycle are common metabolomic features. Glycerophospholipid metabolism in lipid pathways is considered a key metabolic feature, along with fatty acid and carnitine metabolism related to fatty acid oxidation and energy supply pathways. Purine and pyrimidine metabolism in cellular energy systems are also identified as significant metabolic changes, with potential as future exposure biomarkers [99].

Most current studies utilize metabolomics in the blood (serum, plasma, cord blood) and liver to identify related metabolites, with fewer studies on urine and placental metabolite changes. Some studies simultaneously use serum or plasma and liver samples to explore the impact on enterohepatic circulation. Given the greater practicality of detecting metabolites in different samples for clinical translation, we summarize recent studies and advancements in metabolomics from various samples, as detailed in Tables 2 and 3.

4.1.1. Blood

As illustrated in Tables 2 and 3, in recent years there have been 19 studies investigating the effects of PFASs on metabolomic alterations in blood (plasma, serum, or cord blood). In human studies, PFAS exposure detection and metabolomic analyses typically utilize the same sample matrix, encompassing the general population, special populations (pregnant women, fetuses, children), occupational groups, and diseased cohorts. Diseased cohorts primarily include overweight or obese children, adolescents or young adults, adults and children with NAFLD, HCC cases, and individuals at high risk for or diagnosed with type 2 diabetes. Additionally, a few studies have inconsistent samples for PFAS exposure detection and metabolomic analyses, such as examining maternal PFAS exposure and subsequent metabolomic changes in cord blood, newborn, or child blood to investigate prenatal exposure effects on offspring blood metabolomics.

PFAS-related studies using metabolomics have found that PFAS exposure is primarily associated with altered metabolism of amino acids, lipids, glycerophospholipids, fatty acids, and bile acids in blood [69,76,90,100–102]. For example, PFAS exposure has been found to be primarily associated with altered lipid metabolism in the blood, including phospholipid metabolism, long-chain saturated fatty acids, fatty acid metabolism (acylcarnitines, monounsaturated fatty acids), medium-chain fatty acids, and long-chain polyunsaturated fatty acids (n3 and n6), in studies of the male population, pregnant women and in the elderly population [90,100,101]. Two population studies found that increased PFAS exposure was associated with serum amino acids, such as valine, leucine, isoleucine, tryptophan, and alphaketobutyrate, as well as glycerophospholipid phospholipids (PCs) aa C36:1, 1-palmitoyl-GPC (16:0), and Lyso-PC a C18:1 by examining the metabolomics correlation analyses of PFAS exposure [69,101]. In the elderly population, Yang et al. found that the nine metabolites most strongly associated with PFASs mainly included four fatty acyl groups, such as oleoyl-L-carnitine, 7α-hydroxy-3-oxo-4-cholestenoic acid (7-HOCA), 22β-hydroxycholesterol and vitamin D3, three steroids and steroid derivatives, lipids, and lipid-like substances including γ-tocopherol and PG (16:0/18:1) [100]. In addition, two population studies have found that PFAS exposure was primarily associated with the metabolism of lipids and bile acids, among others, involving polyunsaturated omega-6 fatty acids, arachidonic and linoleic acids, cholesteryl esters, LPE, alkylPCs, stearic acid, arachidic acid, monounsaturated palmitoleic acid, linoleic acid, HCA, TaMCA, GCDCA, CA, and many other metabolites [76,102]. Some studies have also found that PFASs were associated with disturbances in purine–pyrimidine metabolism, and retinol and sterol metabolism [51,103]. For example, PFAS exposure in pregnant women predominantly affects fatty acid and retinol metabolism [51]. In children, PFAS exposure is associated with pathways involved in de novo fatty acid biosynthesis, the tricarboxylic acid (TCA) cycle, and pyrimidine and purine metabolism [103].

Table 2. PFAS exposure and associated metabolites in epidemiological studies.

PFAS	Study Object	Sample Size	Sample Matrix	Main Finding(s)	Ref.
PFOA, PFNA, PFDA, PFUdA, PFTrDA, PFHxS, PFOS, 6:2 FTS, etc.	Non-occupationally exposed residents near industrial parks of Shandong Province	91	serum	• PFOA and PFOS were associated with 126 and 117 metabolites, respectively • The most highly correlated metabolites were predominantly lipids and lipoids, including oleoyl-L-carnitine, 5,6-dihydroxy-8Z,11Z,14Z-eicosatrienoic acid (5,6-DHET), dec-anoyl-L-carnitine, 9,10- dihydroxy-12-octadecenoic acid (9,10-DHOME), 7α-hydroxy-3-oxo-4-cholestenoic acid (7-HOCA), 22β-hydroxycholesterol, vitamin D3, γ-tocopherol, and PG (16:0/18:1)	Yang Y et al., 2024 [100]
PFOS, PFHxS, PFOA, PFNA	Pregnant African American Newborns in Atlanta, Georgia	267	serum	• 435, 559, 154, and 1262 metabolites were associated with PFOA, PFOS, PFNA, and PFHxS, respectively • Contains carnitine and bile acids, polyunsaturated omega-6 fatty acids, arachidonic, linoleic acids, indole-3-acetic acid methyl ester, 4-hydroxy-3-methyloxybenzeneglycol, β-nicotinamide adenine dinucleotide, benzoic acid, s -lactic acid, and glutathione, etc.	Taibl KR et al., 2023 [76]
23PFSA, mainly PFHpA, PFOA, PFBS, PFPeS, PFHxS, PFHpS, PFOS, 6:2 FTS	Occupational workers and residents of fluorine chemical plants in Hubei, China	225	urine	• 78 metabolites associated with PFAS exposure	He A et al., 2023 [96]
PFOS, PFHxS, PFHpS, PFOA, PFNA, PFDA	The Study of Latino Adolescents at Risk (SOLAR), Southern California Children's Health Study (CHS)	312, 137	plasma	• 463 and 200 metabolites of amino acids, lipids, prostaglandin formation from arachidonate and de novo fatty acid biosynthesis, and the aromatic amino acid metabolic pathway associated with PFASs in the SOLAR and CHS cohorts, respectively • Thyroxine (T4), L-glutamic acid, equine uric acid, arachidonic acid, and aminoadipic acid were positively associated with PFAS • Triglycerides and acetoacetate were negatively associated with PFASs in the SOLAR cohort but positively associated with PFASs in the CHS cohort	Goodrich JA et al., 2023 [104]

Table 2. *Cont.*

PFAS	Study Object	Sample Size	Sample Matrix	Main Finding(s)	Ref.
PFOA, PFOS, PFHXS, PFDA, PFUdA, PFNA	Cohort investigating pregnant women and children in the US VDAART cohort study	459, 401	plasma	• 43 metabolites of sphingomyelin, lysophospholipid(LPL), long chain polyunsaturated fatty acids were associated with PFASs in the parent, containing 17 amino acid metabolites • 80 metabolites of LPL and phosphatidylcholine (PC) metabolic pathways were associated with PFASs in child, containing 20 amino acid metabolites • The most associations with PFASs were LPL and PC pathway metabolites, with the smallest p value being 1-palmitoyl-GPC (16:0)	Prince N et al., 2023 [101]
PFHxS, PFOA, PFHpS, PFNA, PFOS, 6:2 Cl-PFESA, PFDA, PFUdA, 8:2 Cl—PFESA	Male residents recruited from Guangzhou	278	serum	• 51, 137, 34, 112, 153, 112, 101, 49, and 46 metabolites related to PFHxS, PFOA, PFHpS, PFNA, PFOS, PFDA, PFUdA, 6:2Cl-PFESA, and 8:2Cl-PFESA, respectively • Glycocholic acid was negatively correlated with PFOS; bile acids were negatively correlated with PFNA, PFOS, PFDA, and PFUdA; deoxycholic acid was negatively correlated with PFOS, PFDA, PFDA, and 8:2 Cl- PFESA • glycerophospholipids, aryl alcohol lipids, and fatty acyls metabolites related to Cl-PFESA	Chen Y et al., 2023 [90]
PFHxS, PFNA, PFOA, 2 PFOS isomers	Cohort with NAFLD who underwent laparoscopic bariatric surgery	105	Liver, serum	• PFASs were correlated with hepatic primary BAs (TCA, GCDCA, TCDCA) and hepatic secondary BAs (DCA, GHCA and GUDCA) • PFOA and PFOS were positively correlated with Cers (e.g., Cer(d18:0/16:0), HexCer(d18:1/18:0)), ether phospholipids (e.g., O-PC(40:4)), TGs, and DGs	Sen P et al., 2022 [63]
PFOS, PFHxS, PFOA, PFDA, PFNA, PFUdA	A nested case-control study of HCC	50 pairs	plasma	• 433 metabolites associated with high levels of PFOS exposure	Goodrich JA et al., 2022 [97]
PFOA, PFNA, PFDA, PFUdA, PFHxS, PFHpS, 6:2 Cl-PFESA, 8:2 Cl-PFESA	Matched samples of pregnant women delivering in Beijing's hospitals	84 pairs	serum, cord blood	• 279 and 338 metabolites were associated with PFASs in maternal and cord serum, respectively • PFOA and PFHxS were the two PFASs with high correlation to metabolites, with 190, 218(maternal) and 201, 215 (cord blood) metabolites, respectively • Two Cl-PFESAs were less relevant to metabolites, both having only one in the maternal blood	Li Y et al., 2021 [51]

Table 2. *Cont.*

PFAS	Study Object	Sample Size	Sample Matrix	Main Finding(s)	Ref.
PFOS, PFOA, PFHxS, PFNA, EtFOSAA, MeFOSAA, PFDA, PFOSA	A Diabetes Prevention Program project for a multicenter randomized clinical trial in people at risk for type 2 diabetes	691	plasma	• 17, 17, 10, 30, 23, and 34 metabolites were associated with total PFOS, n-PFOS, Sm-PFOS, total PFOA, n-PFOA, and Sb-PFOA, respectively • Sphingolipids and glycerophospholipids were positively correlated with PFAS, including C34:2 PE, C36:2 PE, C34:1 PC and C36:1 PC, etc. PFOA and PFOS were correlated with phosphatidylethanolamine, triacylglycerols, C16:1 SM, and C18:2 SM • PFASs correlated with branched-chain amino acids (isoleucine, leucine, valine) and glycine	Mitro SD et al., 2021 [105]
PFHxS, PFOS, PFOA, 6:2 Cl-PFESA, PFNA, PFDA, PFUdA, PFHpS	Cord blood samples stored in the Beijing Cord Blood Bank	104	cord blood	• 73 metabolites were associated with PFAS, including lipids, free fatty acids, and bile acids • Mainly correlated with cholesteryl esters, LPE, alkylPCs, stearic acid, arachidic acid, monounsaturated palmitoleic acid, linoleic acid, HCA, TaMCA, GCDCA, and CA	Sinisalu L et al., 2021 [102]
PFOA, PFOA, PFHxS, PFDA, PFNA, PFUdA	A case-control study on T2D	187 pairs	plasma	• 171 metabolites were associated with PFAS • 19, 14, 18, 66, 62, 59 metabolites related to PFOS, PFOA, PFHxS, PFDA, PFNA and PFUdA, respectively	Schillemans T et al., 2020 [91]
PFOS, PFOA, PFNA, PFHxS, PFUdA	The Human Early-Life Exposome project	1105 pairs	serum	• 66 and 34 metabolites were associated with PFASs at high and low risk of liver injury, respectively. • Including amino acids, biogenic amine, glycerophospholipids, sphingolipids, and hexose	Stratakis N et al., 2020 [69]
PFOA, PFOS and PFHxS	NAFLD patients enrolled at Children's Healthcare of Atlanta	74	plasma	• 348, 349, and 662 metabolites were associated with PFOA, PFOS, and PFHxS, respectively	Jin R et al., 2020 [62]
PFOA, PFOS, PFHxS	Overweight and Obese Hispanic Children in Downtown Los Angeles	40	plasma	• 149, 298, and 17 metabolites were associated with PFOA, PFOS, and PFHxS, respectively	Alderete TL et al., 2019 [93]

Table 2. *Cont.*

PFAS	Study Object	Sample Size	Sample Matrix	Main Finding(s)	Ref.
PFOA, PFOS, PFNA, PFHxS	The Health Outcomes and Measures of the Environment Study	114	serum	• Negative ion pattern, 17, 63, 47, and 29 metabolites associated with PFOA, PFOS, PFNA, and PFHxS, respectively • Positive ion mode, 18, 253, 76, and 39 metabolites associated with PFOA, PFOS, PFNA, and PFHxS, respectively	Kingsley SL et al., 2019 [103]

PFOA, perfluorooctanoic acid; PFNA, perfluorononanoic acid; PFDA, perfluorodecanoic acid; PFTrDA, perfluorotridecanoic acid; PFHxS, perfluorohexane sulfonic acid; PFOS, perfluorooctane sulfonic acid; 6:2 FTS, 6:2 fluorotelomer sulfonic acid; PFHpA, perfluoroheptanoic acid; PFBS, perfluorobutane sulfonic acid; PFPeS, perfluoropentanc sulfonic acid; PFHpS, perfluoroheptane sulfonic acid; PFUdA, perfluoroundecanoic acid; 6:2 Cl-PFESA, 6:2 chlorinated polyfluorinated ether sulfonate; 8:2 Cl-PFESA, 8:2 chlorinated polyfluorinated ether sulfonate; EtFOSAA, 2-(N-ethyl-perfluorooctane sulfonamido) acetic acid; MeFOSAA, 2-(N-methyl-perfluorooctane sulfonamido) acetic acid; PFOSA, perfluorooctane sulfonamide.

Table 3. PFAS exposure and associated metabolites in rodent studies.

PFAS	Study Object	Dose	Sample Matrix	Main Finding(s)	Ref.
PFOA	Humanized PPARα mice	8 μM/kg, 6–7 weeks	liver	• Positively correlated with Cer(18:0/23:0), LPE, DG(36:2), TG(18:1/18:1/18:1), TG(50:3), and TG(51:3), and negatively correlated with HexCer(18:1/22:0) and SM(d18:0/24:1)	Sen P et al., 2022 [63]
HFPO-DA or "GenX", NBP2	C57BL/6 mice	0.5, 5, 100 mg/kg, 28 d	liver	• 199 and 123 metabolites associated with GenX and NBP2 group, respectively, and 54 and 86 common differential metabolites in males and females, respectively • The most common dysregulated fatty acids were oleic acid (OA) and dihomo-γ-linoleic acid (DGLA)	Kirkwood-Donelson KI et al., 2024 [106]
FTEOs, PFOA	CD-1 mice	5, 100 ng/L, 17.5 d	placenta	• 14 related metabolites including asparagine, fatty acids, glucose, threonine, lactate, lysine, and threonine • PFOA increased glucose and threonine and decreased creatine. FTEOs decreased asparagine and lysine and increased creatine	Adams H et al., 2024 [107]

Table 3. *Cont.*

PFAS	Study Object	Dose	Sample Matrix	Main Finding(s)	Ref.
PFOS	SD rats	0.03, 0.3 mg/kg, 18 d	liver	• Negative ion mode, 164 and 158 metabolites in low and high dose groups, respectively, including E, 5′-Hydroxy-3′,4′,7-trimethoxyflavan, tragopogonsaponin F, enkephalin L, and tetracosapentaenoic acid (24:5n-3), etc. • Positive ion model, 104 and 56 metabolites in low- and high-dose groups, respectively; the major metabolites were glycerophospholipids, carboxylic acids and their derivatives, fatty acyls, and aryl alcohol lipids	Yu G et al., 2023 [94]
Mixture of PFOA, PFNA, PFDA, PFUdA, PFDoDA, PFTrDA, PFTeDA, PFOS	A/J mice	3 g/week, 10 weeks	liver	• 95 and 72 differential metabolites in male and female mice, respectively, and 54 metabolites were present in both sexes • PC 38:4, PC 38:5, PC 38:6, TG 52:5, TG 56:5, and PE 38:6; PC 34:2 differed in both sexes; TG 54:3 and TG 56:6 differed in females; PC 36:4 and TG 56:8 differed in males • 4 fatty acids, alpha-linolenic acid, DHA, DPA, and FA 20:4, were significantly different in both sexes	Khan EA et al., 2023 [73]
PFOA	C57BL/6 mice	1 ppm/kg, 40 ppb/kg, 4 weeks	Serum, liver	• In serum, 145 and 83 differential metabolites in high and low dose group, respectively, including TAG 62:13 and TAG 62:14, PC (32:2), PC (33:2), PC (35:4), LPC 20:3, and PI 36:4, etc. • In liver, 268 and 117 differential metabolites in the high- and low-dose group, respectively. PFOA increased unsaturated triglycerides and decreased sphingomyelin, saturated phosphatidylcholine, saturated lysophosphatidylcholine, and phospholipids ethers	Gao B et al., 2022 [108]

Table 3. *Cont.*

PFAS	Study Object	Dose	Sample Matrix	Main Finding(s)	Ref.
Mixture of POPs with six PFAS: PFHxS, PFOS, PFOA, PFNA, PFDA, PFUdA	NOD/SHiLtJ mice	0.14, 2.9 μg/kg, 12 weeks	serum	• Bile acid dysregulation, TDCA, and GDCA were negatively correlated with PFAS, LCA was positively correlated with PFAS, and most BAs were negatively correlated • LPC 18:2, LPC 20:4, LPC 22:5, and LPC 22:6 were negatively correlated with PFAS, whereas LPC 20:3 was positively correlated with PFAS; 48 PCs were correlated with PFAS • 3 and 19 polarity-differentiated metabolites in the low- and high-dose groups, respectively, and PFASs were negatively correlated with serine, threonine, glycerol-3-phosphate, palmitic acid, linoleic acid, and cholesterol; it was positively correlated with 3-hydroxybutyrate, fumaric acid, glutamate, and uric acid	Sinioja T et al., 2022 [82]
PFAS mixture, PFOA, PFOS, PFNA, PFHxS, GenX	C57BL/6J mice	10 mg/L, 12 weeks	plasma, liver	• PFAS exposure altered sterol, bile acid, ketone body, and acylcarnitine metabolites in the liver and circulatory system with inconsistent alterations • The hepatic sterol metabolites cholesterol sulfate and 7-HOCA were decreased, and 4-cholesten-3-one was increased; ketone body 3-BHBA was increased; most of the acylcarnitines were increased • Plasma levels of the sterol metabolites cholesterol and 7-HOCA were increased; metabolites of bile acids cholate, chenodeoxycholate, and deoxycholate were increased; ketone bodies 3-BHBA were increased; acylcarnitines C6 and C8 were decreased	Roth K et al., 2021 [71]

Table 3. *Cont.*

PFAS	Study Object	Dose	Sample Matrix	Main Finding(s)	Ref.
PFHxA	ICR mice	50, 200 mg/kg, 2 months	Serum, liver	• 30 and 49 differential metabolites in serum and liver, respectively, mainly interfering with purine and glutathione metabolism • Xanthine and uric acid in liver decreased. In serum, xanthine increased and uric acid level increased slightly with the increase in PFHxA	Jiang L et al., 2021 [66]
PFOS	BALB/c mouse	100, 1000 μg/kg, 2 months	liver	• 54 differential lipid metabolites, mainly glycerophospholipids, sphingolipids, and TGs • Eight PCs, eight LPCs, two ceramides, four phosphatidylinositols (PIs), four lysophosphatidylinositols (LPIs), and one lysophosphatidylethanolamine (LPE) were up-regulated, while nine TGs, eight sphingomyelins (SMs), three cardiolipins (CLs), two phosphatidylethanolamines (PEs), three phosphatidylglycerol (PG), and two coenzymes were downregulated.	Li X et al., 2021 [72]

HFPO-DA or "GenX", hexafluoropropylene oxide dimer acid; NBP2, nafion byproduct 2; 6:2 FTEOs, 6:2 fluorotelomer ethoxylates; PFDoDA, perflurododecanoic acid; PFTeDA, perfluorotetradecanoic acid; PFHxA, perfluorohexanoic acid.

The close association of PFAS exposure with the metabolism of lipids, amino acids, and bile acids is also well documented in metabolome-related studies in disease populations [62,93,97,105]. In studies of PFAS-related metabolites examined in children with NAFLD, in populations at high risk for diabetes, and in overweight and obese populations, it was found that amino acid and glycerophospholipid metabolisms were most relevant after PFAS exposure [62,93,105]. For example, the pathways found to be most affected after PFAS exposure in children with NAFLD include tyrosine metabolism, alanine and aspartate metabolism, glycine, serine, alanine and threonine metabolism, urea cycle metabolism, and glycerophospholipid metabolism [62]. Differential metabolites have been found in overweight and obese children, mainly associated with amino acids such as aspartate, tyrosine, arginine, and proline and lipid metabolic pathways such as sphingolipid metabolism, fatty acid metabolism, de novo lipogenesis, and linoleic acid metabolism [93]. In people at high risk of diabetes, PFASs are mainly associated with amino acid and glycerophospholipid metabolism, and PFASs are closely related to branched-chain amino acids (isoleucine, leucine, and valine) [105]. In addition, metabolites have been found to be closely associated with PFOS in HCC cases, mainly interfering with amino acids, carbohydrates, and glycan biosynthesis, and metabolism in relation to them [97].

Similarly, several animal studies utilizing blood samples have also analyzed blood metabolomics by PFAS exposure and found that PFAS exposure is also primarily associated with lipid and bile acid metabolites [66,71,82,108]. A comprehensive metabolomics study examining lipid, bile acid, and sterol metabolites found that PFAS exposure in both sexes increased plasma levels of sterol metabolites, bile acids, and ketone bodies, while acylcarnitine levels decreased [71]. Among them, bile acids, such as cholate (CA), chenodeoxycholate (CDCA), beta-muricholate, deoxycholate (DCA), taurodeoxycholate (TDCA), ursodeoxycholate (UDCA), tauroursodeoxycholate (TUDCA), and hyocholate, significantly increased, and cholesterol, 7-alpha-hydroxy-3-oxo-4-cholestenoate (7-HOCA), and 4-cholesten-3-one levels increased among sterol metabolites. Additionally, plasma levels of acylcarnitines, such as hexanoylcarnitine (C6) and octanoylcarnitine (C8), decreased. Similarly, PFAS exposure during pregnancy also dysregulated bile acid and lipid metabolism in mice, with taurine deoxycholic acid (TDCA) and glycodeoxycholic acid (GDCA) negatively correlating with PFAS, lithocholic acid (LCA) positively correlating with PFAS, and the majority of the BAs trending towards a negative correlation with PFASs [82]. In addition, a serum lipid metabolome of male mice found that PFOA altered 11 lipids, including PC and PI, at either high or low doses compared to controls [108]. Two other studies found that PFAS exposure also interfered with amino acid, fatty acid, purine, and glutathione metabolic pathways in the blood, with PFASs negatively correlating primarily with serine, threonine, aspartic acid, glycerol-3-phosphate, palmitic acid, linoleic acid, stearic acid, arachidonic acid, cholesterol, and 1,5-nonhydrohexanol [66,82]. PFASs positively correlated with 3-hydroxybutyric acid, fumaric acid, methionine, glutamic acid and gluconic acids, uric acid, and glycerol monostearate [82].

4.1.2. Liver

The impact of PFAS exposure on liver metabolites primarily manifests in disruptions to bile acid and lipid metabolism, as well as effects on amino acid, purine, and glutathione metabolism. Human studies on liver metabolomics are limited, likely due to the difficulty in obtaining liver samples. However, one study involving liver biopsies from individuals with NAFLD undergoing laparoscopic surgery analyzed the relationship between PFAS exposure levels in blood and liver metabolites, finding that PFAS exposure was primarily associated with disruptions in liver bile acid and lipid metabolic pathways [63]. In individuals with NAFLD, studies have demonstrated gender-specific associations between PFAS exposure and liver metabolites. In females, serum concentrations of PFASs (PFNA, PFOA, and PFOS) positively correlated with primary liver bile acids (TCA, GCDCA, TCDCA), while PFOA was positively associated with various secondary bile acids (DCA, GHCA, and GUDCA). Additionally, PFOA and PFOS were positively correlated with ceramides (e.g., Cer(d18:0/16:0), HexCer(d18:1/18:0)), ether phospholipids (e.g., O-PC(40:4)), TGs), and diglycerides (DGs). These associations were not observed in males. For both sexes, shared metabolic impacts included overexpression of primary bile acid biosynthesis, glycerophospholipid metabolism, and alanine, aspartate, and glutamate metabolism. Similar lipid associations were observed in mice, where PFOA exposure was linked to Cer, hexosylceramides(HexCer), LPE, DG, and TG. Most studies linking PFAS exposure and liver metabolomics focus on rodent models, particularly targeting lipid metabolism. Three animal studies have all found that PFAS exposure leads to alterations in the hepatic metabolome and is primarily associated with lipid metabolism [72,73,106]. For example, Kirkwood-Donelson KI et al. studied the effects of NBP2 or GenX exposure on the hepatic metabolome and found that over half of the detected lipids were significantly dysregulated, with oleic acid (OA) and dihomo-γ-linoleic acid (DGLA) being the most commonly affected fatty acids [106]. Khan et al. found that, similar to the blood metabolome, lipid differential metabolites were dominated by eight different lipid major classes, including Cer, DG, PC, PE, PS, SM, sterols, and TG, with common lipid species across genders including PC 38:4, PC 38:5, PC 38:6, TG 52:5, TG 56:5, and PE 38:6 [73]. In addition, further studies on fatty acids revealed that four fatty acids, alpha-linolenic acid, DHA, DPA, and FA 20:4, were significantly different between males and females. Li et al. analyzed PFAS exposure and lipid metabolites and found that 54 lipids had differentiated lipid metabolites and that the major abnormal lipids were glycerophospholipids, sphingolipids, and TG [72]. In addition, in studies analyzing the metabolism of sterols, ketone bodies, and acylcarnitines in the liver, it was found that sterol metabolites in the liver mainly showed a decrease, ketone bodies an increase, and most of the acylcarnitines an increase [71]. In addition, two animal studies have found that PFAS exposure is associated with amino acid, purine, and glutathione metabolism [66,94]. Among them, Yu et al. found that more metabolites were found in the negative ion pattern and more differential metabolites were up-regulated after PFAS exposure. These metabolites were enriched in pathways such as α-linolenic acid metabolism, linoleic acid metabolism, purine metabolism, glycine, serine, and threonine metabolism, steroid biosynthesis, glycolysis/gluconeogenesis, and glucagon signaling pathways [94]. Jiang et al. found that PFHxA exposure was also associated with purine and glutathione metabolism [66].

4.1.3. Urine and Placenta

Limited research has examined metabolite changes in urine and placenta. He et al. found that PFAS concentrations in urine closely correlated with those in serum among workers and residents near a large fluorochemical plant in Hubei Province, China, suggesting that urinary PFAS could serve as a good indicator of serum PFAS levels [96]. Eight potential exposure biomarkers were identified, with differential metabolites associated with amino acids, steroids, fatty acid derivatives, carboxylic acids, isoprenoids, amines, and vitamin pathways. Adams et al. investigated placental metabolites in healthy CD-1 pregnant mice exposed to PFOA and FTEOs, identifying significant changes primarily in amino acids [107]. Fourteen different metabolites, including asparagine, fatty acids, glucose, threonine, lactate, lysine, and creatine, were identified. PFOA exposure led to increased glucose and threonine and decreased creatine in the placenta, whereas FTEO exposure decreased asparagine and lysine while increasing creatine. Pathway analysis indicated alterations in glycine, serine, and threonine metabolism, as well as valine, leucine, and isoleucine biosynthesis following PFOA exposure, with biotin metabolism affected by FTEO exposure.

4.2. *Effect Biomarkers*

There is a substantial body of research on the relationship between PFAS exposure and metabolites; however, studies that connect these metabolites to both PFAS exposure and biological effect indicators are limited, primarily focusing on changes in blood metabolites. To date, metabolomics research has enhanced our understanding of the comprehensive impact of PFAS exposure on human health and has identified associated metabolites and intermediary markers involved in the onset and progression of various diseases (e.g., liver diseases, glucose and lipid abnormalities, cancer). These metabolites may serve as potential biomarkers for the association between PFAS exposure and biological effects. By synthesizing recent metabolomic studies on PFAS exposure and toxicity, we can elucidate consistent metabolic response patterns to PFAS exposure, revealing common pathways and potential biomarkers affected in both human models and laboratory rodents, as detailed in Table 4.

Table 4. PFAS exposure intermediate metabolites associated with biological effects.

PFAS	Study Object	Sample Size	Toxic Effect	Sample Matrix	Main Finding(s)	Ref.
PFHxS, PFHpS, PFOA, PFNA, PFDA, PFOS, PFUdA, 6:2Cl-PFESA, 8:2Cl-PFESA	Residents of Guangzhou, China	278	Liver function indicators (APOB, GGT, DBIL)	serum	• 78, 58 46 metabolites mediate the association between PFASs and APOB, DBIL, and GGT, respectively • The main markers associated with APOB and PFASs were glycerophospholipids; the main marker associated with DBIL was 5-(14-nonadecenyl)-1,3-benzenediol; the main markers associated with GGT were (R)-dihydromaleimide, Ile Leu, (R)-(+)-2-pyrrolidone-5-carboxylic acid, and L-glutamate	Chen Y et al., 2024 [90]
PFOS, PFHxS, PFOA, PFNA	Pregnant African American Newborns in Atlanta, Georgia	267	Premature labor and gestational age at delivery	serum	• 56 metabolites were associated with both PFASs and birth outcomes • Amino acids and protein metabolites were the most abundant, and were associated with PFNA and PFHxS and gestational age and PTB; Carnosine and bile acids were associated with PFASs and gestational age; polyunsaturated omega-6 fatty acids, arachidonic and linoleic acids, and tetradecenoyl levocarnosine were correlated with PFHxS and gestational week; indole-3-acetic acid methyl ester, benzoic acid, and s -lactic acid glutathione were associated with PFASs and premature delivery and gestational age	Taibl KR et al., 2023 [76]
PFHxS, PFOA, PFHpS, PFNA, PFOS, 6:2 Cl-PFESA, PFDA, PFUdA	Male residents of Guangzhou	278	Dyslipidemia (TC and LDL)	plasma	• 105 and 82 metabolites mediated the association between PFASs and TC and LDL, respectively • Most glycerophospholipids were positively associated with PFASs and TC • Important markers were 24-Hydroxycholesterol, 3alpha,7alpha-dihydroxy-5beta-cholestan-26-al, PC(18:0/0:0), PC(22:5/0:0), GPCho(18:1/18:1), LysoPC(22:2(13Z,16Z)), LysoPC(16:0), 9(S)-HODE, 9,10-DHOME, L-glutamate, 4-hydroxybutyric acid, cytosine, PC(14:1(9Z)/18:0), sphinganine, and (S)-beta- aminoisobutyrate	Chen Y et al., 2023 [90]

Table 4. *Cont.*

PFAS	Study Object	Sample Size	Toxic Effect	Sample Matrix	Main Finding(s)	Ref.
PFOA, PFOS, PFNA, PFUdA, PFDA, PFHxS, PFBS, PFDoA, PFHpA, PFOSA	A birth cohort study in Shanghai, China	1671	Obesity trajectory in children by age 4	plasma	• Increased odds of sustained growth trajectories (category 1) among three BMI z-score trajectories were associated with prenatal PFAS exposure • Octanoylcarnitine (C8) was the only mediator found in the sum of PFOS, PFNA, PFDA, a total of seven PFASs (PFOA, PFOS, PFNA, PFDA, PFUdA, PFHxS, PFHpA), and BMI trajectory	Zeng X et al., 2023 [79]
23 PFASs, mainly PFHpA, PFOA, PFBS, PFPeS, PFHxS, PFHpS, PFOS, 6:2FTS	Workers and residents in Hubei, China	225	Renal function indicators (Urea, Creatine, Uric acid)	urine	• 8 differential metabolites were correlated with PFAS and renal function indicators, of which farnesyl pyrophosphate and taurine were negatively correlated with PFASs, and S-Inosyl-L-homocysteine,3α,12α-dihydroxy-5β-chol-6-en-24-oic acid, 4-hydroxyphenylacetic acid, pyridoxine, versiconal, and cholic acid were positively correlated with PFASs • S-Inosyl-L-homocysteine and farnesyl pyrophosphate had the highest correlations	He A et al., 2023 [96]
PFHxS, PFOS, PFOA, PFNA	A birth cohort of African American pregnant women	313	Decreased fetal growth	serum	• 10 metabolites overlapping with PFASs and fetal growth endpoints, including glycine, taurine, uric acid, ferulic acid, 2-hexyl-3-phenyl-2-primary acid, unsaturated fatty acid C18:1, androgen couplers, maternal bile acids, and bile acid-glycine couplers • Uric acid is a potential intermediate biomarker of serum PFOA and PFNA associated with reduced fetal growth	Chang CJ et al., 2022 [84]
PFOS, PFHxS, PFOA, PFDA, PFNA, PFUdA	A nested case-control study	50 pairs	Non-viral HCC risk	plasma	• High levels of PFOS were associated with increased risk of HCC • Four metabolites that were positively associated with both PFOS exposure and HCC risk were glucose, butyric acid, α-ketoisovaleric acid, and 7α-hydroxy-3-oxo-4-cholestenoate.	Goodrich JA et al., 2022 [97]

Table 4. *Cont.*

PFAS	Study Object	Sample Size	Toxic Effect	Sample Matrix	Main Finding(s)	Ref.
PFDA, PFNA, PFUdA	A nested case-control study	187	Cholesterol and triglycerides	plasma	• 50 metabolites were associated with PFASs and triglycerides, and a pattern of metabolites dominated by glycerophospholipids was associated with long-chain PFASs and negatively correlated with triglycerides • PFAS-related metabolites not associated with cholesterol	Schillemans T et al., 2023 [88]
PFOS, PFOA, PFNA, PFHxS, PFUdA	The Human Early-Life Exposome project	1105	Risk of liver injury in children	serum	• Liver injury risk associated with BCAAs (valine, leucine, and isoleucine), AAAs (tryptophan and phenylalanine), biogenic amine acetyl ornithine, and the glycerophospholipid phospholipids (PCs) aa C36:1 and Lyso-PC a C18:1	Stratakis N et al., 2020 [69]
PFOA, PFOS and PFHxS	Cross-sectional cohort of NAFLD	74	Histologic severity of NAFLD	plasma	• Differential metabolites were categorized into two potential clusters, and cluster 2 was associated with an increased incidence of NASH • Cluster 2 was positively associated with PFASs and plasma metabolite patterns alters, including elevated ethanolamine phosphate, tyrosine, phenylalanine, aspartate, and creatine, and decreased betaine	Jin R et al., 2020 [62]
PFOA, PFOA, PFHxS, PFDA, PFNA, PFUdA	a case-control study on T2D nested	187 pairs	Type 2 diabetes risk	plasma	• PFASs were positively associated with two model metabolites, but two model metabolites were inversely associated with T2D risk • 35 differential metabolites such as gamma-butyrobetaine, 3,4,5-trimethoxycinnamic acid, and DHA were associated with T2D, including mainly PC1, which is predominantly diacylglycerol metabolite, and PC2, which is predominantly glycerophospholipid	Schillemans T et al., 2021 [91]

Table 4. *Cont.*

PFAS	Study Object	Sample Size	Toxic Effect	Sample Matrix	Main Finding(s)	Ref.
PFOA, PFOS, PFHxS	Hispanic children in Los Angeles	40	glucose homeostasis	plasma	• Differential metabolites identified in two clusters, with higher concentrations of PFASs in children categorized as "high-risk" ("cluster 2") associated with elevated 2 h blood glucose levels • The "high risk" cluster was positively associated with PFASs and plasma metabolites related to palmitic acid, hydroperoxylated linoleic acid, tyrosine and phenylalanine and arginine, sphingomyelin and linoleic acid, and aspartic acid metabolism	Alderete TL et al., 2019 [93]
PFAS	**Study Object**	**Dose**	**Toxic Effect**	**Sample Matrix**	**Main Finding(s)**	**Ref.**
Eight PFAS mixtures (PFOA, PFNA, PFDA, PFUdA, PFDoDA, PFTrDA, PFTeDA, PFOS),	A/J mice	3 g/mouse/ week, 10 weeks	Hepatosomatic index (HSI)	liver	• HSI was associated with differential metabolites related to PFAS exposure • In males, eight metabolites were correlated with HSI, which was positively correlated with PC 36:4, PC 38:5, and TG 56:5, and negatively correlated with PC 38:6, PE 38:6, TG 52:5, TG 54:7, and TG 56:8; in females, two lipids were significantly correlated, with TG 54:3 and TG 52:5 being positively and negatively correlated with HSI, respectively.	Khan EA et al., 2023 [73]
PFOS	SD rats	0.03 and 0.3 mg/kg, GD1-18d	fasting glucose homeostasis	liver	• Key metabolites, such as glycerol 3-phosphate and lactose ceramide may be associated with changes in fasting blood glucose • Lactosylceramide (d18:1/12:0), glycerol 3-phosphate and 4-Hydroxy-5-(phenyl)-valeric acid-O-glucuronide, 20-Oxo-leukotriene E4, licoagrochalcone B, tobramycin, ajmaline, 4-Methyl-2-pentyl-1,3-dioxolane, 3b,17a, 21-Trihydroxypregnenone, etc., were strongly correlated with FBG	Yu G et al., 2023 [94]

4.2.1. Blood

Research on the association between PFASs and liver and glucose–lipid abnormalities has identified glycerophospholipids as major associated metabolites with potential as effect biomarkers. Four studies have all found that glycerophospholipids may be the mediating metabolite that mediates the association of PFAS exposure with liver function or dyslipidemia [62,70,88,90]. A mediator analysis of metabolites mediating the association of PFAS exposure with abnormalities in APOB, an index of liver function, showed that glycerophospholipids were the main markers for the association of PFASs with APOB [70]. PFASs were also found to be predominantly associated with glycerophospholipid and amino acid metabolism in a study of the histologic severity of nonalcoholic fatty liver disease, in which differential metabolites were divided into two potential clusters, and cluster 2 was found to be associated with an increased incidence of NASH and higher PFAS concentrations [62]. In dyslipidemia studies, most glycerophospholipids were positively associated with PFASs and were associated with an increased risk of TC abnormalities, and the metabolites that played a mediating role mainly involved glycerophospholipid metabolism, linoleic acid metabolism, and primitive bile acid biosynthesis, in which 24-Hydroxycholesterol, 3alpha,7alpha-dihydroxy-5beta-cholestan-26-al, PC(18:0/0:0), PC(22:5/0:0), GPCho(18:1/18:1), LPC(22:2(13Z,16Z)), LPC(16:0), 9(S)-HODE, 9,10-DHOME, L-glutamate, 4-hydroxybutyric acid, cytosine, PC(14:1(9Z)/18:0), sphinganine, and (S)-beta-aminoisobutyrate were important markers [90]. Similarly, in studies related to plasma cholesterol and triglycerides, a metabolite pattern predominantly involving glycerophospholipids mediated the association between PFASs and triglycerides, showing a negative correlation with triglycerides [88]. In HCC risk studies, high PFOS levels were associated with increased HCC risk, with glucose, butyric acid, α-ketoisovaleric acid, and 7α-hydroxy-3-oxo-4-cholestenoate identified as potential mediators of the association between PFOS exposure and increased HCC risk [97]. The possible mechanism linking PFASs to increased liver cancer risk is through alterations in glucose, amino acid, and bile acid metabolism. In type 2 diabetes (T2D) risk studies, PFASs were found to be positively associated with two metabolite patterns, but these patterns had opposite relationships with T2D risk [91]. The metabolite pattern PC2, predominantly involving glycerophospholipids, was associated with reduced T2D risk, whereas the metabolite pattern PC1, primarily involving diacylglycerols, was associated with increased T2D risk. In studies related to changes in glucose homeostasis, children with high PFAS concentrations were found to have elevated 2 h glucose levels at baseline and during follow-up [93]. Further metabolomic analysis divided differential metabolites into two potential clusters, with the "high-risk" cluster positively correlated with PFASs, mainly involving the metabolism of palmitic acid, hydroperoxylinoleic acid, tyrosine, phenylalanine, arginine, sphingolipids, linoleic acid, and aspartic acid.

In the study of intermediary metabolites associated with PFASs and reproductive development, it was found that PFASs are primarily associated with the metabolism of amino acids and lipids and identified Octanoylcarnitine (C8) and uric acid as potential intermediary biomarkers [76,79,84]. In studies examining maternal PFAS exposure and childhood obesity trajectories before age four, three BMI z-score trajectories in early childhood were identified, in which an increased likelihood of a persistently increasing trajectory type was associated with prenatal PFAS exposure [79]. Further analysis identified octanoylcarnitine (C8) as the sole mediator between PFAS exposure and childhood BMI trajectory. In a study of maternal PFAS exposure associated with gestational age and PTB, PFASs and gestational age and PTB were found to be primarily involved in the pathways of glycerophospholipid metabolism, amino acid metabolism in the urea cycle, and tryptophan metabolism [76]. In another study on PFAS exposure and fetal growth in pregnant women, 10 metabolites, such as glycine, taurine, uric acid, ferulic acid, and the unsaturated fatty acid C18:1, were identified as overlapping with the endpoints of PFASs and fetal growth. Further analysis indicated that uric acid may be a potential intermediate biomarker representing an early response to PFAS exposure and predicting reduced fetal growth [84]. In addition, in studies on prenatal PFAS exposure and childhood liver injury risk, an increase in prenatal PFAS exposure was associated with elevated serum levels of branched-chain amino acids (BCAAs: valine, leucine, and isoleucine), aromatic amino acids (AAAs: tryptophan and phenylalanine), biogenic amine acetylornithine, glycerophospholipid PC aa C36:1, and Lyso-PC a C18:1 in children at high risk for liver injury compared to those at low risk [69].

4.2.2. Liver

Metabolomic studies investigating the association between PFAS exposure and biological effect indicators in the liver primarily focus on animal models, revealing that liver damage and disruptions in glucose homeostasis are primarily attributed to alterations in hepatic lipid metabolism. In a study examining liver lipid metabolism in relation to hepatic steatosis index, significant correlations were observed in males, with eight lipids showing notable associations with HSI [73]. Specifically, PC 36:4, PC 38:5, and TG 56:5 exhibited positive correlations with HSI, while PC 38:6, PE 38:6, TG 52:5, TG 54:7, and TG 56:8 were negatively correlated with HSI. In females, only two lipids demonstrated significant associations, with TG 54:3 positively correlated and TG 52:5 negatively correlated with HSI. A study on the impact of liver metabolites on maternal fasting blood glucose levels identified key metabolic products such as glycerol 3-phosphate and lactosylceramide (d18:1/12:0) that may be associated with changes in fasting blood glucose levels [94]. Additionally, metabolites including 20-Oxo-leukotriene E4, licoagrochalcone B, tobramycin, ajmaline, P,P-Dioctyldiphenylamine, 4-Methyl-2-pentyl-1,3-dioxolane, 3b,17a,21-Trihydroxypregnenone, and TG(20:5(5Z,8Z,11Z,14Z,17Z)/22:6(4Z,7Z,10Z,13Z,16Z,19Z)/o-18:0) were found to be associated with fasting blood glucose levels.

4.2.3. Urine

Population studies utilizing metabolomic analysis of urine revealed eight differential metabolites associated with PFAS exposure and biochemical indicators of kidney function, including urea, creatine, and uric acid [96]. Among these metabolites, farnesyl pyrophosphate and taurine were negatively correlated with PFAS exposure, while S-Inosyl-L-homocysteine, 3α,12α-Dihydroxy-5β-chol-6-en-24-oic acid, 4-hydroxyphenylacetic acid, pyridoxine, versiconal, and cholic acid showed positive correlations with PFAS exposure. Notably, S-Inosyl-L-homocysteine (amino acid class) and farnesyl pyrophosphate (isoprenoid class) exhibited the highest correlation.

5. Summary and Outlook

5.1. Summary

Exposure to PFASs has been associated with various biological abnormalities. The advancement of metabolomics technology has provided us with tools to develop an earlier and more comprehensive understanding of the impact of PFAS exposure on disease occurrence and progression mechanisms, as well as facilitating the identification of exposure and effect biomarkers. As illustrated in Figure 1, existing research indicates that PFASs primarily affect liver damage, reproductive toxicity, cardiovascular diseases, abnormal glucose metabolism, kidney diseases, and cancer by influencing the metabolism of amino acids, lipids, and bile acids. Several animal experiments have demonstrated a dose–response relationship of PFASs to alterations in lipid and amino acid metabolites, such as in the metabolome of 0.03 and 0.3 mg/kg PFOS-exposed pregnant rats, where a dose–response relationship was also found in bile secretion, glycine, serine and threonine metabolism, and linoleic acid metabolism, as well as a dose–response relationship in the liver transcriptome in terms of alterations in bile secretion, valine, leucine and isoleucine biosynthesis, and arachidonic acid metabolism [94]. Gao et al. observed an effect of PFOA on lipids, such as unsaturated triglycerides, sphingomyelins, saturated phosphatidylcholines, phospholipid ethers, and other lipids, in a dose–response relationship [108]. In addition, reports of BA metabolism in animal models and in vivo studies have demonstrated that PFAS exposure inhibits de novo synthesis of BA via inhibition of cholesterol 7a-hydroxylase (CYP7A1) and that subsequent inhibition of CYP7A1 leads to the down-regulation of primary BAs (CA, CDCA) [63,71]. Current studies suggest that PFASs primarily induce liver damage by disrupting bile acid, amino acid, and lipid metabolism, reproductive toxicity by perturbing amino acid, lipid, fatty acid, glycerophospholipid, bile acid, uric acid, and carbohydrate metabolism, and lipid and cholesterol abnormalities by disturbing bile acid and cholesterol metabolism. Additionally, PFASs can also influence the occurrence and development of glucose metabolism, kidney diseases, and cancer through metabolic regulation. Although the specific metabolites disrupted by toxicity in different tissues and organs may vary, there are common metabolic disturbances, such as associations with amino acids, lipids, and bile acids observed in almost all abnormalities. Changes in various metabolites in blood, liver, and urine were associated with exposure to PFASs, which offers a reservoir of metabolites that may be associated with PFAS exposure. Some unique metabolites or a combination of multiple metabolites could be potential biomarkers upon further validation. Reviewing the past literature reveals that metabolites associated with PFASs primarily include lipids, amino acids, bile acids, steroids, and acylcarnitines, with changes in blood metabolites dominating the research. Glycerophospholipids show promising potential as effect biomarkers in studies related to PFAS exposure and toxic effects. In addition, we have some limitations with this review in that the alterations in metabolites such as lipids, amino acids, and bile acids identified by this study may not be unique to PFASs alone; however, a comprehensive characterization of the metabolite changes would provide clues to the underlying mechanistic alterations that underlie these adverse outcomes. Further research is needed for a more specific investigation into lipid and amino acid metabolism. Moreover, due to differences in sample matrix and physiological pathological states, different biomarkers may be identified, emphasizing the need for further analysis of specific biomarkers under specific physiological conditions.

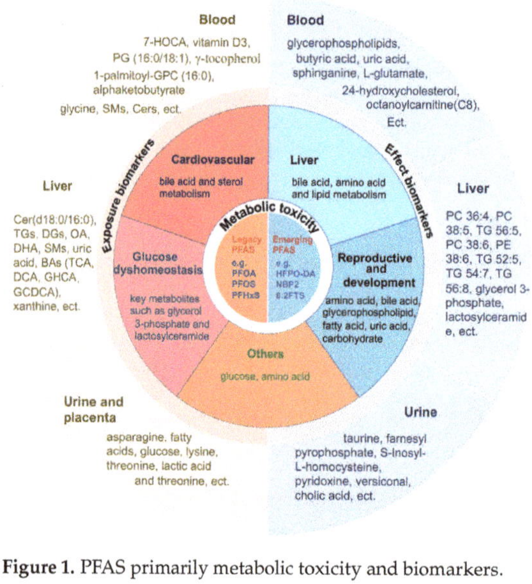

Figure 1. PFAS primarily metabolic toxicity and biomarkers.

5.2. Outlook

Current research on PFAS exposure primarily focuses on legacy PFAS compounds, with a predominant emphasis on individual PFAS studies. Some studies have indicated that disturbances in metabolomic profiles caused by PFAS mixtures may be more pronounced than those induced by single PFAS compounds. However, the composite exposure characteristics of perfluoroalkyl substances and their effects on internal metabolites remain unclear. This significantly hampers the accuracy of risk assessments for populations with high exposure to perfluoroalkyl substances. Therefore, there is a need to fill the data gap in metabolomic studies of novel PFAS substitutes and PFAS mixtures. It has also been observed that the toxicity varies among different monomers. Even with higher accumulation levels in the liver, the disruptive effects of their metabolites may not necessarily be greater [85,106]. Studies have highlighted that the novel PFAS substitute NBP2 exhibits developmental toxicity in rats, with oral toxicity slightly lower than PFOS but stronger than HFPO-DA. Comparing the toxicity of different PFAS monomers is crucial for identifying substitutes with lower toxicity. Thus, further research is needed to compare the toxicity of different novel PFAS compounds.

Existing research on the association between PFAS exposure and metabolomics mainly focuses on changes in metabolites associated with PFAS exposure to explore underlying mechanisms. However, there is a lack of research on exposure biomarkers and metabolomic markers that further mediate these effects.

Research on the biological effects of PFASs primarily targets the general population, with limited studies on the effects of exposure on pregnant women, infants, toddlers, and children. Additionally, studies have found higher detection rates among occupational populations near factories, with differential metabolites associated with oxidative stress, impaired fatty acid β-oxidation, and kidney damage. The impact of occupational exposure on worker health cannot be overlooked, highlighting the need for further research on populations with high exposure risks [100].

With the advancement of omics technologies, transcriptomics, metagenomics, and proteomics have become valuable tools for studying the biological effects of PFAS exposure. However, current research lacks comprehensive multi-level investigations combining metabolomics with other omics approaches to elucidate the mechanisms underlying the occurrence and development of PFAS-related effects. There is a need to integrate various omics techniques to provide a more comprehensive understanding of the impact of PFAS exposure and to validate metabolomics findings across multiple levels using other omics methodologies.

Author Contributions: Conceptualization: X.M., Q.C. and Z.M.; literature search: X.M., D.C., Z.Z., S.Z. and Z.W.; writing—original draft: X.M., D.C., Z.M., H.S., Z.H. and Q.C.; Writing—review and editing: all authors. All authors have read and agreed to the published version of the manuscript.

Funding: This research received no external funding.

Institutional Review Board Statement: Not applicable.

Informed Consent Statement: Not applicable.

Data Availability Statement: No new data were created or analyzed in this study.

Conflicts of Interest: The authors declare no conflicts of interest.

References

1. Buck, R.C.; Franklin, J.; Berger, U.; Conder, J.M.; Cousins, I.T.; de Voogt, P.; Jensen, A.A.; Kannan, K.; Mabury, S.A.; van Leeuwen, S.P. Perfluoroalkyl and polyfluoroalkyl substances in the environment: Terminology, classification, and origins. *Integr. Environ. Assess. Manag.* **2011**, *7*, 513–541. [CrossRef] [PubMed]
2. Gaines, L.G.T. Historical and current usage of per- and polyfluoroalkyl substances (PFAS): A literature review. *Am. J. Ind. Med.* **2023**, *66*, 353–378. [CrossRef] [PubMed]
3. Thijs, M.; Laletas, E.; Quinn, C.M.; Raguraman, S.V.; Carr, B.; Bierganns, P. Total and Class-Specific Determination of Fluorinated Compounds in Consumer and Food Packaging Samples Using Fluorine-19 Solid-State Nuclear Magnetic Resonance Spectroscopy. *Anal. Chem.* **2024**, *96*, 8282–8290. [CrossRef] [PubMed]
4. Casal, P.; Zhang, Y.; Martin, J.W.; Pizarro, M.; Jiménez, B.; Dachs, J. Role of Snow Deposition of Perfluoroalkylated Substances at Coastal Livingston Island (Maritime Antarctica). *Environ. Sci. Technol.* **2017**, *51*, 8460–8470. [CrossRef] [PubMed]
5. Kowalczyk, J.; Ehlers, S.; Oberhausen, A.; Tischer, M.; Fürst, P.; Schafft, H.; Lahrssen-Wiederholt, M. Absorption, distribution, and milk secretion of the perfluoroalkyl acids PFBS, PFHxS, PFOS, and PFOA by dairy cows fed naturally contaminated feed. *J. Agric. Food Chem.* **2013**, *61*, 2903–2912. [CrossRef] [PubMed]

6. Miner, K.R.; Clifford, H.; Taruscio, T.; Potocki, M.; Solomon, G.; Ritari, M.; Napper, I.E.; Gajurel, A.P.; Mayewski, P.A. Deposition of PFAS 'forever chemicals' on Mt. Everest. *Sci. Total Environ.* **2021**, *759*, 144421. [CrossRef] [PubMed]
7. Chen, J.; Zheng, L.; Tian, L.; Wang, N.; Lei, L.; Wang, Y.; Dong, Q.; Huang, C.; Yang, D. Chronic PFOS Exposure Disrupts Thyroid Structure and Function in Zebrafish. *Bull. Environ. Contam. Toxicol.* **2018**, *101*, 75–79. [CrossRef] [PubMed]
8. Butenhoff, J.L.; Chang, S.C.; Olsen, G.W.; Thomford, P.J. Chronic dietary toxicity and carcinogenicity study with potassium perfluorooctanesulfonate in Sprague Dawley rats. *Toxicology* **2012**, *293*, 1–15. [CrossRef]
9. Wang, Z.; DeWitt, J.C.; Higgins, C.P.; Cousins, I.T. A Never-Ending Story of Per- and Polyfluoroalkyl Substances (PFASs)? *Environ. Sci. Technol.* **2017**, *51*, 2508–2518. [CrossRef]
10. Lindstrom, A.B.; Strynar, M.J.; Libelo, E.L. Polyfluorinated compounds: Past, present, and future. *Environ. Sci. Technol.* **2011**, *45*, 7954–7961. [CrossRef]
11. Xu, C.; Jiang, Z.Y.; Liu, Q.; Liu, H.; Gu, A. Estrogen receptor beta mediates hepatotoxicity induced by perfluorooctane sulfonate in mouse. *Environ. Sci. Pollut. Res. Int.* **2017**, *24*, 13414–13423. [CrossRef]
12. Sun, P.; Nie, X.; Chen, X.; Yin, L.; Luo, J.; Sun, L.; Wan, C.; Jiang, S. Nrf2 Signaling Elicits a Neuroprotective Role against PFOS-mediated Oxidative Damage and Apoptosis. *Neurochem. Res.* **2018**, *43*, 2446–2459. [CrossRef]
13. Soloff, A.C.; Wolf, B.J.; White, N.D.; Muir, D.; Courtney, S.; Hardiman, G.; Bossart, G.D.; Fair, P.A. Environmental perfluorooctane sulfonate exposure drives T cell activation in bottlenose dolphins. *J. Appl. Toxicol.* **2017**, *37*, 1108–1116. [CrossRef] [PubMed]
14. Wang, X.; Bai, Y.; Tang, C.; Cao, X.; Chang, F.; Chen, L. Impact of Perfluorooctane Sulfonate on Reproductive Ability of Female Mice through Suppression of Estrogen Receptor α-Activated Kisspeptin Neurons. *Toxicol. Sci.* **2018**, *165*, 475–486. [CrossRef]
15. Liu, C.; Chang, V.W.; Gin, K.Y.; Nguyen, V.T. Genotoxicity of perfluorinated chemicals (PFCs) to the green mussel (*Perna viridis*). *Sci. Total Environ.* **2014**, *487*, 117–122. [CrossRef] [PubMed]
16. Toskos, T.; Panagiotakis, I.; Dermatas, D. Per- and polyfluoroalkyl substances—Challenges associated with a family of ubiquitous emergent contaminants. *Waste Manag. Res.* **2019**, *37*, 449–451. [CrossRef]
17. Pezzatti, J.; Boccard, J.; Codesido, S.; Gagnebin, Y.; Joshi, A.; Picard, D.; González-Ruiz, V.; Rudaz, S. Implementation of liquid chromatography-high resolution mass spectrometry methods for untargeted metabolomic analyses of biological samples: A tutorial. *Anal. Chim. Acta* **2020**, *1105*, 28–44. [CrossRef]
18. Gobelius, L.; Glimstedt, L.; Olsson, J.; Wiberg, K.; Ahrens, L. Mass flow of per- and polyfluoroalkyl substances (PFAS) in a Swedish municipal wastewater network and wastewater treatment plant. *Chemosphere* **2023**, *336*, 139182. [CrossRef] [PubMed]
19. Abafe, O.A.; Macheka, L.R.; Abafe, O.T.; Chokwe, T.B. Concentrations and human exposure assessment of per and polyfluoroalkyl substances in farmed marine shellfish in South Africa. *Chemosphere* **2021**, *281*, 130985. [CrossRef]
20. Gebbink, W.A.; van Leeuwen, S.P.J. Environmental contamination and human exposure to PFASs near a fluorochemical production plant: Review of historic and current PFOA and GenX contamination in the Netherlands. *Environ. Int.* **2020**, *137*, 105583. [CrossRef]
21. Katz, D.R.; Sullivan, J.C.; Rosa, K.; Gardiner, C.L.; Robuck, A.R.; Lohmann, R.; Kincaid, C.; Cantwell, M.G. Transport and fate of aqueous film forming foam in an urban estuary. *Environ. Pollut.* **2022**, *300*, 118963. [CrossRef] [PubMed]
22. Augustsson, A.; Lennqvist, T.; Osbeck, C.M.G.; Tibblin, P.; Glynn, A.; Nguyen, M.A.; Westberg, E.; Vestergren, R. Consumption of freshwater fish: A variable but significant risk factor for PFOS exposure. *Environ. Res.* **2021**, *192*, 110284. [CrossRef] [PubMed]
23. Ragnarsdóttir, O.; Abdallah, M.A.; Harrad, S. Dermal uptake: An important pathway of human exposure to perfluoroalkyl substances? *Environ. Pollut.* **2022**, *307*, 119478. [CrossRef] [PubMed]
24. Eun, H.; Yamazaki, E.; Taniyasu, S.; Miecznikowska, A.; Falandysz, J.; Yamashita, N. Evaluation of perfluoroalkyl substances in field-cultivated vegetables. *Chemosphere* **2020**, *239*, 124750. [CrossRef] [PubMed]
25. Gazzotti, T.; Sirri, F.; Ghelli, E.; Zironi, E.; Zampiga, M.; Pagliuca, G. Perfluoroalkyl contaminants in eggs from backyard chickens reared in Italy. *Food Chem.* **2021**, *362*, 130178. [CrossRef] [PubMed]
26. Sharp, S.; Sardiña, P.; Metzeling, L.; McKenzie, R.; Leahy, P.; Menkhorst, P.; Hinwood, A. Per- and Polyfluoroalkyl Substances in Ducks and the Relationship with Concentrations in Water, Sediment, and Soil. *Environ. Toxicol. Chem.* **2021**, *40*, 846–858. [CrossRef]
27. Macheka, L.R.; Olowoyo, J.O.; Mugivhisa, L.L.; Abafe, O.A. Determination and assessment of human dietary intake of per and polyfluoroalkyl substances in retail dairy milk and infant formula from South Africa. *Sci. Total Environ.* **2021**, *755*, 142697. [CrossRef] [PubMed]
28. Monge Brenes, A.L.; Curtzwiler, G.; Dixon, P.; Harrata, K.; Talbert, J.; Vorst, K. PFOA and PFOS levels in microwave paper packaging between 2005 and 2018. *Food Addit. Contam. Part B Surveill.* **2019**, *12*, 191–198. [CrossRef] [PubMed]
29. Toptancı, İ.; Ketenoglu, O.; Kıralan, M. Assessment of the migration of perfluorinated compounds and primary aromatic amines from PTFE-coated non-stick cookware marketed in Turkey. *Environ. Sci. Pollut. Res. Int.* **2022**, *29*, 38535–38549. [CrossRef]
30. Cara, B.; Lies, T.; Thimo, G.; Robin, L.; Lieven, B. Bioaccumulation and trophic transfer of perfluorinated alkyl substances (PFAS) in marine biota from the Belgian North Sea: Distribution and human health risk implications. *Environ. Pollut.* **2022**, *311*, 119907. [CrossRef]
31. Bolan, N.; Sarkar, B.; Vithanage, M.; Singh, G.; Tsang, D.C.W.; Mukhopadhyay, R.; Ramadass, K.; Vinu, A.; Sun, Y.; Ramanayaka, S.; et al. Distribution, behaviour, bioavailability and remediation of poly- and per-fluoroalkyl substances (PFAS) in solid biowastes and biowaste-treated soil. *Environ. Int.* **2021**, *155*, 106600. [CrossRef] [PubMed]

32. Xu, B.; Qiu, W.; Du, J.; Wan, Z.; Zhou, J.L.; Chen, H.; Liu, R.; Magnuson, J.T.; Zheng, C. Translocation, bioaccumulation, and distribution of perfluoroalkyl and polyfluoroalkyl substances (PFASs) in plants. *iScience* **2022**, *25*, 104061. [CrossRef] [PubMed]
33. Zhou, Y.; Zhou, Z.; Lian, Y.; Sun, X.; Wu, Y.; Qiao, L.; Wang, M. Source, transportation, bioaccumulation, distribution and food risk assessment of perfluorinated alkyl substances in vegetables: A review. *Food Chem.* **2021**, *349*, 129137. [CrossRef]
34. Li, Y.; Yao, J.; Zhang, J.; Pan, Y.; Dai, J.; Ji, C.; Tang, J. First Report on the Bioaccumulation and Trophic Transfer of Perfluoroalkyl Ether Carboxylic Acids in Estuarine Food Web. *Environ. Sci. Technol.* **2022**, *56*, 6046–6055. [CrossRef] [PubMed]
35. Munoz, G.; Mercier, L.; Duy, S.V.; Liu, J.; Sauvé, S.; Houde, M. Bioaccumulation and trophic magnification of emerging and legacy per- and polyfluoroalkyl substances (PFAS) in a St. Lawrence River food web. *Environ. Pollut.* **2022**, *309*, 119739. [CrossRef]
36. Göckener, B.; Eichhorn, M.; Lämmer, R.; Kotthoff, M.; Kowalczyk, J.; Numata, J.; Schafft, H.; Lahrssen-Wiederholt, M.; Bücking, M. Transfer of Per- and Polyfluoroalkyl Substances (PFAS) from Feed into the Eggs of Laying Hens. Part 1: Analytical Results Including a Modified Total Oxidizable Precursor Assay. *J. Agric. Food Chem.* **2020**, *68*, 12527–12538. [CrossRef]
37. Lupton, S.J.; Smith, D.J.; Scholljegerdes, E.; Ivey, S.; Young, W.; Genualdi, S.; DeJager, L.; Snyder, A.; Esteban, E.; Johnston, J.J. Plasma and Skin Per- and Polyfluoroalkyl Substance (PFAS) Levels in Dairy Cattle with Lifetime Exposures to PFAS-Contaminated Drinking Water and Feed. *J. Agric. Food Chem.* **2022**, *70*, 15945–15954. [CrossRef] [PubMed]
38. Aker, A.; Ayotte, P.; Caron-Beaudoin, E.; De Silva, A.; Ricard, S.; Gaudreau, É.; Lemire, M. Plasma concentrations of perfluoroalkyl acids and their determinants in youth and adults from Nunavik, Canada. *Chemosphere* **2023**, *310*, 136797. [CrossRef]
39. Chen, F.; Yin, S.; Kelly, B.C.; Liu, W. Chlorinated Polyfluoroalkyl Ether Sulfonic Acids in Matched Maternal, Cord, and Placenta Samples: A Study of Transplacental Transfer. *Environ. Sci. Technol.* **2017**, *51*, 6387–6394. [CrossRef]
40. Pan, Y.; Zhang, H.; Cui, Q.; Sheng, N.; Yeung, L.W.Y.; Guo, Y.; Sun, Y.; Dai, J. First Report on the Occurrence and Bioaccumulation of Hexafluoropropylene Oxide Trimer Acid: An Emerging Concern. *Environ. Sci. Technol.* **2017**, *51*, 9553–9560. [CrossRef]
41. Pérez, F.; Nadal, M.; Navarro-Ortega, A.; Fàbrega, F.; Domingo, J.L.; Barceló, D.; Farré, M. Accumulation of perfluoroalkyl substances in human tissues. *Environ. Int.* **2013**, *59*, 354–362. [CrossRef] [PubMed]
42. Hall, S.M.; Zhang, S.; Hoffman, K.; Miranda, M.L.; Stapleton, H.M. Concentrations of per- and polyfluoroalkyl substances (PFAS) in human placental tissues and associations with birth outcomes. *Chemosphere* **2022**, *295*, 133873. [CrossRef] [PubMed]
43. Onteeru, M.; Barnes, L.E.; O'Connell, K.; Bhimani, J.; Du, M.; Romano, M.E.; Kantor, E.D. Association between fish oil supplements use and serum per- and polyfluoroalkyl substances (PFAS): Results from the National Health and Nutrition Examination Survey. *Environ. Res.* **2022**, *215*, 114205. [CrossRef]
44. Yu, C.H.; Riker, C.D.; Lu, S.E.; Fan, Z.T. Biomonitoring of emerging contaminants, perfluoroalkyl and polyfluoroalkyl substances (PFAS), in New Jersey adults in 2016–2018. *Int. J. Hyg. Environ. Health* **2020**, *223*, 34–44. [CrossRef]
45. Richterová, D.; Govarts, E.; Fábelová, L.; Rausová, K.; Rodriguez Martin, L.; Gilles, L.; Remy, S.; Colles, A.; Rambaud, L.; Riou, M.; et al. PFAS levels and determinants of variability in exposure in European teenagers—Results from the HBM4EU aligned studies (2014–2021). *Int. J. Hyg. Environ. Health* **2023**, *247*, 114057. [CrossRef] [PubMed]
46. Bartolomé, M.; Gallego-Picó, A.; Cutanda, F.; Huetos, O.; Esteban, M.; Pérez-Gómez, B.; Castaño, A. Perfluorinated alkyl substances in Spanish adults: Geographical distribution and determinants of exposure. *Sci. Total Environ.* **2017**, *603–604*, 352–360. [CrossRef] [PubMed]
47. Li, J.; Cai, D.; Chu, C.; Li, Q.; Zhou, Y.; Hu, L.; Yang, B.; Dong, G.; Zeng, X.; Chen, D. Transplacental Transfer of Per- and Polyfluoroalkyl Substances (PFASs): Differences between Preterm and Full-Term Deliveries and Associations with Placental Transporter mRNA Expression. *Environ. Sci. Technol.* **2020**, *54*, 5062–5070. [CrossRef] [PubMed]
48. Mamsen, L.S.; Jönsson, B.A.G.; Lindh, C.H.; Olesen, R.H.; Larsen, A.; Ernst, E.; Kelsey, T.W.; Andersen, C.Y. Concentration of perfluorinated compounds and cotinine in human foetal organs, placenta, and maternal plasma. *Sci. Total Environ.* **2017**, *596–597*, 97–105. [CrossRef]
49. Liu, Y.; Zhang, Q.; Li, Y.; Hao, Y.; Li, J.; Zhang, L.; Wang, P.; Yin, Y.; Zhang, S.; Li, T.; et al. Occurrence of per- and polyfluoroalkyl substances (PFASs) in raw milk and feed from nine Chinese provinces and human exposure risk assessment. *Chemosphere* **2022**, *300*, 134521. [CrossRef]
50. Norén, E.; Lindh, C.; Glynn, A.; Rylander, L.; Pineda, D.; Nielsen, C. Temporal trends, 2000–2017, of perfluoroalkyl acid (PFAA) concentrations in serum of Swedish adolescents. *Environ. Int.* **2021**, *155*, 106716. [CrossRef]
51. Li, Y.; Lu, X.; Yu, N.; Li, A.; Zhuang, T.; Du, L.; Tang, S.; Shi, W.; Yu, H.; Song, M.; et al. Exposure to legacy and novel perfluoroalkyl substance disturbs the metabolic homeostasis in pregnant women and fetuses: A metabolome-wide association study. *Environ. Int.* **2021**, *156*, 106627. [CrossRef] [PubMed]
52. Chu, C.; Zhou, Y.; Li, Q.Q.; Bloom, M.S.; Lin, S.; Yu, Y.J.; Chen, D.; Yu, H.Y.; Hu, L.W.; Yang, B.Y.; et al. Are perfluorooctane sulfonate alternatives safer? New insights from a birth cohort study. *Environ. Int.* **2020**, *135*, 105365. [CrossRef]
53. Choi, G.W.; Choi, E.J.; Kim, J.H.; Kang, D.W.; Lee, Y.B.; Cho, H.Y. Gender differences in pharmacokinetics of perfluoropentanoic acid using non-linear mixed-effect modeling in rats. *Arch. Toxicol.* **2020**, *94*, 1601–1612. [CrossRef] [PubMed]
54. Li, J.; Wang, L.; Zhang, X.; Liu, P.; Deji, Z.; Xing, Y.; Zhou, Y.; Lin, X.; Huang, Z. Per- and polyfluoroalkyl substances exposure and its influence on the intestinal barrier: An overview on the advances. *Sci. Total Environ.* **2022**, *852*, 158362. [CrossRef] [PubMed]
55. Zhao, L.; Teng, M.; Zhao, X.; Li, Y.; Sun, J.; Zhao, W.; Ruan, Y.; Leung, K.M.Y.; Wu, F. Insight into the binding model of per- and polyfluoroalkyl substances to proteins and membranes. *Environ. Int.* **2023**, *175*, 107951. [CrossRef] [PubMed]

56. Cao, H.; Zhou, Z.; Hu, Z.; Wei, C.; Li, J.; Wang, L.; Liu, G.; Zhang, J.; Wang, Y.; Wang, T.; et al. Effect of Enterohepatic Circulation on the Accumulation of Per- and Polyfluoroalkyl Substances: Evidence from Experimental and Computational Studies. *Environ. Sci. Technol.* **2022**, *56*, 3214–3224. [CrossRef] [PubMed]
57. Kim, S.J.; Choi, E.J.; Choi, G.W.; Lee, Y.B.; Cho, H.Y. Exploring sex differences in human health risk assessment for PFNA and PFDA using a PBPK model. *Arch. Toxicol.* **2019**, *93*, 311–330. [CrossRef] [PubMed]
58. Pizzurro, D.M.; Seeley, M.; Kerper, L.E.; Beck, B.D. Interspecies differences in perfluoroalkyl substances (PFAS) toxicokinetics and application to health-based criteria. *Regul. Toxicol. Pharmacol.* **2019**, *106*, 239–250. [CrossRef] [PubMed]
59. Dzierlenga, A.L.; Robinson, V.G.; Waidyanatha, S.; DeVito, M.J.; Eifrid, M.A.; Gibbs, S.T.; Granville, C.A.; Blystone, C.R. Toxicokinetics of perfluorohexanoic acid (PFHxA), perfluorooctanoic acid (PFOA) and perfluorodecanoic acid (PFDA) in male and female Hsd:Sprague dawley SD rats following intravenous or gavage administration. *Xenobiotica* **2020**, *50*, 722–732. [CrossRef]
60. Chang, S.C.; Das, K.; Ehresman, D.J.; Ellefson, M.E.; Gorman, G.S.; Hart, J.A.; Noker, P.E.; Tan, Y.M.; Lieder, P.H.; Lau, C.; et al. Comparative pharmacokinetics of perfluorobutyrate in rats, mice, monkeys, and humans and relevance to human exposure via drinking water. *Toxicol. Sci.* **2008**, *104*, 40–53. [CrossRef]
61. Attanasio, R. Sex differences in the association between perfluoroalkyl acids and liver function in US adolescents: Analyses of NHANES 2013–2016. *Environ. Pollut.* **2019**, *254*, 113061. [CrossRef] [PubMed]
62. Jin, R.; McConnell, R.; Catherine, C.; Xu, S.; Walker, D.I.; Stratakis, N.; Jones, D.P.; Miller, G.W.; Peng, C.; Conti, D.V.; et al. Perfluoroalkyl substances and severity of nonalcoholic fatty liver in Children: An untargeted metabolomics approach. *Environ. Int.* **2020**, *134*, 105220. [CrossRef] [PubMed]
63. Sen, P.; Qadri, S.; Luukkonen, P.K.; Ragnarsdottir, O.; McGlinchey, A.; Jäntti, S.; Juuti, A.; Arola, J.; Schlezinger, J.J.; Webster, T.F.; et al. Exposure to environmental contaminants is associated with altered hepatic lipid metabolism in non-alcoholic fatty liver disease. *J. Hepatol.* **2022**, *76*, 283–293. [CrossRef] [PubMed]
64. Wu, Z.; Ouyang, T.; Liu, H.; Cao, L.; Chen, W. Perfluoroalkyl substance (PFAS) exposure and risk of nonalcoholic fatty liver disease in the elderly: Results from NHANES 2003–2014. *Environ. Sci. Pollut. Res. Int.* **2023**, *30*, 64342–64351. [CrossRef] [PubMed]
65. Chen, M.; Zhu, L.; Wang, Q.; Shan, G. Tissue distribution and bioaccumulation of legacy and emerging per-and polyfluoroalkyl substances (PFASs) in edible fishes from Taihu Lake, China. *Environ. Pollut.* **2021**, *268*, 115887. [CrossRef] [PubMed]
66. Jiang, L.; Hong, Y.; Xie, G.; Zhang, J.; Zhang, H.; Cai, Z. Comprehensive multi-omics approaches reveal the hepatotoxic mechanism of perfluorohexanoic acid (PFHxA) in mice. *Sci. Total Environ.* **2021**, *790*, 148160. [CrossRef] [PubMed]
67. Wang, G.; Pan, R.; Liang, X.; Wu, X.; Wu, Y.; Zhang, H.; Zhao, J.; Chen, W. Perfluorooctanoic acid-induced liver injury is potentially associated with gut microbiota dysbiosis. *Chemosphere* **2021**, *266*, 129004. [CrossRef]
68. Jiang, L.; Hong, Y.; Xiao, P.; Wang, X.; Zhang, J.; Liu, E.; Li, H.; Cai, Z. The Role of Fecal Microbiota in Liver Toxicity Induced by Perfluorooctane Sulfonate in Male and Female Mice. *Environ. Health Perspect.* **2022**, *130*, 67009. [CrossRef] [PubMed]
69. Stratakis, N.; Conti, D.V.; Jin, R.; Margetaki, K.; Valvi, D.; Siskos, A.P.; Maitre, L.; Garcia, E.; Varo, N.; Zhao, Y.; et al. Prenatal Exposure to Perfluoroalkyl Substances Associated with Increased Susceptibility to Liver Injury in Children. *Hepatology* **2020**, *72*, 1758–1770. [CrossRef]
70. Chen, Y.; Wu, Y.; Lv, J.; Zhou, S.; Lin, S.; Huang, S.; Zheng, L.; Deng, G.; Feng, Y.; Zhang, G.; et al. Overall and individual associations between per- and polyfluoroalkyl substances and liver function indices and the metabolic mechanism. *Environ. Int.* **2024**, *183*, 108405. [CrossRef]
71. Roth, K.; Yang, Z.; Agarwal, M.; Liu, W.; Peng, Z.; Long, Z.; Birbeck, J.; Westrick, J.; Liu, W.; Petriello, M.C. Exposure to a mixture of legacy, alternative, and replacement per- and polyfluoroalkyl substances (PFAS) results in sex-dependent modulation of cholesterol metabolism and liver injury. *Environ. Int.* **2021**, *157*, 106843. [CrossRef] [PubMed]
72. Li, X.; Li, T.; Wang, Z.; Wei, J.; Liu, J.; Zhang, Y.; Zhao, Z. Distribution of perfluorooctane sulfonate in mice and its effect on liver lipidomic. *Talanta* **2021**, *226*, 122150. [CrossRef] [PubMed]
73. Khan, E.A.; Grønnestad, R.; Krøkje, Å.; Bartosov, Z.; Johanson, S.M.; Müller, M.H.B.; Arukwe, A. Alteration of hepato-lipidomic homeostasis in A/J mice fed an environmentally relevant PFAS mixture. *Environ. Int.* **2023**, *173*, 107838. [CrossRef] [PubMed]
74. Hærvig, K.K.; Petersen, K.U.; Hougaard, K.S.; Lindh, C.; Ramlau-Hansen, C.H.; Toft, G.; Giwercman, A.; Høyer, B.B.; Flachs, E.M.; Bonde, J.P.; et al. Maternal Exposure to Per- and Polyfluoroalkyl Substances (PFAS) and Male Reproductive Function in Young Adulthood: Combined Exposure to Seven PFAS. *Environ. Health Perspect.* **2022**, *130*, 107001. [CrossRef]
75. Rickard, B.P.; Rizvi, I.; Fenton, S.E. Per- and poly-fluoroalkyl substances (PFAS) and female reproductive outcomes: PFAS elimination, endocrine-mediated effects, and disease. *Toxicology* **2022**, *465*, 153031. [CrossRef]
76. `Taibl, K.R.; Dunlop, A.L.; Barr, D.B.; Li, Y.Y.; Eick, S.M.; Kannan, K.; Ryan, P.B.; Schroder, M.; Rushing, B.; Fennell, T.; et al. Newborn metabolomic signatures of maternal per- and polyfluoroalkyl substance exposure and reduced length of gestation. *Nat. Commun.* **2023**, *14*, 3120. [CrossRef]
77. Ouidir, M.; Buck Louis, G.M.; Kanner, J.; Grantz, K.L.; Zhang, C.; Sundaram, R.; Rahman, M.L.; Lee, S.; Kannan, K.; Tekola-Ayele, F.; et al. Association of Maternal Exposure to Persistent Organic Pollutants in Early Pregnancy with Fetal Growth. *JAMA Pediatr.* **2020**, *174*, 149–161. [CrossRef] [PubMed]
78. Zheng, T.; Kelsey, K.; Zhu, C.; Pennell, K.D.; Yao, Q.; Manz, K.E.; Zheng, Y.F.; Braun, J.M.; Liu, Y.; Papandonatos, G.; et al. Adverse birth outcomes related to concentrations of per- and polyfluoroalkyl substances (PFAS) in maternal blood collected from pregnant women in 1960–1966. *Environ. Res.* **2024**, *241*, 117010. [CrossRef]

79. Zeng, X.; Chen, T.; Cui, Y.; Zhao, J.; Chen, Q.; Yu, Z.; Zhang, Y.; Han, L.; Chen, Y.; Zhang, J. In utero exposure to perfluoroalkyl substances and early childhood BMI trajectories: A mediation analysis with neonatal metabolic profiles. *Sci. Total Environ.* **2023**, *867*, 161504. [CrossRef]
80. Zhuchen, H.Y.; Wang, J.Y.; Liu, X.S.; Shi, Y.W. Research Progress on Neurodevelopmental Toxicity in Offspring after Indirect Exposure to PFASs in Early Life. *Toxics* **2023**, *11*, 571. [CrossRef]
81. Liu, D.; Yan, S.; Liu, Y.; Chen, Q.; Ren, S. Association of prenatal exposure to perfluorinated and polyfluoroalkyl substances with childhood neurodevelopment: A systematic review and meta-analysis. *Ecotoxicol. Environ. Saf.* **2024**, *271*, 115939. [CrossRef] [PubMed]
82. Sinioja, T.; Bodin, J.; Duberg, D.; Dirven, H.; Berntsen, H.F.; Zimmer, K.; Nygaard, U.C.; Orešič, M.; Hyötyläinen, T. Exposure to persistent organic pollutants alters the serum metabolome in non-obese diabetic mice. *Metabolomics* **2022**, *18*, 87. [CrossRef] [PubMed]
83. Luo, K.; Liu, X.; Nian, M.; Wang, Y.; Qiu, J.; Yu, H.; Chen, X.; Zhang, J. Environmental exposure to per- and polyfluoroalkyl substances mixture and male reproductive hormones. *Environ. Int.* **2021**, *152*, 106496. [CrossRef] [PubMed]
84. Chang, C.J.; Barr, D.B.; Ryan, P.B.; Panuwet, P.; Smarr, M.M.; Liu, K.; Kannan, K.; Yakimavets, V.; Tan, Y.; Ly, V.; et al. Per- and polyfluoroalkyl substance (PFAS) exposure, maternal metabolomic perturbation, and fetal growth in African American women: A meet-in-the-middle approach. *Environ. Int.* **2022**, *158*, 106964. [CrossRef]
85. Conley, J.M.; Lambright, C.S.; Evans, N.; Medlock-Kakaley, E.; Hill, D.; McCord, J.; Strynar, M.J.; Wehmas, L.C.; Hester, S.; MacMillan, D.K.; et al. Developmental toxicity of Nafion byproduct 2 (NBP2) in the Sprague-Dawley rat with comparisons to hexafluoropropylene oxide-dimer acid (HFPO-DA or GenX) and perfluorooctane sulfonate (PFOS). *Environ. Int.* **2022**, *160*, 107056. [CrossRef]
86. Schillemans, T.; Yan, Y.; Ribbenstedt, A.; Donat-Vargas, C.; Lindh, C.H.; Kiviranta, H.; Rantakokko, P.; Wolk, A.; Landberg, R.; Åkesson, A.; et al. OMICs Signatures Linking Persistent Organic Pollutants to Cardiovascular Disease in the Swedish Mammography Cohort. *Environ. Sci. Technol.* **2024**, *58*, 1036–1047. [CrossRef] [PubMed]
87. Mi, X.; Wu, L.Y.; Liu, J.J.; Fang, Q.L.; Qian, Z.M.; Chu, C.; Li, Q.Q.; Su, F.; Zhang, Y.T.; Zhou, P.; et al. The effects of Cl-PFESAs exposure on blood lipids—A community-based large population study in Guangzhou. *Sci. Total Environ.* **2022**, *806*, 150634. [CrossRef] [PubMed]
88. Schillemans, T.; Bergdahl, I.A.; Hanhineva, K.; Shi, L.; Donat-Vargas, C.; Koponen, J.; Kiviranta, H.; Landberg, R.; Åkesson, A.; Brunius, C. Associations of PFAS-related plasma metabolites with cholesterol and triglyceride concentrations. *Environ. Res.* **2023**, *216*, 114570. [CrossRef]
89. Chen, Z.; Yang, T.; Walker, D.I.; Thomas, D.C.; Qiu, C.; Chatzi, L.; Alderete, T.L.; Kim, J.S.; Conti, D.V.; Breton, C.V.; et al. Dysregulated lipid and fatty acid metabolism link perfluoroalkyl substances exposure and impaired glucose metabolism in young adults. *Environ. Int.* **2020**, *145*, 106091. [CrossRef]
90. Chen, Y.; Lv, J.; Fu, L.; Wu, Y.; Zhou, S.; Liu, S.; Zheng, L.; Feng, W.; Zhang, L. Metabolome-wide association study of four groups of persistent organic pollutants and abnormal blood lipids. *Environ. Int.* **2023**, *173*, 107817. [CrossRef]
91. Schillemans, T.; Shi, L.; Donat-Vargas, C.; Hanhineva, K.; Tornevi, A.; Johansson, I.; Koponen, J.; Kiviranta, H.; Rolandsson, O.; Bergdahl, I.A.; et al. Plasma metabolites associated with exposure to perfluoroalkyl substances and risk of type 2 diabetes—A nested case-control study. *Environ. Int.* **2021**, *146*, 106180. [CrossRef] [PubMed]
92. Yang, A.; Tam, C.H.T.; Wong, K.K.; Ozaki, R.; Lowe, W.L., Jr.; Metzger, B.E.; Chow, E.; Tam, W.H.; Wong, C.K.C.; Ma, R.C.W. Epidemic-specific association of maternal exposure to per- and polyfluoroalkyl substances (PFAS) and their components with maternal glucose metabolism: A cross-sectional analysis in a birth cohort from Hong Kong. *Sci. Total Environ.* **2024**, *917*, 170220. [CrossRef] [PubMed]
93. Alderete, T.L.; Jin, R.; Walker, D.I.; Valvi, D.; Chen, Z.; Jones, D.P.; Peng, C.; Gilliland, F.D.; Berhane, K.; Conti, D.V.; et al. Perfluoroalkyl substances, metabolomic profiling, and alterations in glucose homeostasis among overweight and obese Hispanic children: A proof-of-concept analysis. *Environ. Int.* **2019**, *126*, 445–453. [CrossRef] [PubMed]
94. Yu, G.; Wang, J.; Liu, Y.; Luo, T.; Meng, X.; Zhang, R.; Huang, B.; Sun, Y.; Zhang, J. Metabolic perturbations in pregnant rats exposed to low-dose perfluorooctanesulfonic acid: An integrated multi-omics analysis. *Environ. Int.* **2023**, *173*, 107851. [CrossRef] [PubMed]
95. Stanifer, J.W.; Stapleton, H.M.; Souma, T.; Wittmer, A.; Zhao, X.; Boulware, L.E. Perfluorinated Chemicals as Emerging Environmental Threats to Kidney Health: A Scoping Review. *Clin. J. Am. Soc. Nephrol.* **2018**, *13*, 1479–1492. [CrossRef] [PubMed]
96. He, A.; Li, J.; Li, Z.; Lu, Y.; Liang, Y.; Zhou, Z.; Man, Z.; Lv, J.; Wang, Y.; Jiang, G. Novel Insights into the Adverse Health Effects of per- and Polyfluoroalkyl Substances on the Kidney via Human Urine Metabolomics. *Environ. Sci. Technol.* **2023**, *57*, 16244–16254. [CrossRef] [PubMed]
97. Goodrich, J.A.; Walker, D.; Lin, X.; Wang, H.; Lim, T.; McConnell, R.; Conti, D.V.; Chatzi, L.; Setiawan, V.W. Exposure to perfluoroalkyl substances and risk of hepatocellular carcinoma in a multiethnic cohort. *JHEP Rep.* **2022**, *4*, 100550. [CrossRef] [PubMed]
98. Yu, M.; Teitelbaum, S.L.; Dolios, G.; Dang, L.T.; Tu, P.; Wolff, M.S.; Petrick, L.M. Molecular Gatekeeper Discovery: Workflow for Linking Multiple Exposure Biomarkers to Metabolomics. *Environ. Sci. Technol.* **2022**, *56*, 6162–6171. [CrossRef] [PubMed]

99. Guo, P.; Furnary, T.; Vasiliou, V.; Yan, Q.; Nyhan, K.; Jones, D.P.; Johnson, C.H.; Liew, Z. Non-targeted metabolomics and associations with per- and polyfluoroalkyl substances (PFAS) exposure in humans: A scoping review. *Environ. Int.* **2022**, *162*, 107159. [CrossRef]
100. Yang, Y.; Wang, X.; Yang, M.; Wei, S.; Li, Y. Integrated Analysis of Per- and Polyfluoroalkyl Substance Exposure and Metabolic Profiling of Elderly Residents Living near Industrial Plants. *Environ. Sci. Technol.* **2024**, *58*, 4104–4114. [CrossRef]
101. Prince, N.; Begum, S.; Mínguez-Alarcón, L.; Génard-Walton, M.; Huang, M.; Soeteman, D.I.; Wheelock, C.; Litonjua, A.A.; Weiss, S.T.; Kelly, R.S.; et al. Plasma concentrations of per- and polyfluoroalkyl substances are associated with perturbations in lipid and amino acid metabolism. *Chemosphere* **2023**, *324*, 138228. [CrossRef] [PubMed]
102. Sinisalu, L.; Yeung, L.W.Y.; Wang, J.; Pan, Y.; Dai, J.; Hyötyläinen, T. Prenatal exposure to poly-/per-fluoroalkyl substances is associated with alteration of lipid profiles in cord-blood. *Metabolomics* **2021**, *17*, 103. [CrossRef]
103. Kingsley, S.L.; Walker, D.I.; Calafat, A.M.; Chen, A.; Papandonatos, G.D.; Xu, Y.; Jones, D.P.; Lanphear, B.P.; Pennell, K.D.; Braun, J.M. Metabolomics of childhood exposure to perfluoroalkyl substances: A cross-sectional study. *Metabolomics* **2019**, *15*, 95. [CrossRef] [PubMed]
104. Goodrich, J.A.; Walker, D.I.; He, J.; Lin, X.; Baumert, B.O.; Hu, X.; Alderete, T.L.; Chen, Z.; Valvi, D.; Fuentes, Z.C.; et al. Metabolic Signatures of Youth Exposure to Mixtures of Per- and Polyfluoroalkyl Substances: A Multi-Cohort Study. *Environ. Health Perspect.* **2023**, *131*, 27005. [CrossRef]
105. Mitro, S.D.; Liu, J.; Jaacks, L.M.; Fleisch, A.F.; Williams, P.L.; Knowler, W.C.; Laferrère, B.; Perng, W.; Bray, G.A.; Wallia, A.; et al. Per- and polyfluoroalkyl substance plasma concentrations and metabolomic markers of type 2 diabetes in the Diabetes Prevention Program trial. *Int. J. Hyg. Environ. Health* **2021**, *232*, 113680. [CrossRef]
106. Kirkwood-Donelson, K.I.; Chappel, J.; Tobin, E.; Dodds, J.N.; Reif, D.M.; DeWitt, J.C.; Baker, E.S. Investigating mouse hepatic lipidome dysregulation following exposure to emerging per- and polyfluoroalkyl substances (PFAS). *Chemosphere* **2024**, *354*, 141654. [CrossRef] [PubMed]
107. Adams, H.; Hanrahan, J.; Kiefte, S.; O'Brien, T.; Mercer, G.V.; Steeves, K.L.; Schneider, C.M.; Jobst, K.J.; Cahill, L.S. Differential impact of perfluorooctanoic acid and fluorotelomer ethoxylates on placental metabolism in mice. *Chemosphere* **2024**, *356*, 141923. [CrossRef]
108. Gao, B.; Tu, P.; Chi, L.; Shen, W.; Gao, N. Perfluorooctanoic Acid-Disturbed Serum and Liver Lipidome in C57BL/6 Mice. *Chem. Res. Toxicol.* **2022**, *35*, 2252–2259. [CrossRef]

metabolites

Review

Deciphering the Role of the Gut Microbiota in Exposure to Emerging Contaminants and Diabetes: A Review

Xueqing Li [1,†], Huixia Niu [1,†], Zhengliang Huang [2], Man Zhang [3], Mingluan Xing [1], Zhijian Chen [1], Lizhi Wu [1] and Peiwei Xu [1,*]

1 Zhejiang Provincial Center for Disease Control and Prevention, 3399 Bin Sheng Rd., Binjiang District, Hangzhou 310051, China; xqli@cdc.zj.cn (X.L.); 2111101061@nbu.edu.cn (H.N.); mlxing@cdc.zj.cn (M.X.); zhjchen@cdc.zj.cn (Z.C.); lzhwu@cdc.zj.cn (L.W.)
2 Disease Prevention and Control Center of Jingning She Autonomous County, Lishui 323500, China; hzljmx12688@163.com
3 School of Public Health, Zhejiang Chinese Medical University, Hangzhou 310053, China; ggwszm15726626270@163.com
* Correspondence: pwxu@cdc.zj.cn
† These authors contributed equally to this work.

Abstract: Emerging pollutants, a category of compounds currently not regulated or inadequately regulated by law, have recently become a focal point of research due to their potential toxic effects on human health. The gut microbiota plays a pivotal role in human health; it is particularly susceptible to disruption and alteration upon exposure to a range of toxic environmental chemicals, including emerging contaminants. The disturbance of the gut microbiome caused by environmental pollutants may represent a mechanism through which environmental chemicals exert their toxic effects, a mechanism that is garnering increasing attention. However, the discussion on the toxic link between emerging pollutants and glucose metabolism remains insufficiently explored. This review aims to establish a connection between emerging pollutants and glucose metabolism through the gut microbiota, delving into the toxic impacts of these pollutants on glucose metabolism and the potential role played by the gut microbiota.

Keywords: emerging contaminants; emerging pollutants; gut microbiome; gut microbiota; glucose metabolism; diabetes

1. Introduction

The gut microbiota, due to performing myriad vital functions within human health and being closely intertwined with human health and disease, has attracted increasing attention over the past decades. With deeper research into the gut microbiota, a broad range of functions have been recognized, encompassing carbohydrate digestion, the synthesis of vitamins and other nutrients, and the regulation of the immune system [1,2]. Given its crucial role in human health, the gut microbiota is considered a novel organ within the human body [3]. However, the structure and composition of the gut microbiota are highly susceptible to external compounds, leading to gut microbiota dysbiosis [4,5]. For example, exposure to 1–4 μm polystyrene microplastics for seven days led to significant differences between the microplastic-treated group and the control group in the Shannon and Simpson indices, with notable changes in the abundance of Bacteroidetes, Firmicutes, Proteobacteria, and Verrucomicrobia [6]. Alterations in the gut microbiota may impact host health through metabolic changes. The disruption of the gut microbiota caused by exogenous pollutants has been termed 'gut microbiome toxicity' [7]. Therefore, changes in the abundance and functionality of the gut microbiota following exposure to exogenous pollutants may represent a potential mechanism for pollutant-induced toxicity.

Emerging pollutants, defined as compounds that naturally occur or are synthetically produced and detectable and potentially harmful to the environment, flora, fauna, and humans, yet are currently not or inadequately regulated by law [6,8,9], have become a hot topic in toxicological research in recent years. Importantly, numerous studies have explored the link between emerging pollutants and the gut microbiota, demonstrating that exposure to emerging pollutants can impact the structure and functionality of the gut microbiota, thereby posing potential health risks to the host. Concurrently, the gut microbiota has a strong relationship with host metabolism, especially glucose metabolism. For example, Wang et al. studied the relationship between gestational diabetes mellitus and intestinal flora by recruiting pregnant women with gestational diabetes mellitus and studying the changes in their intestinal flora, and they found that the changes in certain intestinal bacteria were significantly correlated with the oral glucose tolerance test [10].

It is well known that diabetes has become a serious public health problem. Diabetes remains the fifth leading cause of death globally, although scientists have made great efforts to treat diabetes and prolong the life of patients. Compared with people without diabetes, the risk of premature death is increased by 15% in patients with type 1 and type 2 diabetes, and life expectancy is reduced by 10 and 20 years, respectively [11]. According to a systematic analysis, diabetes is the sixth leading cause of disability [12], and the multiple serious complications during the course of the disease bring huge economic pressure to patients and society. According to the International Diabetes Federation statistics in 2017, the number of global diabetics has reached 425 million, and it is estimated that by 2045, the number of diabetics in the world will reach 783 million [13]. Diabetes is mainly a metabolic disease characterized by high blood sugar caused by genetic factors, environmental factors, and unhealthy lifestyles, including type 1 diabetes, type 2 diabetes, and gestational diabetes mellitus. Among them, type 1 diabetes is caused by the immune-mediated destruction of pancreatic β cells and absolute insulin deficiency, which occurs mostly in adolescents. Type 2 diabetes is mainly caused by insulin resistance, and studies have found that Chinese patients with type 2 diabetes account for more than 90% of all diabetics, ranking first in the world and showing a rapid upward trend [14]. When hyperglycemia occurs during pregnancy, it is called gestational diabetes mellitus. It is estimated that in 2019, there were 20.4 million women with hyperglycemia during pregnancy, and the incidence of gestational diabetes mellitus in China was 14.8% [15]. In addition, more studies have found that environmental factors play an important role in the occurrence and development of diabetes [16,17], and new pollutants are also being considered [18].Therefore, the connection among exposure to emerging pollutants, glucose metabolism, and the development of diabetes is an area worthy of further exploration. However, the relationship between emerging pollutant exposure and glucose metabolism has yet to be systematically summarized and discussed.

Therefore, this review aims to explore the role of the gut microbiota in the toxic effects of emerging pollutants on glucose metabolism via systematically reviewing the existing literature. On the one hand, it offers new perspectives for understanding the hazards associated with emerging pollutant exposure, providing new research directions, data support, and scientific rationale for addressing the public health issue of diabetes. On the other hand, this review can provide evidence support for future research into the toxicity of emerging pollutants on glucose metabolism, drawing attention to the relationship between emerging pollutants and diabetes, and laying a foundation for the prevention and understanding of the toxicity of emerging pollutants.

2. Emerging Pollutants

Emerging pollutants in the environment exhibit a broad spectrum and diverse origins. It is estimated that over 3000 different emerging pollutants have been detected in the environment [19]. The current literature categorizes these into four main groups: endocrine disruptors, perfluorinated compounds, microplastics, and antibiotics [20,21]. Compared to traditional environmental pollutants, emerging pollutants are characterized by (i) a degree of risk obscurity, due to their relatively recent identification and low environmental concentrations, making their short-term impacts less evident [22]; (ii) persistence in the environment, as they are difficult to metabolize and degrade and prone to bioaccumulation [23]; (iii) high toxicity to biological organisms, including humans, with many having endocrine-disrupting, carcinogenic, teratogenic, and mutagenic effects [24]; and (iv) challenges in management, owing to their vast variety and low concentrations, which complicates their detection and the understanding of their environmental and biological impacts [23,25]. The paucity of research on their environmental and biological harm, migration, and transformation mechanisms adds to the complexity of managing these pollutants. These characteristics have increasingly drawn attention to the need for in-depth research on emerging pollutants, providing a scientific foundation for their regulation and management (Figure 1).

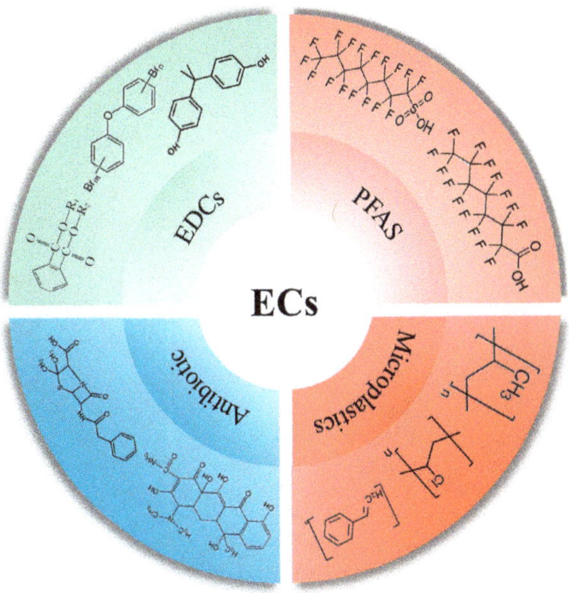

Figure 1. Typical emerging contaminants and representative compounds.

2.1. Sources of Exposure to Emerging Pollutants

Scientists generally agree that dietary exposure is a major source of human exposure to emerging pollutants, with a wide range of pollution sources that cannot be ignored. Dietary pollutants mainly enter higher trophic levels through food chain transmission and nutritional transfer. For instance, studies on microplastics have shown that they are initially absorbed by plants (algae) in the environment, then ingested by consumers (like water fleas and freshwater fish), and ultimately ingested by apex consumers, including humans. It is shown that consuming 250 g of wet-weight mussels can result in an intake of approximately 90 particles of microplastics, and microplastics have been found to accumulate in human tissues such as the blood, placenta, and heart [26–30]. Wang et al. [31] detected 11 types of perfluorinated compounds in various consumer products like pork loin, pig heart, liver, kidney, chicken breast, and liver, with pig liver showing the highest average concentration of 3.438 ng/g, followed by pig kidney (0.508 ng/g). These findings indicate that emerging pollutants can enter and accumulate in biological organisms, including humans, through food chains and nutritional transfer, posing potential health hazards.

Studies measuring the concentration of polychlorinated biphenyls in Crassostrea tulipa (oysters) and Anadara senilis (mussels) found concentrations of 2.95–11.41 mg/kg and 5.55–6.37 mg/kg wet weight, respectively [32]. Apart from marine life, polychlorinated biphenyls and polybrominated diphenyl ethers have also been detected in commonly consumed milk samples [33], indicating potential endocrine-disrupting hazards from everyday dietary exposure. Studies on perfluorinated compounds found that younger populations have higher levels of contact with PFOS and PFOA, with dietary exposure primarily from fish, meat, eggs, and products containing these ingredients, with eggs and dairy products as major sources [34]. PFOS has been detected in meat, fish, shellfish, and fast food [35], and Ericson et al. estimated that an adult male's dietary intake of PFOS could reach 62.5 ng/d [36]. In contrast, research on microplastics found that oral exposure is mainly concentrated in commercial fish, table salt, honey, and bottled water [37]. For instance, researchers found microplastics in various edible salts used by humans, with the highest concentrations in sea salt (550–681 particles/kg) [38]. Mason's research also found that bottled water contains an average concentration of microplastics larger than 100 μm and smaller than 100 μm, at 10.4 particles/L and 325 particles/L, respectively. Cauwenberghe et al. also found microplastics in mussels, estimating that the maximum exposure to microplastics for adults through mussel consumption could reach 11,000 microplastics [28]. Antibiotics, long established in human food supplies, have gradually contaminated food products, including livestock, aquatic products, and vegetables [39]. Known for their use in treating and preventing diseases, as well as additives for promoting growth and improving feed efficiency, antibiotics are likely major contributors to contamination in livestock and aquatic products [40,41]. For example, amoxicillin and penicillin were detected in 81% and 27% of fresh milk samples, respectively [42]. Furthermore, an investigative study involving the random sampling of chicken and beef from supermarkets revealed a considerable detection rate of antibiotics. Quinolone drugs were identified in 45.7% of chicken samples and 57.7% of beef samples, with concentrations reaching 30.8 ± 0.45 μg/kg and 6.64 ± 1.11 μg/kg, respectively [43].

2.2. Connection between Emerging Pollutants and the Gut Microbiome

Contemporary research findings have demonstrated that exposure to emerging environmental pollutants alters the structure and composition of the gut microbiota. For instance, scientists, in a study examining the impact of nanoplastics on the gut microbiome, exposed Eriocheir Sinensis to polystyrene nanoplastics. Compared to the control group, the exposed group exhibited significant changes in the gut microbiota's structure and composition, with a marked decrease in the relative abundance of Firmicutes and Bacteroidetes and an increase in Fusobacteria and Proteobacteria [44]. Another study found that mice exposed to 0.5 μm and 50 μm polystyrene microplastics showed significant alterations in the composition and structure of their gut microbiota, with a notable reduction in diversity and a decrease in the relative abundance of Firmicutes and α-Proteobacteria [45]. Additionally, research on PCB-126 revealed significant changes in the gut microbiota of exposed mice compared to the control group, with notable shifts in the proportion of Bacteroides, Parabacteroides, Romboutsia, and the ratio of Firmicutes to Bacteroidetes [46]. Recent studies in several species have found that Di-(2-ethylhexyl) phthalate (DEHP) exposure during development alters the structure and composition of the gut microbiota and reduces the diversity of the gut microbiota: an increase in the relative abundance of Firmicutes and Akkermansia was found, along with a decrease in the relative abundance of Bacteroidetes and Actinobacteria [47]. Furthermore, Yang et al. found that exposure to DEHP during infancy also altered the composition and diversity of the intestinal flora, with a decrease in the number of Rothia species and Bifidobacterium longum [48]. In their study of the effects of BPA on intestinal flora, Lai et al. found that the growth of TM7 and Proteobacteria was promoted in BPA-exposed mice, while the number of Clostridia was reduced [49]. Furthermore, studies have shown that gut bacteria such as Firmicutes and Bacteroides are significantly associated with type 2 diabetes [50].

3. Gut Microbiome, Diabetes and Potential Mechanisms

Diabetes, a metabolic disease characterized by abnormally elevated blood glucose levels, has become a significant global health issue due to its high prevalence and associated disability and mortality rates. In 2017, the estimated prevalence of diabetes in China was 12.8% [51], with the global diabetic population reaching 425 million people [52], over 90% of whom have type 2 diabetes [14], ranking it as the most prevalent and rapidly increasing trend worldwide. Genetic predisposition, environmental factors, and unhealthy lifestyles are closely linked to the onset and progression of diabetes. Moreover, recent studies have revealed an interesting phenomenon: the gut microbiome plays a crucial role in the development and progression of obesity and type 2 diabetes, with approximately 3.8 $\pm$ 0.2% of gut microbiota relative abundance associated with type 2 diabetes and obesity [50].

3.1. Diabetes and Gut Microbiota

In past diabetes research, the gut bacteria most commonly reported to be negatively associated with diabetes are Bifidobacterium and Bacteroides. The role of Bifidobacterium in type 2 diabetes appears to be consistently supported by the literature: Bifidobacterium potentially exerts a protective effect against type 2 diabetes [53,54]. For example, Gao et al. studied whether the composition and structure of the gut microbiota differed among healthy, overweight, and obese volunteers, finding a significant reduction in gut bacteria including Bifidobacterium, anti-inflammatory Faecalibacterium, and butyrate-producing Ruminococcaceae in the obese population compared to healthy individuals [55]. This phenomenon was validated in animal experiments: transplanting Bifidobacterium into mice on a high-fat diet, researchers found that Bifidobacterium not only reduced weight gain in mice but also significantly improved glucose–insulin disorder and hepatic steatosis, shifting the gut microbiota structure of high-fat diet mice towards that of normal-diet mice. Moreover, numerous cross-sectional studies have discovered a negative correlation between Bacteroides and type 2 diabetes. For instance, an analysis of the gut microbiota of 121 type 2 diabetes patients by Zhang et al. revealed dysbiosis and changes in alpha diversity, with a significant reduction in Bacteroides, only half the amount found in non-diabetic and pre-diabetic patients [56]. When mice were exposed to Bacteroides orally, researchers found that in high-fat-diet mice, not only were serum cholesterol, triglycerides, blood sugar, insulin, and leptin levels reduced, but their oral glucose tolerance was also improved [57].

Additionally, a few articles have reported gut bacteria positively correlated with diabetes or high blood sugar. Specifically, many studies have reported a positive correlation between Firmicutes, Ruminococcus, Lachnospiraceae, and Blautia with type 2 diabetes. For example, numerous population studies have indicated that both pre-diabetic and diabetic patients have relatively higher Operational Taxonomic Units (OTUs) of Ruminococcus. Furthermore, increases in the abundance of Sutterella, Streptococcus, Lachnospiraceae, Clostridiales, Eubacterium, Sporobacter, Abiotrophia, Firmicutes, and Subdoligranulum have been observed [58,59]. Moreover, type 2 diabetic patients have shown a notable decrease in the quantity of Ruminococcus and Lachnospiraceae following metformin treatment [60]. Some cross-sectional studies have found that compared with control groups, case groups have a marked increase in the quantity of Blautia, which diminishes after metformin treatment, thus affirming the significant role of Blautia in the development and progression of diabetes [56,61,62]. Additionally, researchers have discovered that compared with non-diabetic patients, the gut microbiome of diabetic patients is predominantly composed of opportunistic pathogens such as *Bacteroides caccae, Clostridium hathewayi, Clostridium ramosum, Clostridium symbiosum, Eggerthella lenta*, and *Escherichia coli* [50].

There is an evident link between emerging pollutants and abnormalities in glucose metabolism, with changes in the gut microbiota playing a significant role. However, it is important to note that not all studies on gut microbiota yield consistent conclusions. For instance, contrary to the studies mentioned above, Diamante, in his research on the impact of Bisphenol A on metabolic diseases, exposed pregnant mice to Bisphenol A and observed the metabolic phenotype of the offspring. The results indicated that male mice in the Bisphenol A exposure group had significantly lower insulin levels than those in the control group, and the area under the curve in the intraperitoneal glucose tolerance test was reduced. Diamante's correlation analysis of differentially abundant amplicon sequence variants (ASVs) with the metabolic phenotype of mice found 18 ASVs related to body weight and two related to the area under the glucose tolerance curve [63]. Additionally, population studies have found an increase in the relative abundance of Firmicutes in both pre-diabetic and diabetic patients [59], whereas a decrease in Firmicutes was observed in high-fat-diet-induced obese mice [49]. The inconsistencies across different studies could be attributed to the use of varied animal models, which might affect metabolic levels and microbiota. Furthermore, the structure and composition of the gut microbiota are closely related to numerous factors, and studying the impact of just one exogenous chemical exposure is quite limited. In addition, the experimental conditions and environmental exposure during the experiment might affect the results and the composition of the gut microbiota.

3.2. Potential Mechanisms of Gut Microbiota-Induced Glucose Metabolic Abnormalities

The gut microbiota impacts host metabolism through various pathways, including inflammatory responses, intestinal permeability, glucose metabolism, and the collective action of the microbiome. It is well established that certain gut bacteria and their metabolites can alter levels of pro-inflammatory and anti-inflammatory factors, as well as lipopolysaccharides (LPS), in the host. Given that inflammation and inflammatory mediators are closely linked with the development of type 2 diabetes, the influence of the gut microbiota on glucose metabolism through inflammatory responses has garnered widespread attention. For example, several studies have reported that patients with type 2 diabetes exhibit elevated serum endotoxin levels, a decrease in butyrate-producing bacteria and Firmicutes abundance, and an increase in Lactobacillus and Betaproteobacteria abundance. These findings suggest that certain gut bacteria may induce type 2 diabetes through endotoxin-induced inflammatory responses [50,64,65]. Chen et al., in their research on the role of Lactobacillus in diabetes progression, found that, besides significantly reduced fasting blood glucose and postprandial 2 h blood glucose levels, exposed mice showed decreased levels of pro-inflammatory cytokines such as TNF-α and LPS and increased levels of anti-inflammatory cytokines like IL-10 [66]. Another critical feature of type 2 diabetes is increased intestinal permeability, which allows gut microbiota and their metabolites to enter the bloodstream. Chelakkot et al. [67] found that, compared to diabetic patients, healthy individuals had a higher number of extracellular vesicles from *Akkermansia muciniphila* (AmEVs) in their fecal samples, which also enhanced the function of tight junction proteins in the intestines of diabetic mice. To verify their direct effect, Chelakkot applied AmEVs to lipopolysaccharide-treated Caco-2 cells and observed improved cellular permeability and increased the expression of the occludin protein. These findings suggest that AmEVs can regulate the expression of key proteins and affect intestinal permeability functions. Importantly, the gut microbiota might also alter blood glucose levels by affecting glucose homeostasis and insulin resistance in primary metabolic organs like the liver, muscles, and adipose tissue. Dang et al. found that after treating diabetic mice with *Lactobacillus paracasei*, not only were the mice's fasting blood glucose, postprandial blood glucose, and glucose tolerance adjusted, but the expression of genes associated with gluconeogenesis, such as G-6-Pase and PEPCK, was inhibited, and levels of IRS-2, PI3K, and Akt were increased to normal. Additionally, the treatment of diabetic mice with L. casei CCFM419 effectively improved the downregulated mRNA expression levels of PI3K and GS and significantly decreased the expression of the GSK3β gene [68]. Besides these potential mechanisms, some gut bacteria might also affect host physiology through interactions with other bacteria [69,70] (Figure 2).

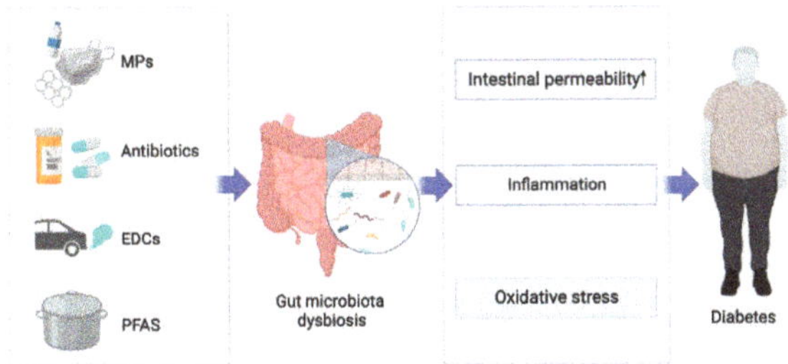

Figure 2. Possible mechanisms of exposure to emerging pollutants affecting diabetes through gut microbiota dysbiosis.

4. Emerging Pollutants, Gut Microbiome, and Diabetes

Given the known role of the gut microbiome in the development and progression of diabetes, coupled with the association between emerging pollutant exposure and changes in the gut microbiome, it is plausible that the disruption of glucose metabolism via the gut microbiome may be a potential mechanism by which emerging pollutants contribute to diabetes. This section will discuss, category by category—microplastics, antibiotics, endocrine disruptors, and perfluorinated compounds—their impacts on glucose metabolism and the role and function of the gut microbiome therein.

4.1. Microplastics

The concept of microplastics, defined as plastic particles with a diameter of $\leq$5 mm, was formally introduced by Thompson et al. in 2004 [71]. Since then, scientific research on microplastics present in the environment has become increasingly extensive. Numerous studies have found that plastics in the environment, upon degradation through physical, chemical, and biological processes into microplastic particles, pose toxicological threats to the natural environment, ecosystems, and flora and fauna [15,72–74]. Many studies focus on the significant impacts of microplastics entering the human body through various pathways on glucose metabolism and their potential mechanisms.

In recent years, more scientists have begun to focus on changes in the gut microbiota of diabetic mice following microplastic exposure, attempting to elucidate the relationship between the toxicity of microplastics to the gut microbiota and sugar–lipid metabolism (Table 1). For instance, Shi et al. [75] exposed mice to 1 μm polystyrene microplastics and found an increase in fasting blood glucose and insulin levels, hypothesizing that this may be due to a disruption of the gut–liver axis prompted by changes in the structure and composition of the gut microbiota. An analysis of the gut microbiome revealed a marked reduction in its diversity and significant changes at the phylum level, with a decrease in the abundance of Bacteroidetes and Verrucomicrobia and an increase in Firmicutes, Deferribacteres, and Actinobacteria. Similarly, another study arrived at consistent results, with a decreased abundance of Bacteroidetes in mice fed a high-fat diet containing microplastics, also noting reductions in the Chao1, Shannon, and Gini–Simpson indices [76]. Huang et al. [77] also found that after microplastic exposure in high-fat-diet mice, there was a reduction in microbial richness and diversity, with a relative increase in the abundance of Gram-negative rods (such as Prevotellaceae and Enterobacteriaceae). However, intriguingly, a different conclusion was reached in another study. Liu et al. [78] examined the varying responses of healthy mice and diabetic mice to exposure to polystyrene microplastics, finding that the gut microbiota in both healthy and diabetic mice changed post exposure. The difference was that in healthy mice, the proportion of probiotics (Alloprevotella, Bacteroides, Dubosiella, Lachnospiraceae_NK4A136_group, Lactobacillus, Weissella) declined while the proportion of pathogens (Helicobacter, Parabacteroides, Candidatus_Saccharimonas, Lachnoclostridium) increased; in contrast, diabetic mice exhibited an increased proportion of probiotics and a decreased proportion of pathogens.

Table 1. Effects of microplastic exposure on glucose metabolism and gut microbiota.

Species	Microplastics	Altered Glucose Metabolism	Altered Gut Microbiota	Reference
ICR mice	Polystyrene microplastics (1 μm)	• Fasting blood glucose, fasting insulin, and HOMA-IR levels were significantly elevated in mice after microplastics exposure.	• pCoA analysis showed that the diversity of the gut microbiota was significantly reduced in the exposure group mice compared to the control mice. • At the phylum level, the relative abundance of Bacteroidetes and Verrucomicrobia was significantly reduced in the exposure group mice, whereas the relative abundance of Firmicute, Deferribacteres, and Actinobacteria was significantly increased. • At the genus level, the relative abundance of Oscillospira, Akkermansia, and Desulfovibrio was decreased, and the relative abundance of Lactobacillus and Bifidobacterium was increased in the exposure group mice. • Further analysis showed that the relative abundance of Bacillales, Staphylococcaceae, Jeotgalicoccus, and Rikenella was significantly altered in the exposure group mice.	[75]
High-fat-diet mice (C57BL/6J)	Polystyrene microplastics	• In the IPGTT (intraperitoneal glucose tolerance test), ND and HFD mice fed added microplastics had higher blood glucose levels than both ND and HFD mice. • In the ITT (insulin tolerance test), HFD mice with added microplastics had higher blood glucose levels than HFD mice.	• At the gate level, the relative abundance of Bacteroidetes was decreased, and the relative abundance of Proteobactria was increased in the exposure group mice. • The Chao 1 index, Shannon index, and GiniSimpson index of gut microbiota abundance decreased in the exposure group mice.	[76]

Table 1. *Cont.*

Species	Microplastics	Altered Glucose Metabolism	Altered Gut Microbiota	Reference
ICR mice	Polystyrene microplastics (5 μm, 50 μm, 100 μm, 200 μm)	• Fasting glucose and fasting insulin levels were significantly elevated in mice, and the AUC of OGTT and ITT was elevated.	• Significant changes in the structure and composition of the gut microbiota. • At the phylum level, the relative abundance of Bacteroidetes was elevated, and the relative abundance of Firmicutes decreased, causing a decrease in the Firmicutes-to-Bacteroidetes ratio. • At the family level, the relative abundance of Muribaculaceae and Helicobacteraceae decreased, and the relative abundance of Prevotellaceae, Enterobacteriaceae, Desulfovibrionaceae, and Rikenellaceae increased. • The relative abundance of Prevotellaceae and Enterobacteriaceae increased.	[77]
db/db mice	Polystyrene microplastics (100 nm)	• Elevated fasting blood glucose levels in mice after microplastic exposure suggested impaired glucose control. • Mice in the exposure group had significantly higher blood glucose levels at 90 min and 120 min after oral glucose administration in the OGTT test.	• Microplastic exposure increased gut microbiota diversity in healthy mice and decreased gut microbiota diversity in diabetic mice. • Microplastic exposure resulted in a decrease in the proportion of probiotic bacteria and an increase in the proportion of pathogenic bacteria in healthy mice gut microbiota, whereas diabetic mice gut microbiota showed an increase in the proportion of probiotic bacteria and a decrease in the proportion of pathogenic bacteria.	[78]

4.2. Antibiotics

Developed in the late 1940s for treating bacterial pathogen-induced infections, antibiotics are known to selectively target potential pathogens within microbial populations while significantly disrupting the human gut microbiome, with effects lasting for several months or even longer [79,80]. Recent studies have begun to explore the impact of antibiotic treatment on the gut microbiome and human metabolism.

Vrieze et al. [81] conducted a study to investigate the effect of vancomycin on insulin sensitivity and the gut microbiota. Patients with metabolic syndrome were randomly assigned to different treatment groups. They found that after one week of treatment, patients in the vancomycin group showed a significant decrease in insulin sensitivity and noticeable changes in fecal microbiome diversity, including a reduction in the relative abundance of Gram-positive bacteria and a compensatory increase in Gram-negative bacteria. These results highlight the influential role of the gut microbiome in how antibiotics affect glucose metabolism. Additionally, Hwang's animal study showed that antibiotics alter the structure and composition of the gut microbiome, enhancing the microbes' capacity to collect and store energy, thereby changing insulin sensitivity and glucose tolerance in mice [82,83]. In addition, in an experiment that treated ob/ob mice with Ampicillin and neomycin, it was found that compared with mice that did not receive antibiotic treatment, the treated mice not only showed better glucose tolerance but also showed significant changes in the structure and composition of their gut microbiota, which was only 22% similar to that of the mice on the high-fat diet before the treatment [84]. Another study found that antibiotic therapy improved fasting glucose, glucose tolerance, and gut microbiota in mice [85]. It is well known that gut bacteria ferment carbohydrates in the intestine into short-chain fatty acids [86], which, in turn, affect metabolism and energy balance by altering the expression and secretion of intestinal hormones. For instance, Livanos used antibiotics to treat mice, finding a reduction in the diversity of the gut microbiome and the selection of unique microbial community structures and a significant increase in diabetes incidence [87].

4.3. Endocrine Disruptors

Beyond the aforementioned microplastics and antibiotics, endocrine disruptors represent another significant category of emerging pollutants in the environment, one that is gradually garnering scientific attention. Environmental endocrine disruptors primarily include phenolic compounds, pesticides, and persistent organic pollutants. These disruptors enter organisms through various exposure pathways and gradually accumulate in tissues and organs [88–90], interfering with normal hormone synthesis and secretion, leading to hormonal imbalances and consequent endocrine diseases like obesity and diabetes [91].

Yan et al. studied endosulfan sulfate (ES) exposure in pregnant mice and found that ES inhibits high-fat-diet-induced adipogenesis, reduces glucose tolerance, and affects glucose homeostasis by promoting lipolytic metabolism and fatty acid oxidation and altering the composition of the intestinal flora [92]. Meanwhile, Fan et al. showed that pregnant mice exposed to DEHP had offspring with abnormalities in adipogenesis, energy expenditure, glucose tolerance, and dysbiosis of the intestinal flora, and linear discriminant analysis (LDA) showed significant differences in 16 characteristic flora at the phylum and genus levels between exposed and control mice. Meanwhile, Fan et al. showed that the offspring of mice exposed to DEHP during gestation had abnormal lipogenesis, energy expenditure, and glucose tolerance and had dysbiosis in the gut microbiota. LDA analysis showed that the 16 characteristic flora of exposed mice and control mice were significantly different at the phylum and genus levels [93]. However, relatively few studies have been conducted on the disruption of glucose metabolism by affecting the structure and composition of the intestinal flora after exposure to EDCs. Nevertheless, numerous studies have found that exposure to endocrine disruptors affects glucose metabolism, leading to elevated blood glucose levels. Marmugi et al. exposed CD-1 mice to bisphenol A and observed that compared to the control group, the exposed mice had significantly higher blood glucose and plasma cholesterol levels. Moreover, mice exposed to a dose of 5000 mg/kg/d showed decreased glucose tolerance and a significant increase in the area under the curve [94]. Many studies have drawn similar conclusions, finding that C57BL/6 mice exposed to bisphenol A, even on a normal diet, showed an increase in body weight, elevated insulin levels, and impaired glucose tolerance, with bisphenol A exacerbating high-fat-diet-induced weight gain and insulin resistance in mice [95,96]. Interestingly, Lai et al. exposed CD-1 mice to bisphenol A and found that it induced gut microbiome community structures similar to those induced by a high-fat diet, with an increased relative abundance of Proteobacteria and Helicobacteraceae and a decrease in Firmicutes and Clostridiapopulations [49]. Consequently, the role of the gut microbiome in endocrine disruptor-induced glucose metabolic disorders is increasingly recognized. Tian et al. [46] exposed C57BL/6 high-fat-diet mice to polychlorinated biphenyl-126 and found that early-life exposure to PCB-126 in mice led to decreased glucose tolerance and a significant increase in metabolites involved in the tricarboxylic acid cycle (e.g., pyruvate, succinate, citrate). Tian concluded from these results that early exposure to PCB-126 exacerbates glucose homeostasis impairment characterized by abnormal glucose tolerance and increased tricarboxylic acid cycle flux in high-fat-diet mice. Additionally, researchers have found that PCB-126 affects the structure and composition of the mouse gut microbiome. For example, compared to control mice, high-fat-diet mice exposed to PCB-126 in early life showed a significant decrease in the relative abundance of Muribaculum, Duncaniella, Bacteroides, Parabacteroides, and Prevotella in the cecum, while the ratio of Firmicutes/Bacteroidetes, Romboutsia, and Adlercreutzia significantly increased. Qin et al.'s study found that the aforementioned groups, such as Firmicutes, Clostridium, Bacteroides, and Parabacteroides, are significantly associated with type 2 diabetes [50]. Li et al.'s research on the impact of TCDD exposure during pregnancy and lactation in mice found significant changes in the structure and composition of the gut microbiome, characterized by an upregulation of Firmicutes, Bacteroidetes, Clostridia, and Lachnospiraceae. Moreover, Pearson correlation coefficients suggested that affected tryptophan metabolism (positively correlated with type 2 diabetes) was positively correlated with harmful bacteria and negatively correlated with beneficial bacteria [97,98].

4.4. Perfluorinated Compounds

Similar to endocrine disruptors, perfluorinated compounds such as Perfluorooctane sulfonate (PFOS) and Perfluorooctanoic acid (PFOA) are characterized by unique structures and stability, allowing them to persist in the environment and accumulate in organisms following exposure, thereby posing health risks [99–103].

Perfluorinated compounds are also ubiquitously present in the environment and pose significant threats to the organisms living in it, particularly regarding metabolic effects. Consequently, scientists are focusing on the impact of perfluorinated compound exposure on glucose metabolism by disrupting flora metabolism. Wei et al. exposed adult male mice to 25 mg/kg/d DEHP (Di-(2-ethylhexyl) phthalate) via continuous oral exposure, finding that the mice developed elevated fasting blood glucose levels and hepatic fat accumulation. Interestingly, research on the intestinal flora of mice revealed significant differences in the community structure of the gut microbiota of exposed mice compared to the control group. Moreover, LDA showed that 29 features were significantly different between control and exposed mice from the gate level to the genus level. Compared to control mice, the relative abundance of cyanobacteria in the intestinal flora of exposed mice was significantly increased at the phylum level, whereas at the genus level, the relative abundance of Bacteroides was decreased, and the relative abundance of Allobaculumin was increased [104]. However, relatively few studies have been conducted in this area, and current articles have focused on the effects on glucose metabolism following exposure to PFAS or changes in gut microbiota caused by exposure to PFAS. Rats exposed to PFOS during gestation exhibited pre-diabetic symptoms in their offspring, with elevated fasting insulin and leptin levels and impaired glucose tolerance compared to the control group. Lv inferred from these results that exposure to PFOS during development could lead to glucose metabolism disorders in adulthood in rats [105]. Yan's study arrived at a similar conclusion: mice exposed to PFOA showed higher insulin sensitivity and glucose tolerance and reduced hepatic glycogen synthesis compared to the control group [106]. Additionally, researchers found that mice exposed to PFOS exhibited disorders in fat and glucose metabolism [107]. Interestingly, upon analyzing the gut microbiota of mice, researchers found a significant increase in the relative abundance of Turicibacterales and Turicibacteraceae in the exposed group; the glucose metabolism disorder in mice was notably positively correlated with the relative abundance of Turicibacteraceae [61,107]. Furthermore, an increase in the relative abundance of Allobaculum, which contributes to insulin resistance and obesity in mice, was found in the exposed mice. Significant changes in the relative abundance of Turicibacter, Allobaculum, B. acidifaciens, and Dehalbacteriaceae, which are considered related to dysregulation of sugar and lipid metabolism, were observed [107]. In studies on OBS (Sodium ρ-perfluorous nonenoxybenzene sulfonate, a PFASs substitute), it was found that OBS exposure in zebrafish led to a decrease in cytoplasmic phosphoenolpyruvate carboxykinase gene levels in the liver (related to glucose metabolism levels), with a decrease in the relative abundance of β-Proteobacteria, Bacteroidetes and Actinobacteria, α-Proteobacteria, γ-Proteobacteria, and Verrucomicrobia [108]. Moreover, exposure to F-53B (a PFOS substitute) also caused an increase in the relative abundance of Verrucomicrobia and a decrease in Firmicutes in the gut microbiome of mice, with significant changes in Akkermansia, Bacteroides, and Ruminococcus (significantly related to type 2 diabetes) [50,109].

However, unlike microplastics and antibiotics, few studies on EDCs and PFAS have addressed the effects of gut microbiota disruption on glucose metabolism. Nevertheless, we reviewed the effects of EDCs and PFAS on the gut microbiota and analyzed the possible correlation between altered gut microbiota and glucose metabolism (Table 2).

Table 2. Effects of EDCs and PFAS on glucose metabolism and gut microbiota.

Species	Chemical	Altered Gut Microbiota	Gut Microbiota Associated with Glucose Metabolism	Reference
Pregnant CD-1 mice	Endosulfan sulfate	• Alleviation of obesity and liver triglyceride accumulation due to high-fat diet. • Elevated fasting blood glucose and reduced glucose tolerance.	• There was a reduced α diversity of gut microbiota in the exposure group of mice. • ES treatment alleviated high-fat-diet-induced increases in the relative abundance of Actinobacteria and Proteobacteria. • ES treatment alleviated high-fat-diet-induced increases in the relative abundance of Enterorhabdus and Bifidobacterium. • Chronic exposure to ES caused an increase in the relative abundance of Bacteroides.	[92]
Pregnant mice	Di-(2-ethylhexyl)-phthalate	• Reduced glucose tolerance and disturbed glucose metabolism in offspring mice. • Lifelong metabolic consequences for offspring in a gender-dependent manner.	• LDA analysis showed that mice in the offspring of the exposed group were significantly different at the gate level for 16 features. • There was an increased α diversity of gut microbiota in offspring mice of the exposure group.	[93]
CD-1 mice	Bisphenol A	• Compared with the control mice, the exposure mice showed a significant reduction in the diversity of the gut flora. • α-diversity and β-diversity analyses suggest that BPA leads to a gut microbiota community structure similar to that induced by high-fat diets. • Proteobacteria were significantly elevated in both bisphenol A exposure group mice and high-fat dietary mice.	• Elevated Helicobacteraceae and Proteobacteria and reduced Firmicutes and Clostridiapopulations were observed in both BPA-exposed mice and high-fat-diet mice.	[49]

Table 2. *Cont.*

Species	Chemical	Altered Gut Microbiota	Gut Microbiota Associated with Glucose Metabolism	Reference
High-fat-diet mice	Polychlorinated Biphenyl 126	• High-fat-diet mice exposed to PCB126 early in life showed a significant decrease in the relative abundance of Muribaculum, Duncaniella, Bacteroides, Parabacteroides, and Prevotella and an increase in the relative abundance of Romboutsia, Akkermansia, and Adlercreutzia.	• Early exposure to PCB126 significantly decreased the abundance of Bacteroidetes and increased the ratio of Firmicutes/Bacteroidetes in mice on a high-fat diet, whereas the ratio of Firmicutes/Bacteroidetes was not significantly altered, and the relative abundance of Firmicutes and Verrucomicrobia was significantly altered in mice on a normal diet.	[46]
C57BL/6 mice (pregnant and lactating mice)	2,3,7,8-tetrachlorodibenzo-p-dioxin	• Alterations in the structure and composition of the gut microbiota of both mothers and offspring, reflected in an upregulation of harmful bacteria and a downregulation of beneficial bacteria.	• The dominant phylum in mothers and offspring of mice in the exposure group was Firmicutes and Bacteroidetes; the dominant class was Bacteroidia and Clostridia; the dominant order was Bacteroidales and Clostridiales; and the dominant families were S24-7 and Lachnospiraceae.	[97]
CD-1 mice	Di-(2-ethylhexyl) phthalate	• Significant increase in blood glucose levels and hepatic fat accumulation in mice in the exposed group.	• Significant reduction in the alpha diversity of the gut microbiota of mice in the exposure group. • Significant increase in Cyanobacteria relative abundance at the gate level. • Significant increase in the relative abundance of Allobaculumin and a decrease in the relative abundance of Bacteroides at the genus level.	[104]
CD-1 mice	Perfluorooctane sulfonic acid	• Metabolic disorders, especially fat and glucose metabolism, occurred in exposure group mice. • The relative abundance of Firmicutes, Bacteroidetes, Proteobacteria and Cyanobacteria were altered.	• Disturbed glucose metabolism was positively associated with Turicibacteraceae, similar to the increase in Turicibacterales in previously hypercholesterolemic fed mice. • An increased abundance of Allobaculum (a hypothesized short-chain fatty acid-producing bacterium) in PFOS-exposed mice contributed to insulin resistance and obesity. • There were significant changes in the abundance of Turicibacter, Allobaculum, B. acidifaciens, and Dehalbacteriaceae, which are thought to be associated with disturbed glycolipid metabolism.	[107]

5. Conclusions and Outlook

In recent years, more studies have begun to focus on the impact of exposure to emerging pollutants in the environment on metabolic disorders, particularly glucose metabolism disorders. Given the significant role of the gut microbiota in glucose metabolism, scientists are starting to pay attention to the structural and compositional disorders of the gut microbiota caused by pollutants, as well as changes in microbial diversity. Although an increasing number of studies are focusing on the role of the gut microbiota in glucose metabolism disorders caused by emerging pollutants, current research is just the tip of the iceberg, and our understanding in this area remains very limited:

(1) Many studies have shown that exposure to emerging pollutants can cause pre-diabetic symptoms or exacerbate existing glucose metabolism disorders in organisms. Interestingly, different results have been found for the same compound, possibly due to different diabetes animal models or exposure periods used in the studies. Therefore, there is an urgent need for more in-depth research to standardize diabetes animal models or exposure forms.

(2) It is well known that the gut microbiota plays a crucial role in the development of diabetes. However, studies on whether exposure to emerging pollutants affects the glucose metabolism process by altering the structure and composition of the gut microbiota are relatively scarce. Additionally, there is still controversy over changes in certain specific gut bacteria like Akkermansia, Parabacteroides, and Verrucomicrobia after exposure to emerging pollutants or during glucose metabolism disorders. Therefore, more research is needed to explore the changes in the gut microbiota after exposure to emerging pollutants and its relationship with glucose metabolism.

(3) Most current research on the impact of emerging pollutants on the gut microbiota and glucose metabolism focuses on animal models. Due to interspecies differences, studies on the impact of emerging pollutants on human populations are very limited. Therefore, large-scale population studies are needed to elucidate the impact of emerging pollutant exposure on human glucose metabolism and the role of the gut microbiota in this process.

(4) Currently, most research on emerging contaminants focuses on the effects of exposure on glucose metabolism or on a single aspect of the gut microbiota, while relatively few studies have been conducted on whether they affect the development of diabetes by altering biological glucose metabolism through the gut microbiota, especially with regard to endocrine disruptors and perfluorinated compounds. Therefore, large-scale and more in-depth studies are needed to elucidate whether exposure to emerging contaminants causes glucose metabolism disorders through the gut microbiota and its specific mechanisms.

Author Contributions: Conceptualization, X.L., H.N. and P.X.; literature search, X.L., H.N., M.Z., M.X., Z.H. and Z.C.; writing—original draft preparation, X.L., H.N., Z.H. and L.W.; writing—review and editing, X.L., H.N., Z.C. and P.X.; supervision, P.X. All authors have read and agreed to the published version of the manuscript.

Funding: This research was funded by Zhejiang Provincial Project for Medical Research and Health Sciences (Grant No. 2024KY910, 2023KY644, 2021PY042, and 2020KY515).

Institutional Review Board Statement: Not applicable.

Informed Consent Statement: Not applicable.

Data Availability Statement: Not applicable.

Conflicts of Interest: The authors declare no conflict of interest.

References

1. Buffie, C.G.; Bucci, V.; Stein, R.R.; McKenney, P.T.; Ling, L.; Gobourne, A.; No, D.; Liu, H.; Kinnebrew, M.; Viale, A.; et al. Precision microbiome reconstitution restores bile acid mediated resistance to Clostridium difficile. *Nature* **2015**, *517*, 205–208. [CrossRef] [PubMed]
2. Sharon, G.; Garg, N.; Debelius, J.; Knight, R.; Dorrestein, P.C.; Mazmanian, S.K. Specialized metabolites from the microbiome in health and disease. *Cell Metab.* **2014**, *20*, 719–730. [CrossRef] [PubMed]
3. O'Hara, A.M.; Shanahan, F. The gut flora as a forgotten organ. *EMBO Rep.* **2006**, *7*, 688–693. [CrossRef] [PubMed]
4. Claus, S.P.; Guillou, H.; Ellero-Simatos, S. The gut microbiota: A major player in the toxicity of environmental pollutants? *NPJ Biofilms Microbiomes* **2016**, *2*, 16003. [CrossRef] [PubMed]
5. Lu, K.; Abo, R.P.; Schlieper, K.A.; Graffam, M.E.; Levine, S.; Wishnok, J.S.; Swenberg, J.A.; Tannenbaum, S.R.; Fox, J.G. Arsenic exposure perturbs the gut microbiome and its metabolic profile in mice: An integrated metagenomics and metabolomics analysis. *Environ. Health Perspect.* **2014**, *122*, 284–291. [CrossRef]
6. Bao, L.-J.; Wei, Y.-L.; Yao, Y.; Ruan, Q.-Q.; Zeng, E.Y. Global trends of research on emerging contaminants in the environment and humans: A literature assimilation. *Environ. Sci. Pollut. Res. Int.* **2015**, *22*, 1635–1643. [CrossRef]
7. Tu, P.; Chi, L.; Bodnar, W.; Zhang, Z.; Gao, B.; Bian, X.; Stewart, J.; Fry, R.; Lu, K. Gut Microbiome Toxicity: Connecting the Environment and Gut Microbiome-Associated Diseases. *Toxics* **2020**, *8*, 19. [CrossRef]
8. Dey, S.; Bano, F.; Malik, A. Pharmaceuticals and personal care product (PPCP) contamination—A global discharge inventory. In *Pharmaceuticals and Personal Care Products: Waste Management and Treatment Technology*; Elsevier: Amsterdam, The Netherlands, 2019; pp. 1–26.
9. Naidu, R.; Arias Espana, V.A.; Liu, Y.; Jit, J. Emerging contaminants in the environment: Risk-based analysis for better management. *Chemosphere* **2016**, *154*, 350–357. [CrossRef]
10. Wang, J.; Zheng, J.; Shi, W.; Du, N.; Xu, X.; Zhang, Y.; Ji, P.; Zhang, F.; Jia, Z.; Wang, Y.; et al. Dysbiosis of maternal and neonatal microbiota associated with gestational diabetes mellitus. *Gut* **2018**, *67*, 1614–1625. [CrossRef]
11. Byers, T. Excess Mortality among Persons with Type 2 Diabetes. *N. Engl. J. Med.* **2016**, *374*, 788. [CrossRef]
12. Vos, T.; Allen, C.; Arora, M.; Barber, R.M.; Bhutta, Z.A.; Brown, A.; Carter, A.; Casey, D.C.; Charlson, F.J.; Chen, A.Z.; et al. Global, regional, and national incidence, prevalence, and years lived with disability for 310 diseases and injuries, 1990–2015: A systematic analysis for the Global Burden of Disease Study 2015. *Lancet* **2016**, *388*, 1545–1602. [CrossRef] [PubMed]
13. Zheng, Y.; Ley, S.H.; Hu, F.B. Global aetiology and epidemiology of type 2 diabetes mellitus and its complications. *Nat. Rev. Endocrinol.* **2018**, *14*, 88–98. [CrossRef] [PubMed]
14. Zhao, L.; Shi, W.; Hu, F.; Song, X.; Cheng, Z.; Zhou, J. Prolonged oral ingestion of microplastics induced inflammation in the liver tissues of C57BL/6J mice through polarization of macrophages and increased infiltration of natural killer cells. *Ecotoxicol. Environ. Saf.* **2021**, *227*, 112882. [CrossRef] [PubMed]
15. Gao, C.; Sun, X.; Lu, L.; Liu, F.; Yuan, J. Prevalence of gestational diabetes mellitus in mainland China: A systematic review and meta-analysis. *J. Diabetes Investig.* **2019**, *10*, 154–162. [CrossRef] [PubMed]
16. Yao, X.; Geng, S.; Zhu, L.; Jiang, H.; Wen, J. Environmental pollutants exposure and gestational diabetes mellitus: Evidence from epidemiological and experimental studies. *Chemosphere* **2023**, *332*, 138866. [CrossRef]
17. Kolb, H.; Martin, S. Environmental/lifestyle factors in the pathogenesis and prevention of type 2 diabetes. *BMC Med.* **2017**, *15*, 131. [CrossRef] [PubMed]
18. Taylor, K.W.; Novak, R.F.; Anderson, H.A.; Birnbaum, L.S.; Blystone, C.; Devito, M.; Jacobs, D.; Köhrle, J.; Lee, D.-H.; Rylander, L.; et al. Evaluation of the association between persistent organic pollutants (POPs) and diabetes in epidemiological studies: A national toxicology program workshop review. *Environ. Health Perspect.* **2013**, *121*, 774–783. [CrossRef]
19. Puri, M.; Gandhi, K.; Kumar, M.S. Emerging environmental contaminants: A global perspective on policies and regulations. *J. Environ. Manag.* **2023**, *332*, 117344. [CrossRef]
20. Bodus, B.; O'Malley, K.; Dieter, G.; Gunawardana, C.; McDonald, W. Review of emerging contaminants in green stormwater infrastructure: Antibiotic resistance genes, microplastics, tire wear particles, PFAS, and temperature. *Sci. Total Environ.* **2023**, *906*, 167195. [CrossRef]
21. Chen, X.; Wang, S.; Mao, X.; Xiang, X.; Ye, S.; Chen, J.; Zhu, A.; Meng, Y.; Yang, X.; Peng, S.; et al. Adverse health effects of emerging contaminants on inflammatory bowel disease. *Front. Public. Health* **2023**, *11*, 1140786. [CrossRef]
22. Ouda, M.; Kadadou, D.; Swaidan, B.; Al-Othman, A.; Al-Asheh, S.; Banat, F.; Hasan, S.W. Emerging contaminants in the water bodies of the Middle East and North Africa (MENA): A critical review. *Sci. Total Environ.* **2021**, *754*, 142177. [CrossRef] [PubMed]
23. Khan, S.; Naushad, M.; Govarthanan, M.; Iqbal, J.; Alfadul, S.M. Emerging contaminants of high concern for the environment: Current trends and future research. *Environ. Res.* **2022**, *207*, 112609. [CrossRef]
24. Mohammadi, A.; Dobaradaran, S.; Schmidt, T.C.; Malakootian, M.; Spitz, J. Emerging contaminants migration from pipes used in drinking water distribution systems: A review of the scientific literature. *Environ. Sci. Pollut. Res. Int.* **2022**, *29*, 75134–75160. [CrossRef]
25. Kumar, N.; Shukla, P. Microalgal-based bioremediation of emerging contaminants: Mechanisms and challenges. *Environ. Pollut.* **2023**, *337*, 122591. [CrossRef]
26. Ragusa, A.; Svelato, A.; Santacroce, C.; Catalano, P.; Notarstefano, V.; Carnevali, O.; Papa, F.; Rongioletti, M.C.A.; Baiocco, F.; Draghi, S.; et al. Plasticenta: First evidence of microplastics in human placenta. *Environ. Int.* **2021**, *146*, 106274. [CrossRef]

27. Leslie, H.A.; van Velzen, M.J.M.; Brandsma, S.H.; Vethaak, A.D.; Garcia-Vallejo, J.J.; Lamoree, M.H. Discovery and quantification of plastic particle pollution in human blood. *Environ. Int.* **2022**, *163*, 107199. [CrossRef] [PubMed]
28. Van Cauwenberghe, L.; Janssen, C.R. Microplastics in bivalves cultured for human consumption. *Environ. Pollut.* **2014**, *193*, 65–70. [CrossRef] [PubMed]
29. Mattsson, K.; Ekvall, M.T.; Hansson, L.-A.; Linse, S.; Malmendal, A.; Cedervall, T. Altered behavior, physiology, and metabolism in fish exposed to polystyrene nanoparticles. *Environ. Sci. Technol.* **2015**, *49*, 553–561. [CrossRef]
30. Yang, Y.; Xie, E.; Du, Z.; Peng, Z.; Han, Z.; Li, L.; Zhao, R.; Qin, Y.; Xue, M.; Li, F.; et al. Detection of Various Microplastics in Patients Undergoing Cardiac Surgery. *Environ. Sci. Technol.* **2023**, *57*, 10911–10918. [CrossRef]
31. Li, X.; Yin Yeung, L.W.; Xu, M.; Taniyasu, S.; Lam, P.K.S.; Yamashita, N.; Dai, J. Perfluorooctane sulfonate (PFOS) and other fluorochemicals in fish blood collected near the outfall of wastewater treatment plant (WWTP) in Beijing. *Environ. Pollut.* **2008**, *156*, 1298–1303. [CrossRef]
32. Dodoo, D.K.; Essumang, D.K.; Jonathan, J.W.A. Accumulation profile and seasonal variations of polychlorinated biphenyls (PCBs) in bivalves Crassostrea tulipa (oysters) and Anadara senilis (mussels) at three different aquatic habitats in two seasons in Ghana. *Ecotoxicol. Environ. Saf.* **2013**, *88*, 26–34. [CrossRef]
33. Asante, K.; Sudaryanto, A.; Gnanasekaran, D.; Bello, M.; Takahashi, S.; Isobe, T.; Tajima, Y. Polybrominated Diphenyl Ethers and Polychlorinated Biphenyls in Cow Milk Samples from Ghana. *Interdiscip. Stud. Environ. Chem. Environ. Specimen Bank.* **2010**, *4*, 191–198.
34. Knutsen, H.K.; Alexander, J.; Barregård, L.; Bignami, M.; Brüschweiler, B.; Ceccatelli, S.; Cottrill, B.; Dinovi, M.; Edler, L.; Grasl-Kraupp, B.; et al. Risk to human health related to the presence of perfluorooctane sulfonic acid and perfluorooctanoic acid in food. *EFSA J.* **2018**, *16*, e05194. [CrossRef]
35. Tittlemier, S.A.; Pepper, K.; Seymour, C.; Moisey, J.; Bronson, R.; Cao, X.-L.; Dabeka, R.W. Dietary exposure of Canadians to perfluorinated carboxylates and perfluorooctane sulfonate via consumption of meat, fish, fast foods, and food items prepared in their packaging. *J. Agric. Food Chem.* **2007**, *55*, 3203–3210. [CrossRef]
36. Ericson, I.; Martí-Cid, R.; Nadal, M.; Van Bavel, B.; Lindström, G.; Domingo, J.L. Human exposure to perfluorinated chemicals through the diet: Intake of perfluorinated compounds in foods from the Catalan (Spain) market. *J. Agric. Food Chem.* **2008**, *56*, 1787–1794. [CrossRef]
37. Niu, H.; Liu, S.; Jiang, Y.; Hu, Y.; Li, Y.; He, L.; Xing, M.; Li, X.; Wu, L.; Chen, Z.; et al. Are Microplastics Toxic? A Review from Eco-Toxicity to Effects on the Gut Microbiota. *Metabolites* **2023**, *13*, 739. [CrossRef]
38. Yang, D.; Shi, H.; Li, L.; Li, J.; Jabeen, K.; Kolandhasamy, P. Microplastic Pollution in Table Salts from China. *Environ. Sci. Technol.* **2015**, *49*, 13622–13627. [CrossRef] [PubMed]
39. Kirchhelle, C. Pharming animals: A global history of antibiotics in food production (1935–2017). *Palgrave Commun.* **2018**, *4*, 96. [CrossRef]
40. Cabello, F.C. Heavy use of prophylactic antibiotics in aquaculture: A growing problem for human and animal health and for the environment. *Environ. Microbiol.* **2006**, *8*, 1137–1144. [CrossRef]
41. Liu, S.; Bekele, T.-G.; Zhao, H.; Cai, X.; Chen, J. Bioaccumulation and tissue distribution of antibiotics in wild marine fish from Laizhou Bay, North China. *Sci. Total Environ.* **2018**, *631–632*, 1398–1405. [CrossRef] [PubMed]
42. Khanal, B.K.S.; Sadiq, M.B.; Singh, M.; Anal, A.K. Screening of antibiotic residues in fresh milk of Kathmandu Valley, Nepal. *J. Environ. Sci. Health B* **2018**, *53*, 57–86. [CrossRef] [PubMed]
43. Er, B.; Onurdag, F.K.; Demirhan, B.; Ozgacar, S.Ö.; Oktem, A.B.; Abbasoglu, U. Screening of quinolone antibiotic residues in chicken meat and beef sold in the markets of Ankara, Turkey. *Poult. Sci.* **2013**, *92*, 2212–2215. [CrossRef] [PubMed]
44. Liu, Z.; Yu, P.; Cai, M.; Wu, D.; Zhang, M.; Chen, M.; Zhao, Y. Effects of microplastics on the innate immunity and intestinal microflora of juvenile Eriocheir sinensis. *Sci. Total Environ.* **2019**, *685*, 836–846. [CrossRef] [PubMed]
45. Lu, L.; Wan, Z.; Luo, T.; Fu, Z.; Jin, Y. Polystyrene microplastics induce gut microbiota dysbiosis and hepatic lipid metabolism disorder in mice. *Sci. Total Environ.* **2018**, *631*, 449–458. [CrossRef] [PubMed]
46. Tian, Y.; Rimal, B.; Gui, W.; Koo, I.; Smith, P.B.; Yokoyama, S.; Patterson, A.D. Early Life Polychlorinated Biphenyl 126 Exposure Disrupts Gut Microbiota and Metabolic Homeostasis in Mice Fed with High-Fat Diet in Adulthood. *Metabolites* **2022**, *12*, 894. [CrossRef] [PubMed]
47. Fu, X.; Han, H.; Li, Y.; Xu, B.; Dai, W.; Zhang, Y.; Zhou, F.; Ma, H.; Pei, X. Di-(2-ethylhexyl) phthalate exposure induces female reproductive toxicity and alters the intestinal microbiota community structure and fecal metabolite profile in mice. *Environ. Toxicol.* **2021**, *36*, 1226–1242. [CrossRef]
48. Yang, Y.-N.; Yang, Y.-C.S.H.; Lin, I.H.; Chen, Y.-Y.; Lin, H.-Y.; Wu, C.-Y.; Su, Y.-T.; Yang, Y.-J.; Yang, S.-N.; Suen, J.-L. Phthalate exposure alters gut microbiota composition and IgM vaccine response in human newborns. *Food Chem. Toxicol.* **2019**, *132*, 110700. [CrossRef]
49. Lai, K.-P.; Chung, Y.-T.; Li, R.; Wan, H.-T.; Wong, C.K.-C. Bisphenol A alters gut microbiome: Comparative metagenomics analysis. *Environ. Pollut.* **2016**, *218*, 923–930. [CrossRef]
50. Qin, J.; Li, Y.; Cai, Z.; Li, S.; Zhu, J.; Zhang, F.; Liang, S.; Zhang, W.; Guan, Y.; Shen, D.; et al. A metagenome-wide association study of gut microbiota in type 2 diabetes. *Nature* **2012**, *490*, 55–60. [CrossRef]

51. Li, Y.; Teng, D.; Shi, X.; Qin, G.; Qin, Y.; Quan, H.; Shi, B.; Sun, H.; Ba, J.; Chen, B.; et al. Prevalence of diabetes recorded in mainland China using 2018 diagnostic criteria from the American Diabetes Association: National cross sectional study. *BMJ* **2020**, *369*, m997. [CrossRef]
52. Zhao, Y.; Li, Y.; Zhuang, Z.; Song, Z.; Wang, W.; Huang, N.; Dong, X.; Xiao, W.; Jia, J.; Liu, Z.; et al. Associations of polysocial risk score, lifestyle and genetic factors with incident type 2 diabetes: A prospective cohort study. *Diabetologia* **2022**, *65*, 2056–2065. [CrossRef] [PubMed]
53. Sedighi, M.; Razavi, S.; Navab-Moghadam, F.; Khamseh, M.E.; Alaei-Shahmiri, F.; Mehrtash, A.; Amirmozafari, N. Comparison of gut microbiota in adult patients with type 2 diabetes and healthy individuals. *Microb. Pathog.* **2017**, *111*, 362–369. [CrossRef]
54. Le, T.K.C.; Hosaka, T.; Nguyen, T.T.; Kassu, A.; Dang, T.O.; Tran, H.B.; Pham, T.P.; Tran, Q.B.; Le, T.H.H.; Pham, X.D. *Bifidobacterium* species lower serum glucose, increase expressions of insulin signaling proteins, and improve adipokine profile in diabetic mice. *Biomed. Res.* **2015**, *36*, 63–70. [CrossRef] [PubMed]
55. Gao, R.; Zhu, C.; Li, H.; Yin, M.; Pan, C.; Huang, L.; Kong, C.; Wang, X.; Zhang, Y.; Qu, S.; et al. Dysbiosis Signatures of Gut Microbiota Along the Sequence from Healthy, Young Patients to Those with Overweight and Obesity. *Obesity* **2018**, *26*, 351–361. [CrossRef]
56. Zhang, X.; Shen, D.; Fang, Z.; Jie, Z.; Qiu, X.; Zhang, C.; Chen, Y.; Ji, L. Human gut microbiota changes reveal the progression of glucose intolerance. *PLoS ONE* **2013**, *8*, e71108. [CrossRef]
57. Gauffin Cano, P.; Santacruz, A.; Moya, Á.; Sanz, Y. *Bacteroides* uniformis CECT 7771 ameliorates metabolic and immunological dysfunction in mice with high-fat-diet induced obesity. *PLoS ONE* **2012**, *7*, e41079. [CrossRef]
58. Candela, M.; Biagi, E.; Soverini, M.; Consolandi, C.; Quercia, S.; Severgnini, M.; Peano, C.; Turroni, S.; Rampelli, S.; Pozzilli, P.; et al. Modulation of gut microbiota dysbioses in type 2 diabetic patients by macrobiotic Ma-Pi 2 diet. *Br. J. Nutr.* **2016**, *116*, 80–93. [CrossRef]
59. Allin, K.H.; Tremaroli, V.; Caesar, R.; Jensen, B.A.H.; Damgaard, M.T.F.; Bahl, M.I.; Licht, T.R.; Hansen, T.H.; Nielsen, T.; Dantoft, T.M.; et al. Aberrant intestinal microbiota in individuals with prediabetes. *Diabetologia* **2018**, *61*, 810–820. [CrossRef]
60. Wu, H.; Esteve, E.; Tremaroli, V.; Khan, M.T.; Caesar, R.; Mannerås-Holm, L.; Ståhlman, M.; Olsson, L.M.; Serino, M.; Planas-Fèlix, M.; et al. Metformin alters the gut microbiome of individuals with treatment-naive type 2 diabetes, contributing to the therapeutic effects of the drug. *Nat. Med.* **2017**, *23*, 850–858. [CrossRef]
61. Lippert, K.; Kedenko, L.; Antonielli, L.; Kedenko, I.; Gemeier, C.; Leitner, M.; Kautzky-Willer, A.; Paulweber, B.; Hackl, E. Gut microbiota dysbiosis associated with glucose metabolism disorders and the metabolic syndrome in older adults. *Benef. Microbes* **2017**, *8*, 545–556. [CrossRef] [PubMed]
62. Patrone, V.; Vajana, E.; Minuti, A.; Callegari, M.L.; Federico, A.; Loguercio, C.; Dallio, M.; Tolone, S.; Docimo, L.; Morelli, L. Postoperative Changes in Fecal Bacterial Communities and Fermentation Products in Obese Patients Undergoing Bilio-Intestinal Bypass. *Front. Microbiol.* **2016**, *7*, 200. [CrossRef] [PubMed]
63. Diamante, G.; Cely, I.; Zamora, Z.; Ding, J.; Blencowe, M.; Lang, J.; Bline, A.; Singh, M.; Lusis, A.J.; Yang, X. Systems toxicogenomics of prenatal low-dose BPA exposure on liver metabolic pathways, gut microbiota, and metabolic health in mice. *Environ. Int.* **2021**, *146*, 106260. [CrossRef] [PubMed]
64. Larsen, N.; Vogensen, F.K.; van den Berg, F.W.J.; Nielsen, D.S.; Andreasen, A.S.; Pedersen, B.K.; Al-Soud, W.A.; Sørensen, S.J.; Hansen, L.H.; Jakobsen, M. Gut microbiota in human adults with type 2 diabetes differs from non-diabetic adults. *PLoS ONE* **2010**, *5*, e9085. [CrossRef] [PubMed]
65. Karlsson, F.H.; Tremaroli, V.; Nookaew, I.; Bergström, G.; Behre, C.J.; Fagerberg, B.; Nielsen, J.; Bäckhed, F. Gut metagenome in European women with normal, impaired and diabetic glucose control. *Nature* **2013**, *498*, 99–103. [CrossRef]
66. Chen, P.; Zhang, Q.; Dang, H.; Liu, X.; Tian, F.; Zhao, J.; Chen, Y.; Zhang, H.; Chen, W. Antidiabetic effect of *Lactobacillus* casei CCFM0412 on mice with type 2 diabetes induced by a high-fat diet and streptozotocin. *Nutrition* **2014**, *30*, 1061–1068. [CrossRef]
67. Chelakkot, C.; Choi, Y.; Kim, D.-K.; Park, H.T.; Ghim, J.; Kwon, Y.; Jeon, J.; Kim, M.-S.; Jee, Y.-K.; Gho, Y.S.; et al. *Akkermansia* muciniphila-derived extracellular vesicles influence gut permeability through the regulation of tight junctions. *Exp. Mol. Med.* **2018**, *50*, e450. [CrossRef] [PubMed]
68. Li, X.; Wang, E.; Yin, B.; Fang, D.; Chen, P.; Wang, G.; Zhao, J.; Zhang, H.; Chen, W. Effects of *Lactobacillus* casei CCFM419 on insulin resistance and gut microbiota in type 2 diabetic mice. *Benef. Microbes* **2017**, *8*, 421–432. [CrossRef]
69. Singh, S.; Sharma, R.K.; Malhotra, S.; Pothuraju, R.; Shandilya, U.K. *Lactobacillus* rhamnosus NCDC17 ameliorates type-2 diabetes by improving gut function, oxidative stress and inflammation in high-fat-diet fed and streptozotocintreated rats. *Benef. Microbes* **2017**, *8*, 243–255. [CrossRef]
70. Moens, F.; Weckx, S.; De Vuyst, L. Bifidobacterial inulin-type fructan degradation capacity determines cross-feeding interactions between bifidobacteria and *Faecalibacterium* prausnitzii. *Int. J. Food Microbiol.* **2016**, *231*, 76–85. [CrossRef]
71. Thompson, R.C.; Olsen, Y.; Mitchell, R.P.; Davis, A.; Rowland, S.J.; John, A.W.G.; McGonigle, D.; Russell, A.E. Lost at Sea: Where Is All the Plastic? *Science* **2004**, *304*, 838. [CrossRef]
72. Wang, J.; Lv, S.; Zhang, M.; Chen, G.; Zhu, T.; Zhang, S.; Teng, Y.; Christie, P.; Luo, Y. Effects of plastic film residues on occurrence of phthalates and microbial activity in soils. *Chemosphere* **2016**, *151*, 171–177. [CrossRef]
73. Nolte, T.M.; Hartmann, N.B.; Kleijn, J.M.; Garnæs, J.; van de Meent, D.; Jan Hendriks, A.; Baun, A. The toxicity of plastic nanoparticles to green algae as influenced by surface modification, medium hardness and cellular adsorption. *Aquat. Toxicol.* **2017**, *183*, 11–20. [CrossRef]

74. Qiang, L.; Cheng, J. Exposure to microplastics decreases swimming competence in larval zebrafish (Danio rerio). *Ecotoxicol. Environ. Saf.* **2019**, *176*, 226–233. [CrossRef]
75. Shi, C.; Han, X.; Guo, W.; Wu, Q.; Yang, X.; Wang, Y.; Tang, G.; Wang, S.; Wang, Z.; Liu, Y.; et al. Disturbed Gut-Liver axis indicating oral exposure to polystyrene microplastic potentially increases the risk of insulin resistance. *Environ. Int.* **2022**, *164*, 107273. [CrossRef]
76. Okamura, T.; Hamaguchi, M.; Hasegawa, Y.; Hashimoto, Y.; Majima, S.; Senmaru, T.; Ushigome, E.; Nakanishi, N.; Asano, M.; Yamazaki, M.; et al. Oral Exposure to Polystyrene Microplastics of Mice on a Normal or High-Fat Diet and Intestinal and Metabolic Outcomes. *Environ. Health Perspect.* **2023**, *131*, 27006. [CrossRef] [PubMed]
77. Huang, D.; Zhang, Y.; Long, J.; Yang, X.; Bao, L.; Yang, Z.; Wu, B.; Si, R.; Zhao, W.; Peng, C.; et al. Polystyrene microplastic exposure induces insulin resistance in mice via dysbacteriosis and pro-inflammation. *Sci. Total Environ.* **2022**, *838*, 155937. [CrossRef] [PubMed]
78. Liu, S.; Wang, Z.; Xiang, Q.; Wu, B.; Lv, W.; Xu, S. A comparative study in healthy and diabetic mice followed the exposure of polystyrene microplastics: Differential lipid metabolism and inflammation reaction. *Ecotoxicol. Environ. Saf.* **2022**, *244*, 114031. [CrossRef] [PubMed]
79. Dethlefsen, L.; Relman, D.A. Incomplete recovery and individualized responses of the human distal gut microbiota to repeated antibiotic perturbation. *Proc. Natl. Acad. Sci. USA* **2011**, *108* (Suppl. S1), 4554–4561. [CrossRef] [PubMed]
80. Korpela, K.; Salonen, A.; Virta, L.J.; Kekkonen, R.A.; Forslund, K.; Bork, P.; de Vos, W.M. Intestinal microbiome is related to lifetime antibiotic use in Finnish pre-school children. *Nat. Commun.* **2016**, *7*, 10410. [CrossRef] [PubMed]
81. Vrieze, A.; Out, C.; Fuentes, S.; Jonker, L.; Reuling, I.; Kootte, R.S.; van Nood, E.; Holleman, F.; Knaapen, M.; Romijn, J.A.; et al. Impact of oral vancomycin on gut microbiota, bile acid metabolism, and insulin sensitivity. *J. Hepatol.* **2014**, *60*, 824–831. [CrossRef] [PubMed]
82. Hwang, I.; Park, Y.J.; Kim, Y.-R.; Kim, Y.N.; Ka, S.; Lee, H.Y.; Seong, J.K.; Seok, Y.-J.; Kim, J.B. Alteration of gut microbiota by vancomycin and bacitracin improves insulin resistance via glucagon-like peptide 1 in diet-induced obesity. *FASEB J.* **2015**, *29*, 2397–2411. [CrossRef]
83. Turnbaugh, P.J.; Ley, R.E.; Mahowald, M.A.; Magrini, V.; Mardis, E.R.; Gordon, J.I. An obesity-associated gut microbiome with increased capacity for energy harvest. *Nature* **2006**, *444*, 1027–1031. [CrossRef]
84. Cani, P.D.; Bibiloni, R.; Knauf, C.; Waget, A.; Neyrinck, A.M.; Delzenne, N.M.; Burcelin, R. Changes in gut microbiota control metabolic endotoxemia-induced inflammation in high-fat diet-induced obesity and diabetes in mice. *Diabetes* **2008**, *57*, 1470–1481. [CrossRef]
85. Chou, C.J.; Membrez, M.; Blancher, F. Gut decontamination with norfloxacin and ampicillin enhances insulin sensitivity in mice. *Nestle Nutr. Workshop Ser. Pediatr. Program.* **2008**, *62*, 127–140. [CrossRef] [PubMed]
86. Crawford, P.A.; Crowley, J.R.; Sambandam, N.; Muegge, B.D.; Costello, E.K.; Hamady, M.; Knight, R.; Gordon, J.I. Regulation of myocardial ketone body metabolism by the gut microbiota during nutrient deprivation. *Proc. Natl. Acad. Sci. USA* **2009**, *106*, 11276–11281. [CrossRef] [PubMed]
87. Livanos, A.E.; Greiner, T.U.; Vangay, P.; Pathmasiri, W.; Stewart, D.; McRitchie, S.; Li, H.; Chung, J.; Sohn, J.; Kim, S.; et al. Antibiotic-mediated gut microbiome perturbation accelerates development of type 1 diabetes in mice. *Nat. Microbiol.* **2016**, *1*, 16140. [CrossRef] [PubMed]
88. Lind, P.M.; Lind, L. Endocrine-disrupting chemicals and risk of diabetes: An evidence-based review. *Diabetologia* **2018**, *61*, 1495–1502. [CrossRef]
89. Porta, M.; Gasull, M.; Puigdomènech, E.; Garí, M.; Bosch de Basea, M.; Guillén, M.; López, T.; Bigas, E.; Pumarega, J.; Llebaria, X.; et al. Distribution of blood concentrations of persistent organic pollutants in a representative sample of the population of Catalonia. *Environ. Int.* **2010**, *36*, 655–664. [CrossRef]
90. Lee, Y.M.; Kim, K.S.; Jacobs, D.R.; Lee, D.H. Persistent organic pollutants in adipose tissue should be considered in obesity research. *Obes. Rev.* **2017**, *18*, 129–139. [CrossRef]
91. Kassotis, C.D.; Vandenberg, L.N.; Demeneix, B.A.; Porta, M.; Slama, R.; Trasande, L. Endocrine-disrupting chemicals: Economic, regulatory, and policy implications. *Lancet Diabetes Endocrinol.* **2020**, *8*, 719–730. [CrossRef]
92. Yan, J.; Wang, D.; Meng, Z.; Yan, S.; Teng, M.; Jia, M.; Li, R.; Tian, S.; Weiss, C.; Zhou, Z.; et al. Effects of incremental endosulfan sulfate exposure and high fat diet on lipid metabolism, glucose homeostasis and gut microbiota in mice. *Environ. Pollut.* **2021**, *268*, 115697. [CrossRef] [PubMed]
93. Fan, Y.; Qin, Y.; Chen, M.; Li, X.; Wang, R.; Huang, Z.; Xu, Q.; Yu, M.; Zhang, Y.; Han, X.; et al. Prenatal low-dose DEHP exposure induces metabolic adaptation and obesity: Role of hepatic thiamine metabolism. *J. Hazard. Mater.* **2020**, *385*, 121534. [CrossRef] [PubMed]
94. Marmugi, A.; Lasserre, F.; Beuzelin, D.; Ducheix, S.; Huc, L.; Polizzi, A.; Chetivaux, M.; Pineau, T.; Martin, P.; Guillou, H.; et al. Adverse effects of long-term exposure to bisphenol A during adulthood leading to hyperglycaemia and hypercholesterolemia in mice. *Toxicology* **2014**, *325*, 133–143. [CrossRef] [PubMed]
95. Ma, Q.; Deng, P.; Lin, M.; Yang, L.; Li, L.; Guo, L.; Zhang, L.; He, M.; Lu, Y.; Pi, H.; et al. Long-term bisphenol A exposure exacerbates diet-induced prediabetes via TLR4-dependent hypothalamic inflammation. *J. Hazard. Mater.* **2021**, *402*, 123926. [CrossRef] [PubMed]

96. Moon, M.K.; Jeong, I.-K.; Jung Oh, T.; Ahn, H.Y.; Kim, H.H.; Park, Y.J.; Jang, H.C.; Park, K.S. Long-term oral exposure to bisphenol A induces glucose intolerance and insulin resistance. *J. Endocrinol.* **2015**, *226*, 35–42. [CrossRef] [PubMed]
97. Li, J.; Li, Y.; Sha, R.; Zheng, L.; Xu, L.; Xie, H.Q.; Zhao, B. Effects of perinatal TCDD exposure on colonic microbiota and metabolism in offspring and mother mice. *Sci. Total Environ.* **2022**, *832*, 154762. [CrossRef]
98. Qi, Q.; Li, J.; Yu, B.; Moon, J.-Y.; Chai, J.C.; Merino, J.; Hu, J.; Ruiz-Canela, M.; Rebholz, C.; Wang, Z.; et al. Host and gut microbial tryptophan metabolism and type 2 diabetes: An integrative analysis of host genetics, diet, gut microbiome and circulating metabolites in cohort studies. *Gut* **2022**, *71*, 1095–1105. [CrossRef]
99. Wang, Z.; DeWitt, J.C.; Higgins, C.P.; Cousins, I.T. A Never-Ending Story of Per- and Polyfluoroalkyl Substances (PFASs)? *Environ. Sci. Technol.* **2017**, *51*, 2508–2518. [CrossRef] [PubMed]
100. Wan, H.T.; Zhao, Y.G.; Wei, X.; Hui, K.Y.; Giesy, J.P.; Wong, C.K.C. PFOS-induced hepatic steatosis, the mechanistic actions on β-oxidation and lipid transport. *Biochim. Biophys. Acta* **2012**, *1820*, 1092–1101. [CrossRef]
101. Lee, J.-W.; Lee, H.-K.; Lim, J.-E.; Moon, H.-B. Legacy and emerging per- and polyfluoroalkyl substances (PFASs) in the coastal environment of Korea: Occurrence, spatial distribution, and bioaccumulation potential. *Chemosphere* **2020**, *251*, 126633. [CrossRef]
102. Sant, K.E.; Jacobs, H.M.; Borofski, K.A.; Moss, J.B.; Timme-Laragy, A.R. Embryonic exposures to perfluorooctanesulfonic acid (PFOS) disrupt pancreatic organogenesis in the zebrafish, Danio rerio. *Environ. Pollut.* **2017**, *220*, 807–817. [CrossRef] [PubMed]
103. Zeeshan, M.; Zhang, Y.-T.; Yu, S.; Huang, W.-Z.; Zhou, Y.; Vinothkumar, R.; Chu, C.; Li, Q.-Q.; Wu, Q.-Z.; Ye, W.-L.; et al. Exposure to isomers of per- and polyfluoroalkyl substances increases the risk of diabetes and impairs glucose-homeostasis in Chinese adults: Isomers of C8 health project. *Chemosphere* **2021**, *278*, 130486. [CrossRef] [PubMed]
104. Wei, X.; Yang, D.; Zhang, B.; Fan, X.; Du, H.; Zhu, R.; Sun, X.; Zhao, M.; Gu, N. Di-(2-ethylhexyl) phthalate increases plasma glucose and induces lipid metabolic disorders via FoxO1 in adult mice. *Sci. Total Environ.* **2022**, *842*, 156815. [CrossRef] [PubMed]
105. Lv, Z.; Li, G.; Li, Y.; Ying, C.; Chen, J.; Chen, T.; Wei, J.; Lin, Y.; Jiang, Y.; Wang, Y.; et al. Glucose and lipid homeostasis in adult rat is impaired by early-life exposure to perfluorooctane sulfonate. *Environ. Toxicol.* **2013**, *28*, 532–542. [CrossRef] [PubMed]
106. Yan, S.; Zhang, H.; Zheng, F.; Sheng, N.; Guo, X.; Dai, J. Perfluorooctanoic acid exposure for 28 days affects glucose homeostasis and induces insulin hypersensitivity in mice. *Sci. Rep.* **2015**, *5*, 11029. [CrossRef]
107. Lai, K.P.; Ng, A.H.-M.; Wan, H.T.; Wong, A.Y.-M.; Leung, C.C.-T.; Li, R.; Wong, C.K.-C. Dietary Exposure to the Environmental Chemical, PFOS on the Diversity of Gut Microbiota, Associated With the Development of Metabolic Syndrome. *Front. Microbiol.* **2018**, *9*, 2552. [CrossRef]
108. Wang, C.; Zhao, Y.; Jin, Y. The emerging PFOS alternative OBS exposure induced gut microbiota dysbiosis and hepatic metabolism disorder in adult zebrafish. *Comp. Biochem. Physiol. C Toxicol. Pharmacol.* **2020**, *230*, 108703. [CrossRef]
109. Pan, Z.; Yuan, X.; Tu, W.; Fu, Z.; Jin, Y. Subchronic exposure of environmentally relevant concentrations of F-53B in mice resulted in gut barrier dysfunction and colonic inflammation in a sex-independent manner. *Environ. Pollut.* **2019**, *253*, 268–277. [CrossRef]

metabolites MDPI

Review

Are Microplastics Toxic? A Review from Eco-Toxicity to Effects on the Gut Microbiota

Huixia Niu [1,†], Shaojie Liu [2,†], Yujie Jiang [1], Yang Hu [1], Yahui Li [3], Luyang He [3], Mingluan Xing [3], Xueqing Li [3], Lizhi Wu [3], Zhijian Chen [3], Xiaofeng Wang [3] and Xiaoming Lou [3,*]

[1] Health Science Center, Ningbo University, Ningbo 315000, China; 2111101061@nbu.edu.cn (H.N.); 2111101059@nbu.edu.cn (Y.J.); 2011101047@nbu.edu.cn (Y.H.)
[2] Department of Urology, Xijing Hospital, Air Force Medical University, Xi'an 710032, China; liusj1013@fmmu.edu.cn
[3] Zhejiang Provincial Center for Disease Control and Prevention, 3399 Binsheng Road, Hangzhou 310051, China; 881012022129@hmc.edu.cn (Y.L.); 202211125811023@zcmu.edu.cn (L.H.); mlxing@cdc.zj.cn (M.X.); xqli@cdc.zj.cn (X.L.); lzhwu@cdc.zj.cn (L.W.); zhjchen@cdc.zj.cn (Z.C.); xfwang@cdc.zj.cn (X.W.)
* Correspondence: xmlou@cdc.zj.cn
† These authors contributed equally to this work.

Abstract: Emerging studies have presented an initial picture of the toxic effects of exposure to environmental micro- and nanoplastics. They have indicated that micro- and nanoplastics may induce toxicity by leading to oxidative stress, energy metabolism disorders, gene damage, and so forth in environmental organisms, marine invertebrates and vertebrates, and laboratory mouse models. In recent years, micro- and nanoplastics have been discovered in human fecal samples, placentas, lung tissue, and even blood; thus, micro- and nanoplastics pose an alarming and ever-increasing threat to global public health. However, current research on the health effects of micro- and nanoplastics and the possible adverse outcomes in humans has only presented the tip of the iceberg. More robust clinical data and basic experiments are still warranted to elucidate the specific relationships and mechanisms. In this paper, we review studies on micro- and nanoplastic toxicity from the perspectives of eco-toxicity, the adverse effects on invertebrates and vertebrates, and the impact of micro- and nanoplastics on the gut microbiota and its metabolites. In addition, we evaluate the toxicological role of micro- and nanoplastic exposure and its potential implications in respect to human health. We also summarize studies regarding preventive strategies. Overall, this review provides insights on micro- and nanoplastic toxicity and its underlying mechanisms, opening up scientific avenues for future in-depth studies.

Keywords: micro- and nanoplastics; microplastics; polystyrene; toxicity; gut microbiota

1. Background

Micro- and nanoplastics pollution has attracted considerable attention from the international scientific community, and its impact on human health and its associated mechanisms have emerged as leading research frontiers. In 2016, the United Nations Environment Assembly identified marine plastic debris and microplastics as significant global environmental issues. Coinciding with the widespread production and utilization of plastic over the past century, the global annual production of plastic in 2017 exceeded 8.3 billion tons [1]. Nearly half of this plastic was used for disposable packaging, leading to immense generation of plastic waste. By 2050, global plastic waste is estimated to reach 12 billion tons [2]. The limited recycling and reuse of plastics result in the continuous accumulation of plastic waste in the environment [3]. Over time, plastic debris undergoes physical, chemical, and biological processes, fragmenting into micrometer-sized (<5 mm) (called microplastics) and nanometer-sized ($\leq$1 μm) (called nanoplastics) particles [4]. These particles, together with industrially produced plastic particles that are released into the environment, are the major sources of microplastics. In 2004, Richard C. Thompson first introduced the concept of microplastics, which has since garnered increasing research interest and global attention [5]. Based on their origin, microplastics are categorized as primary or secondary microplastics. Primary microplastics are intentionally produced plastic particles with diameters smaller than 5 mm, and they are directly incorporated into products such as cosmetics, exfoliants, and toothpaste; secondary microplastics form via the degradation of environmental plastic [6]. The predominant types of environmental microplastics include polystyrene, polyethylene, polyvinyl chloride, and polypropylene. Notably, polystyrene microplastics, which are frequently detected in the environment, have become the primary focus of microplastic research.

Micro- and nanoplastics are pervasive in the environment, with numerous studies reporting their presence in water, soil, and air [7–12]. They have even been detected in remote areas, such as in the deep sea and the Arctic. Research on Arctic ice has revealed approximately 38–234 microplastic particles per cubic meter of ice [13]. Micro- and nanoplastics have also been detected in everyday items, including cosmetics, exfoliants, and toothpaste. Moreover, micro- and nanoplastics can enter organisms through various pathways, such as through the food chain via commercial fish, as well as through canned food and bottled water [14–16]. Micro- and nanoplastics have been detected in human feces, placentas, lung tissue, and blood, confirming their ability to enter the human body through multiple routes and thus underscoring the importance of addressing their potential hazards [17–20].

Current evidence suggests that micro- and nanoplastics are not easily excreted from organisms once ingested, leading to accumulation in organs and tissues [21]. Micro- and nanoplastics have been found to accumulate in the pancreas and gallbladder of zebrafish, as well as in the intestines, liver, and kidneys of mice [22–24]. Micro- and nanoplastic accumulation would have potentially toxic effects. For instance, microplastics have been shown to decrease catalase activity in zebrafish and to disrupt energy homeostasis in mice [24,25]. Micro- and nanoplastic-associated adverse effects have been observed in a range of organisms; however, the concentration of micro- and nanoplastics utilized in experimental research far exceeds that found in the environment; therefore, the toxic effects of micro- and nanoplastics at environmental concentration levels remain unclear. Furthermore, research on the adverse effects of micro- and nanoplastics on mammals is limited; thus, definitive conclusions regarding micro- and nanoplastic toxicity cannot be drawn based on existing studies. Additionally, the gut microbiota plays a critical role in digestion and absorption, vitamin synthesis, immune response, and gut barrier function, and it is closely related to host health. Upon entering the intestines, micro- and nanoplastics first interact with the gut microbiota, potentially affecting its composition and function [26]. Consequently, the impact of micro- and nanoplastics on the gut microbiota after entering organisms has become a focus of research.

Academic reviews of micro- and nanoplastics have focused primarily on the presence of micro- and nanoplastics in the environment, the biological exposure pathways, and toxic hazards caused to aquatic organisms, and a few articles have focused on the effects of micro- and nanoplastics on mammals. However, neither the history of research on the detection of micro- and nanoplastics that accumulate in tissues and organs nor the relationship between micro- and nanoplastics and populations has been reviewed in the literature, so the threat posed by micro- and nanoplastics to humans has not been highlighted. In addition, there does not appear to be a scientific summary of the alteration of the metabolites of the gut microbiota and metabolic pathways of the organism caused by micro- and nanoplastics. Furthermore, it is known that micro- and nanoplastics can cause various toxic reactions and affect the function of an organism when they enter the organism. Therefore, this article also summarizes the measures to prevent and alleviate the toxic reactions caused by micro- and nanoplastics after entering an organism. In this article, we review the history of micro- and nanoplastic discovery, the potential pathways through which micro- and nanoplastics enter the human body, ecological toxicity, the toxic effects on marine invertebrates and vertebrates and in mouse models, and alterations in the gut microbiota that are caused by micro- and nanoplastics. Additionally, mitigation of the toxic hazards posed by micro- and nanoplastics through bioactive food components and microbiota transplantation is discussed. The objective of this review is to systematically explore the toxicity of micro- and nanoplastics, raise public awareness of their potential hazards, and provide a reference for future micro- and nanoplastic research. Furthermore, this review is intended to offer data support and a scientific basis for the management of micro- and nanoplastic pollution and its associated environmental policies.

2. The History of Micro- and Nanoplastic Discovery

Since microplastics were first introduced by Thompson in 2004 [5], microplastics have received increasing scientific attention worldwide. Furthermore, in recent years, many scientists have also been concerned about the presence of smaller plastic particles (nanoplastics) in the environment and their environmental and biological toxicity. However, current studies on micro- and nanoplastics are mainly focused on the toxicity of micro- and nanoplastics to marine invertebrates and vertebrates. Thus, there is a lack of studies on mammals and humans, even though scientists have recently found microplastics in human tissues. Therefore, it is of significance to pay attention to the potential toxicities caused by micro- and nanoplastics, and this is also why we need to discuss if micro- and nanoplastics are toxic and hence proposed this review. Furthermore, to our knowledge, this is the first review of the history of microplastics discovery (Figure 1).

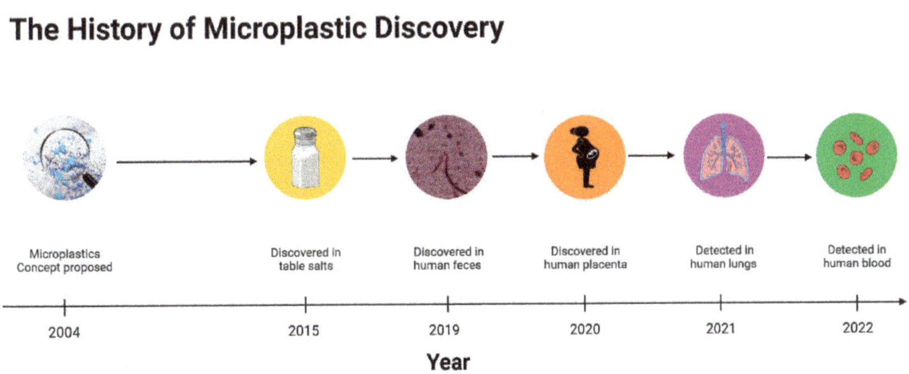

Figure 1. Advances in microplastic discovery. In recent years, scientists have gradually detected microplastics in human tissues.

In 2015, Yang et al. speculated that microplastics might be present in table salt due to its close contact with seawater and lake water [7]. Yang's team proceeded to analyze microplastics in salt samples and found that there were 550–681 microplastic particles/kg in sea salt, 43–364 microplastic particles/kg in lake salt, and 7–204 microplastic particles/kg in rock/well salt. Because table salt is prevalent in the human diet, microplastics may enter the body and threaten human health.

In a prospective cohort study in 2019, 8 healthy adult volunteers, aged 33–65 years, from different countries were invited to provide fecal samples without any interventions [17]. When the fecal samples were examined, researchers found that 3–7 types of microplastic, with an average of 20 pieces of microplastic per 10 g, were detected in each fecal sample, which demonstrates that microplastics can indeed enter the human body through oral exposure.

In 2020, scientists began to find microplastics in human tissue. Moreover, for the first time, Ragusa et al. detected microplastics in human placenta [18]. Ragusa recruited six healthy pregnant women to assess the levels of microplastics in the placenta using Raman Microspectroscopy analysis and found twelve 5–10 μm spherical and irregularly shaped microplastic fragments in four placenta samples, including five on the fetal side, four on the maternal side, and three on the chorionic villi. This suggests that microplastics may potentially be harmful to humans across generations. In 2021, Amato-Lourenço analyzed the lung tissue of 20 nonsmoking adults (lung tissue samples were collected after death via a routine autopsy to verify the cause of death) [19]; in 13 of these samples, microplastic particles and fibers were detected, all of which were less than 5.5 μm in size and constituted an average of approximately 0.56 microplastic particles per gram of lung tissue. In 2022, Leslie developed a double-shot pyrolysis–gas chromatography/mass spectrometry method and applied it to determine the level of microplastic content in the whole blood of humans [20]. The investigators measured the microplastic levels in the whole blood of 22 healthy adult volunteers and found that the mean quantifiable total microplastic concentration in the blood was 1.6 μg/mL. These data indicate that at least some of the microplastics absorbed into the body are bioavailable. However, the manner of entry for the microplastics into the blood and the cells involved remains to be studied.

The abovementioned studies show that microplastics are not only ubiquitous in the environment but also detected in humans. As such, the problem of microplastic pollution is not only an environmental issue but also a global public health issue.

3. Micro- and Nanoplastic Exposure Routes

The non-degradable and ubiquitous nature of micro- and nanoplastics makes it inevitable that organisms, especially humans, are exposed to microplastics in the environment. Studies have shown that humans are exposed to environmental micro- and nanoplastics mainly through routes such as the ingestion of micro- and nanoplastics through food and food packaging materials, the inhalation of microplastic particles and fibers floating in the air, and daily skin contact with micro- and nanoplastics in cosmetics and skin-cleansing products (Figure 2).

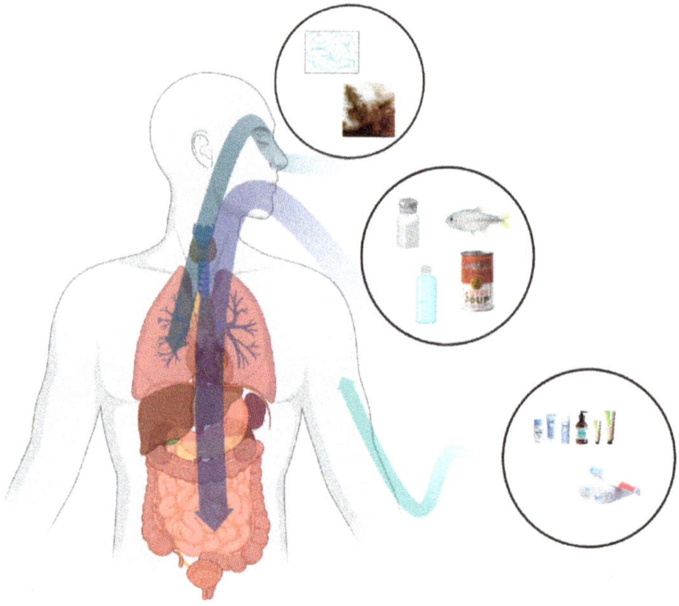

Figure 2. Pathways of human exposure to microplastics mainly include ingestion, inhalation, and dermal contact. Ingestion is the most important route of exposure, and dermal contact is considered the least important route of exposure. For example, micro- and nanoplastics in salt, commercial fish, bottled water, and canned food can enter the human body through ingestion, those in floating fibers and dust in the air can enter the human body through inhalation, and those in toothpaste and skin-cleaning products can enter the human body through skin contact.

3.1. Oral Exposure

Oral exposure is considered to be the predominant microplastic exposure route [27]. Researchers have found micro- and nanoplastics in honey, sugar, bottled water, canned foods, edible salt, and commercial fish [7,14–16,28]. Particulate matter enters organisms orally and reaches the gastrointestinal tract, where it may cause inflammatory responses, oxidative stress, alterations in intestinal permeability, and changes in the composition and function of the gut microbiota [26].

First, micro- and nanoplastics in surface water, groundwater, tap water, and bottled water have gradually been detected by scientists. Surface water, the main source of drinking water, has been found to contain high levels of microplastics, and the maximum concentration of microplastics could reach 44,435 items/km^2 [8]. In addition, Mintenig et al. analyzed the presence of microplastics in groundwater and groundwater-purified drinking water by collecting samples from different locations in the drinking water supply chain; they found that the concentration of microplastics ranged from 0 to 7 particles/m^3, and the particle size was between 50 and 150 μm [29]. In general, however, the level of microplastic contamination in tap water is low and the concentration of microplastics entering human body is negligible. Therefore, bottled water in plastic packaging has received greater attention. Mason tested bottled water from 9 countries and found an average concentration of 10.4 microplastic particles/L in 259 selected bottles with particle sizes that were greater than 100 μm; the average concentration in the particle size range 6.5–100 μm was 325 microplastic particles/L [16]. In addition, the researchers found that these microplastics matched the common plastics used to make bottle caps, so the contamination may have come from the bottling process and the packaging itself. Schymanski also found microplastics in bottled water, where the concentration in recyclable bottles was 118 ± 88 microplastic particles/L, and in single-use bottles, it was 14 ± 14 microplastic particles/L [30].

It is well known that microplastics are transferred through the food chain into higher-trophic-level organisms. Mattsson's study found the transportation of microplastics in the algae–daphnia–freshwater fish food chain [14]. Ultimately, microplastics may enter the human body through the food chain. For example, Cauwenberghe used mussels as a vehicle to study the potential threats of microplastics in marine products to humans through the food chain [31]. It was found that when consuming an average serving (e.g., 250 g wet weight) of mussels, a person can consume roughly 90 microplastic particles. Based on this, it can be calculated that the largest consumer of mollusks in Europe will consume up to 11,000 microplastic particles per year. Moreover, as microplastics accumulate, they are biomagnified in higher trophic levels of organisms. Similar to studies on bottled water, those on canned foods have also concluded that incorrect handling during processing can increase the concentration of microplastics in foods [15]. In such a case, Li et al. compared live mussels with pre-treated (frozen or further processed) mussels and found that live mussels contained 0.9 items/g of microplastics, while processed mussels contained 1.4 items/g [32].

Food packaging materials are also the source of microplastics in terms of food pollution. Substances such as the compound monomers and additives left in food packaging materials may migrate into the food with which they come into contact [33]. To extend the shelf life and freshness of food, and to improve the properties of packaging materials, nanoparticles are increasingly being used in food packaging and can also migrate into food [34,35]. Scientists speculate that microplastics in packaging materials may also contaminate food products that come into contact with them. However, there is a relative lack of research on contamination due to microplastic particles in packaging materials and the threat these microplastics pose to the environment and human health.

3.2. Respiratory Exposure

Micro- and nanoplastics that are suspended in the air are mainly from synthetic fibers, material wear, and the resuspension of surface microplastics [36]. It is well known that the annual production and use of synthetic fibers have been increasing year-by-year in recent years [37]. Browne's sampling of household washing machines found that 1 load of laundry can produce > 1900 fibers per wash [38]. In addition, tire wear contributes significantly to the flow of micro- and nanoplastics into the air. The per capita emissions of particulate matter from car tire wear range from 0.23 to 4.7 kg/year [39]. It is estimated that, in the Netherlands, roughly 17,000 t of tire micro- and nanoparticles are created and released into the environment each year [40]. Micro- and nanoplastics enter the air and are deposited together with dust on the surfaces of roads or objects. Low-density polymers are easily resuspended in the air due to the wind or air movement caused by vehicles, and they can enter the human body through the respiratory system.

In addition, researchers have demonstrated that the human body is exposed to micro- and nanoplastics through the respiratory system by measuring microplastic levels in air and human lung tissue, simulating human exposure to microplastics in the air. Liao selected 13 sites by which to sample both indoor and outdoor air and found that the concentration of microplastics in indoor air (1583 ± 1180 n/m^3) was significantly higher than that in outdoor air (189 ± 85 n/m^3) [41]. Europeans spend around 90% of their time indoors for work and living, resulting in the majority of human exposure to airborne microplastics occurring indoors [42]. Vianello et al. mimicked human exposure to microplastics in indoor air via a breathing thermal manikin [43]. Their sample analysis showed that all the samples were contaminated by microplastics, and the concentrations ranged from 1.7 to 16.2 microplastic particles/m^3. More interestingly, scientists have found microplastics in human lung tissue, whereby 33 microplastic particles and 4 fibers were detected in 13 of 20 samples, as well as 39 particles in 11 of the 13 samples, with an average concentration of 0.69 ± 0.84 microplastic per gram of lung tissue, thereby revealing the respiratory exposure pathway to microplastics in humans [19,44].

3.3. Dermal Exposure

Skin contact is considered to be the least important but most common exposure route due to the use of microplastics in personal care products (e.g., cosmetics, toothpaste, skin-cleansing products). For purposes such as exfoliation, viscosity adjustment, and emulsification, microplastics are widely used as additives in cosmetic and skin-cleansing products [45]. Depending on the functions, different types and sizes of micro- and nanoplastics are selected. Sun et al. state that the particle size of microplastics that are used in cosmetics ranges from 24 μm to 2 mm, and more than 95% were found to be smaller than 350 μm [46]. Praveena surveyed 214 volunteers from Malaysia on cosmetic and personal care products and found that among the selected products, the particle size of microplastics in face wash/scrubs ranged from 10 to 178 μm [47]. Hernandez et al. drew the same conclusion when they examined three commercial face washes containing polyethylene microbeads, whereby they found nanoplastics with particle sizes that ranged from 24 ± 6 nm to 52 ± 14 nm [48].

There is no study to prove that nanoplastics can cross the skin barrier and enter the organism. However, many studies have shown that when humans are exposed to nanoparticles through the dermal contact route, the nanoparticles can enter the body through the skin barrier and cause toxic reactions [49]. Therefore, scientists have speculated that the direct contact between nanoplastics and human skin during the use of cosmetics and personal care products allows nanoplastics to enter the human body through the skin barrier.

4. Toxic Effects of Micro- and Nanoplastics

In recent years, more attention has been paid to the underlying toxic effects of micro- and nanoplastics on the environment and organisms. Studies have revealed that micro- and nanoplastics interacting with the environment can cause eco-toxicity. Micro- and nanoplastics are absorbed by environmental organisms and can enter into consumers through food chain transfer and nutrient transfer, thus causing cytotoxic reactions such as oxidative stress and inflammatory reactions. The normal function of the nervous system and immune system can also be affected by micro- and nanoplastics, causing neurodegenerative diseases and immune system dysfunction.

4.1. Eco-Toxicity

Micro- and nanoplastics accumulate when they enter the environment, where they first cause eco-toxicity in environmental organisms such as plants, earthworms, and oysters (Table S1).

Micro- and nanoplastics act on microorganisms in the environment firstly, affecting their activity, metabolism and ability to break down organic matter, thus affecting ecosystem function. Machado et al. focused on the effects of microplastics on the soil microbiota and illustrated that the increase in microplastic concentration drove an improvement in microbiota activity [50]. The same conclusion was obtained in Liu's study, where high concentrations of polypropylene accelerated the degradation of organic matter in the soil, leading to different metabolite distributions after 7 and 30 d of exposure [10]. However, Lopez-Rojo, when studying the effects of different concentrations of polystyrene microplastics on the decomposition process of dead leaves, demonstrated that the decomposition of leaf litter gradually decreased with increasing microplastic concentrations. There was a significant trend only with the co-existence of harmful substances, and microbiota mediated decomposition as well [51]. Another study suggested that the presence of microplastics in water may affect water ecosystem function, and there is a positive correlation between the concentration of microplastics and the hazards they pose [52]. Additionally, Wang found that microplastic residues caused by the use of mulch in agricultural production contributed to a significant decrease in microbial C, N, and enzyme activities, as well as a significant decrease in microbiota diversity in soil [53].

In addition to affecting the composition of microbiota and their activity, microplastics that are present in the environment are absorbed by plants, affecting seed germination and plant growth. Bosker used a 72 h bioassay to study the effects of microplastic particles of different sizes on cress (*Lepidium sativum*) growth [54]. It was found that, after 8 h of exposure, the germination rate of all microplastic-treated seeds was significantly reduced with the increase in microplastic particle size. In that study, the germination rate of the cress in the 4800 nm nanoplastic group decreased from 78% to 1.7% when compared to the control group. However, more interestingly, the root growth significantly increased after 24 h of exposure to 50 nm nanoplastic, but this decreased significantly after 24 h of exposure to 500 nm nanoplastic. Furthermore, Nolte's and Zhang's studies both elucidated that micro- and nanoplastic particles can reduce photosynthetic efficiency by directly altering the chloroplast fatty acid content, as well as changing the structure of the photosynthetic complex and reducing the chlorophyll content of microalgae (*Skeletonema costatum*), thus resulting in the growth inhibition of microalgae [55,56]. Similarly, Green's study on the effects of microplastics on freshwater ecosystems showed that microplastics significantly reduce the root length and biomass of floating duckweed [52].

Invertebrates can also respond to the eco-toxicity of micro- and nanoplastics. Earthworms (*E. Florida*) were used to study the toxic response to microplastics, and it was concluded that the growth rate of earthworms decreased while the mortality rate increased in the group exposed to high concentrations of microplastics; increasing glutathione (GSH) levels were observed in the earthworms in a dose- and exposure-time-dependent manner [57]. The same conclusion was obtained in Jiang's experiments, i.e., that nanoplastic exposure may cause oxidative stress, a significant increase in GSH content, and a decrease in superoxide dismutase (SOD) activity in earthworms. In addition, the oxidative damage and DNA damage caused by 14 d exposure to 1300 nm nanoplastics was significantly higher than those for 100 nm nanoplastic [58]. Tlili found that microplastic exposure significantly inhibited the acetylcholinesterase (AChE) activity of wedge clams (*Donax trunculus*), suggesting that the accumulation of microplastics was related to the potential neurotoxicity of wedge clams [59]. Sussarellu's study showed that microplastic exposure can cause reproductive toxicity via reducing the oocyte diameter and sperm velocity in oysters after 2 months of exposure to polystyrene microplastics [60]. Larval production and the development of offspring in the exposure group were reduced by 41% and 18%, respectively, when compared to the control group. These data suggest that microplastics not only produce toxic responses in the parents but also present transgenerational hazards that affect the growth and development of offspring.

4.2. Marine Invertebrates and Vertebrates

Micro- and nanoplastics are absorbed by plants and then pass through the food chain. Then, they are transferred via nutrients into animals, thus causing toxic reactions. Some fish and birds are a better basis for speculating on the harmful effects of micro- and nanoplastics on humans than ecosystems. Therefore, the effects of micro- and nanoplastics on marine invertebrates and vertebrates should be of wide concern. Although an increasing number of scientists are concerned about the toxicity of micro- and nanoplastics, the specific mechanisms of toxicity remain unclear (Table S2).

After entering animals' bodies through various routes, micro- and nanoplastics are not simply excreted from the body but accumulate in the tissues. In zebrafish embryos, nanoplastics were found in the yolk sac of fertilized eggs 12 h after fertilization, following their exposure to polystyrene nanoplastics 6 h after fertilization. Furthermore, nanoplastics migrate and accumulate in the gastrointestinal tract, liver, pancreas, bile, heart, and even brain of zebrafish during their growth and development, and they also exist in the maternal–infant transmission process [21,61]. Microplastics have been found in terrestrial animals, and seabirds are thought to be the mediators of these pollutants, i.e., they provide a transmission route from the marine environment to the terrestrial environment [62]. Provencher examined 186 thick-billed murres from the eastern Canadian Arctic and found that 11% of thick-billed murres had microplastics in their gastrointestinal tracts with an average of 0.2 ± 0.8 microplastic particles per bird [63].

The accumulation of micro- and nanoplastics in the tissues of organisms first causes histopathological changes. Xia et al. studied the effect of polyvinyl chloride microplastics on the growth of carp and found that, when compared with the normal radiolucent arrangement of carp hepatocytes in the control group, the hepatocytes in the exposure group were visibly loosened and the cell vacuolation was increased [64]. Moreover, micro- and nanoplastics can cause dilated hepatic sinusoids and hypertrophy and necrosis of hepatocytes, and they can increase the extracellular matrix in liver tissues [23,65]. Additionally, exposed to polystyrene microplastics, the zebrafish intestine showed not only thinning of the intestinal wall and congestive inflammation but also impairments and ruptures in the villi and epithelium, lysis of enterocytes, and a disruption of the integrity of the epithelial barrier at high levels [66,67]. In marine medaka (*Oryzias melastigma*), Wang found that microplastics caused damage to gills and testes. Aside from the liver and intestinal tissues, the gills underwent physical changes, such as a loss of gill lamellae and the loosening of gill filaments after 60 days of microplastic exposure. In addition, the testes of marine medaka showed blurring of the spermatophore structure and lysis of the basement membrane when exposed to high concentrations of microplastics [65].

Micro- and nanoplastics generally enter into the organism and cause cytotoxicity due to oxidative stress and inflammatory responses [36]. In a zebrafish study, catalase (CAT) and SOD activities were significantly higher when compared to the control group. In addition, the expression of the inflammatory gene IL-1β and oxidative stress-associated genes was also higher, thus indicating the occurrence of oxidative stress [23,25,64,66,68,69]. It has also been discovered that exposure to micro- and nanoplastics disrupts the feeding behavior of organisms, thus reducing energy intake by affecting the nervous system (i.e., increased feeding times, decreased feeding behavior, etc.) [70,71]. One study showed that microplastics contribute to changes in neurotransmitter activity; specifically, AChE activity was decreased in Amazonian cichlid heads, thus inhibiting their cholinergic neurotransmission [71]. Further, micro- and nanoplastics have been shown to disrupt immune system function, causing the degranulation of primary neutrophil granules and the release of neutrophil extracellular traps (NETs), thus suggesting impaired innate immune function [72]. Similarly, Karami's study found that microplastics cause decreased globulin levels in fish, which indicates an immunosuppressive response [73].

4.3. Mouse Models

Although rarer in these studies, terrestrial mammals are more representative of humans for speculating on the harm of micro- and nanoplastics to humans. Most of the current studies on the effects of micro- and nanoplastics on mammals use mouse models, and the toxicities induced by exposure to micro- and nanoplastics are listed in Table 1.

Similar to those in marine invertebrates and vertebrates, micro- and nanoplastics first give rise to accumulation in the tissues such as liver, kidney, and gastrointestinal tract tissue, and then, they lead to toxic reactions [24,74]. The long-term interaction of micro- and nanoplastics with tissues will result in histopathological changes. For instance, in some studies, the liver index (liver weight/body weight) of the mice in the microplastic-exposed group increased when compared to the control group, and the microplastics induced severe vacuolar degeneration of the liver tissue and caused hepatocyte edema [75,76]. In mice, Jin and Lu also found that micro- and nanoplastics cause a decrease in the amount of intestinal mucin and mucus secretion, as well as inducing intestinal barrier dysfunction [26,77].

Table 1. Toxic effects of microplastic exposure on laboratory mice.

Properties of Microplastics Used	Toxicity	Reference
Polystyrene microplastics (5–20 μm)	• Accumulated in the kidney, liver, and intestine of mice, and the highest bioaccumulation factor was found in the intestine	[74]
Polystyrene microplastics (5 μm)	• Accumulated in the intestinal tissues and also reduced intestinal mucus secretion and impaired intestinal barrier function • Changes in the intestinal flora diversity with a significant decrease in actinomycetes • Significant changes in metabolic pathways such as pyruvate metabolism, tyrosine metabolism, and fatty acid biosynthesis	[26]
Polystyrene nanoplastics (0.5 μm)	• Increased liver weight, liver index, and liver function indicators • Up-regulation of interferon-γ, TNF-α, IL-1β, IL-6, and IL-33 mRNA expression in non-parenchymal hepatocytes • Down-regulation of IL-4, IL-5, IL-10, IL-18, and transforming growth factor-β1 expression	[75]
Polystyrene microplastics (5 μm)	• Accumulated in the liver tissue accompanied by tissue vacuolar degeneration, chronic inflammatory cell infiltration, and hepatocellular edema • Decreased T-SOD, CAT, and GSH activities and increased MDA levels • L02 hepatocyte rate of apoptosis increased	[76]
Polystyrene micro- and nanoplastics (0.5 μm and 50 μm)	• Reduced body weight, liver weight, and lipid weight in the mice • Decreased intestinal mucus secretion • Relative abundance of *Firmicutes* and *α-Protebacteri* were reduced • Lower liver triglyceride and total cholesterol levels • Decreased mRNA levels of certain key genes associated with adipogenesis and triglyceride synthesis	[77]
Polystyrene microplastics (10–150 μm)	• Increased abundance of *Staphylococcus* and decreased abundance of *Paramecium* • Elevated IL-1α levels • The small intestine showed a significant inflammatory response, as well as increased expression of TLR4, AP-1, and IRF526	[78]
Polystyrene microplastics (5 μm and 20 μm)	• Accumulation in both the kidney and intestine, with tissue accumulation kinetics and distribution patterns dependent on microplastic particle size • Caused disturbances in energy and fat metabolism, as well as causing oxidative stress and neurotoxic reactions	[24]
Polyethylene micro- and nanoplastics (3–16 μm, 100 nm, and 600 nm) Polystyrene micro- and nanoplastics (10 μm, 40 nm, and 250 nm)	• Increased ROS generation	[79]

Table 1. *Cont.*

Properties of Microplastics Used	Toxicity	Reference
Polypropylene microplastics (<200 μm)	• Elevated IL-6 and TNF-α levels • Elevated ROS levels • Caused erythrocyte hemolysis in a concentration-dependent manner • Increased the secretion of histamine, which induces allergic reactions at the cellular level	[80]
Polystyrene microplastics (1 μm, 4 μm, and 10 μm)	• Accumulation in the intestinal tract • Causes a decrease in cell viability at higher concentrations • Macrophage uptake of microplastics was followed by polarization	[81]
Polystyrene nanoplastics (23–26 nm)	• Accumulation in the mouse brain • Changes in anxiety-like behavior and anti-predator defense responses in the face of predators • Reduced DPPH radical scavenging activity and reduced total GSH content • Appearance of DNA damage	[82]
Polystyrene microplastics (5 μm)	• Caused metabolic disorders, intestinal flora dysbiosis, and intestinal barrier dysfunction in the mother • Caused intergenerational effects with long-term metabolic consequences in the F1 and F2 mice • The possibility of hepatic lipid accumulation in the F1 generation mouse in adulthood	[83]

Micro- and nanoplastics frequently cause inflammatory responses and oxidative stress when entering organisms. In an experiment investigating the effect of polyethylene microplastics on the development of inflammation in mice, Li demonstrated that the serum levels of IL-1α were significantly higher in the microplastic-exposed group at different concentrations [78]. Additionally, the small intestine of mice fed with high concentrations of nanoplastics exhibited a pronounced inflammatory response with increased expressions of TLR4, AP-1, and RF526. Similarly, Zhao found that interferon-γ, TNF-α, IL-1β, IL-6, and IL-33 mRNA expressions were up-regulated and IL-4, IL-5, IL-10, IL-18, and transforming growth factor-β1 expressions were down-regulated in liver non-parenchymal cells after nanoplastic exposure, indicating that nanoplastics disrupt the inflammatory process in liver tissues [75]. The increased level of oxidative stress that was induced by micro- and nanoplastics in mice was reflected by the decreased activities of antioxidant enzymes containing T-SOD, CAT, and GSH, as well as by the increased reactive oxygen species (ROS) levels [24,76,84]. In addition to the aforementioned, the ROS levels and inflammatory factors IL-6 and TNF-α were reproducible when human-derived cells were exposed to micro- and nanoplastics [79,80].

Interestingly, micro- and nanoplastic accumulation in tissues also causes metabolic disorders in mice. It was found that the continuous accumulation of microplastics in tissues led to a significant decrease in T-CHO and TG, which also led to lipid droplet accumulation and the relative mRNA levels of the key genes related to adipogenesis and triglyceride synthesis being reduced in liver and epididymal lipids, thus causing hepatic lipid metabolism disorders [24,74,81]. The energy metabolism of mice is also disturbed by microplastics, causing lower ATP levels, higher LDH activity, and alterations in energy-related metabolites including creatine, 2-oxoglutarate, and citric acid [24]. In the nervous system, researchers observed a decrease in AChE activity and cholinergic neurotransmission efficiency in mice treated with nanoplastics. In addition, some of these mice showed increased levels of neurotransmitters such as threonine, aspartate, and taurine, suggesting that nanoplastics can induce neurotoxicity in mice. This could also be reflected by nanoplastic-induced behaviors that are similar to anxiety disorders and anti-predator defense responses that occur when mice are confronted with potential predators in field experiments [82]. Furthermore, micro- and nanoplastic exposure induces DNA damage and intergenerational effects in mouse experiments [82,83] (Figure 3).

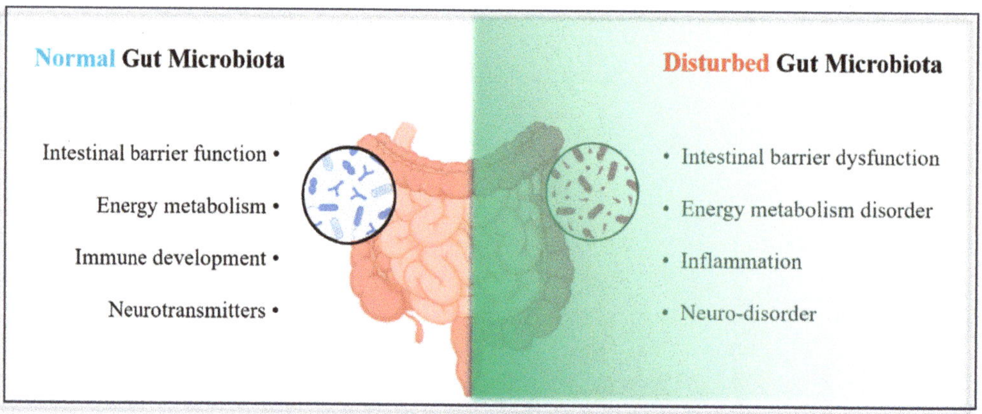

Figure 3. Normal gut microbiota versus micro- and nanoplastic-disturbed gut microbiota.

4.4. Impact of Micro- and Nanoplastic Exposure on the Gut Microbiota

The gastrointestinal tract is a particularly complex system that plays an important role in the intake and absorption of nutrients in humans. Micro- and nanoplastics enter and accumulate in the intestine and interact with tissues, affecting the intestinal barrier's functioning [85]. The intestinal barrier can be divided into the physical barrier, chemical barrier, immune barrier, and microbiological barrier, which consists of a microbiota that maintains intercellular connections and promotes epithelial cell damage repair [86,87]. In their review, Ley et al. noted that there are approximately 10^{14} microbiota present in the human gut, and they emphasized the important role that microbiota play in regard to human health [88]. It has been demonstrated that the host's health is significantly influenced by the gut microbiota [89]. The gut microbiota is involved in the regulation of host physiological functions, such as the regulation of intestinal motility and secretion, the decomposition of macromolecular polysaccharide compounds in food, participation in the biosynthesis of vitamins and nutrients, participation in the digestion and absorption of nutrients, maintenance of the integrity of the epithelial barrier, and promotion and maintenance of the normal development of the immune system and its activities [90–92]. Nevertheless, the accumulation of micro- and nanoplastic particles in the intestine can result in dysbiosis of the gut microbiota, which is closely linked to several illnesses, including diabetes, cardiovascular diseases, hypertension, and others [90–92].

4.4.1. Changes at the Compositional Level

Many studies have found that the entry of micro- and nanoplastics into the intestine causes disturbances in the gut bacteria, in terms of both composition and function (Table 2).

Significant changes in the diversity and composition of the gut microbiota are present. In Chinese mitten crab (*Eriocheir Sinensis*), after exposure to 40 mg/L of polystyrene microplastics (with a particle size of 5 μm), the relative abundance of *Firmicutes* and *Bacteroidetes* decreased, whereas the relative abundance of *Fusobacteria* and *Proteobacteria* increased [93]. Further, marine medaka (*Oryzias melastigma*) exposed to polystyrene micro- and nanoplastics at 45 μm and 50 nm showed increased α-diversity of the gut microbiota and a decreased relative abundance of *Bacteroidetes* [94]. Similarly, Zhao et al., who studied the effects of microplastics on zebrafish, found that the exposure of zebrafish to 1–4 μm of polyethylene microplastics for 7 days aroused significant changes in the abundance of *Bacteroidetes*, *Firmicutes*, *Proteobacteria*, and *Verrucomicrobia* in the intestine [95]. In addition, at the genus level, the abundance of *Aeromonas*, *Shewanella*, *Microbacterium*, *Nevskia*, and *Methyloversatilis* increased significantly, while the abundance of *Pseudomonas*, *Ralstonia*, and *Stenotrophomonas* decreased significantly. Soil collembolans exposed to 80–250 μm polyvinyl chloride microplastics for 56 days showed a significant increase in gut microbiota diversity, with a significant change in the microbiota and the appearance of a large number of unique OUTs [96]. In a mouse model, the diversity, amount of bacteria, and *Staphylococcus* abundance were increased, while the *Parabacteroides* decreased significantly, in mice exposed to different concentrations of microplastics (6, 15, 60, and 600 μg/d) [78]. Conversely, in another study, it was found that when mice were exposed to 50 μm and 0.5 μm polystyrene microplastics, their gut micro- and nanobiota diversity was significantly reduced; in the 0.5 μm exposure group, 310 OUT intestinal microbiota were changed, and at the gate level, the relative abundance of *Firmicutes* and *α-Protebacteri* in feces was significantly decreased [77]. Similarly, Jin found that 5 μm polystyrene microplastics led to dysbiosis of the intestinal microbiota, as well as significant changes in 15 bacterial species at the genus level in mice [26].

Table 2. Gut microbiota composition changes after microplastic exposure.

Species	Properties of Microplastics Used	Changes in Intestinal Microbiota	Reference
Chinese mitten crab (*Eriocheir sinesis*)	Polystyrene microplastics (5 μm)	• Reduced diversity of gut microbiota • The relative abundance of *Firmicutes* and *Bacteroidetes* decreased, and the relative abundance of *Fusobacteria* and *Proteobacteria* increased	[93]
Marine medaka (*Oryzias melastigma*)	Polystyrene micro- and nanoplastics (45 μm and 50 nm)	• Increased α-diversity of gut microbiota in the exposed group • The relative abundance of *Bacteroidetes* and *Vicingus* decreased • The relative abundance of *Lewinella*, *Pseudomonas*, *Thalassospira*, and *Parahaliea* increased	[94]
Larval zebrafish	Polystyrene microplastics (1–4 μm)	• Firmicutes, Bacteroidetes, Proteobacteria, and Verrucomicrobia changed significantly at the gate level • At the genus level, the relative abundance of Aeromonas, Shewanella, Microbacterium, Nevskia, and Methyloversatilis increased significantly, while the relative abundance of Pseudomonas, Ralstonia, and Stenotrophomonas decreased significantly	[95]
Collembolans	Polyvinyl chloride microplastics	• Increased gut microbiota diversity • The relative abundance of *Bacteroidetes* decreased, and the relative abundance of *Firmicutes* increased	[96]
Mouse	Polyethylene microplastics (10–150 μm)	• Increased diversity of gut microbiota • The relative abundance of *Staphylococcus* increased, and the relative abundance of *Parabacteroides* decreased	[78]
Mouse	Polystyrene micro- and nanoplastics (50 μm and 0.5 μm)	• At the gate level, the relative abundance of *Firmicutes* and *α-Protebacteri* was reduced • A total of 6 and 8 bacterial species were altered by exposure to 0.5 and 50 μm micro- and nanoplastics, respectively, at the genus level	[77]
Mouse	Polystyrene microplastics (5 μm)	• Changes in α-diversity and β-diversity occurred • A total of 15 species were significantly altered at the genus level	[26]

4.4.2. Changes at the Metabolite Level

It is not enough to study the structure and composition of microbiota; the function of the microbiota should also be considered when assessing whether micro- and nanoplastics are affecting the homeostasis of the gut microbiota. It is acknowledged that gut microbiota participate in regulating the body's metabolism, intestinal motility, intestinal barrier function, and the digestion and absorption of nutrients, as well as maintenance of the normal function of the immune and nervous systems. Meanwhile, the functions that could interfere with the toxicity of microplastics are mainly a result of the metabolic reprogramming of the gut microbiota (Table 3).

Microplastics cause disruptions in the composition and structure of the gut microbiota and consequently in its metabolic functions: carbohydrate metabolism, lipid metabolism, amino acid, energy metabolism, etc. Microplastic-exposure-induced dysbiosis of the gut microbiota has been reported to significantly alter carbohydrate metabolism; the expression of the pentose phosphate metabolic pathway (i.e., a glucose catabolic pathway that is prevalent in microbiota) was significantly reduced [97]. In addition, fructose and mannose metabolism in human Caco-2 cells was inhibited with down-regulated mannose metabolism gene expression. Inhibition of carbohydrate metabolism also occurred in marine medaka (*Oryzias melastigma*) and in mealworms (*Tenebrio molitor*) [98,99]. In addition, normal lipid metabolism was disturbed after long-term exposure to polystyrene microplastics, both in the gut microbiota and in the host [98,100]. In a recent study, microplastics were demonstrated to reduce the level of taurocholic acid (TCA) via increasing the ratio of *Bifidobacteria* to *Bacteroide* in the intestine, thereby reducing lipid absorption in mice [101]. Microplastic exposure also alters the amino acid and energy metabolism of chicken gut microbiota; this is accompanied by a significant decrease in L-serine and ornithine levels, which have a variety of important functions in the developmental stages, such as the synthesis of neurotransmitters, proteins, and sphingolipids [102]. Peng's study showed that polystyrene microplastics interfere with histidine metabolism in the mealworm gut microbiota, which is followed by alterations in metabolic pathways, such as pentose phosphate metabolism, alanine metabolism, and glutamate metabolism [99]. In another study, in mouse gut microbiota, microplastic exposure for 6 weeks resulted in significant differences in the metabolic pathways of tyrosine-functional genes [26].

Moreover, microplastics affect the synthesis of bile acids by intestinal flora, the production of short-chain fatty acids and neurotransmitters, and the normal function of the body. It is known that Gram-positive bacteria promote the synthesis of bile acids in the intestine. In an experiment on the effect of polyethylene microplastics on the intestinal microbiota, it was found that the proportion of *Firmicutes* decreased, resulting in a reduction in secondary bile acid synthesis [97]. Qiao's study showed that exposing mouse intestinal microbiota to 5 μm and 70 nm polystyrene micro- and nanoplastics resulted in a decrease in short-chain-fatty-acid-producing bacteria and an increase in Gram-negative bacteria [103]. Because short-chain fatty acids and lipopolysaccharide production by Gram-negative bacteria are closely related to intestinal barrier function, the investigators suggested that micro- and nanoplastics acting on intestinal tissues may indirectly disrupt barrier function by modulating the microbial structure. In addition, decreases in *Lactobacillus* and *Bifidobacterium*, which are the main γ-aminobutyric acid (GABA)-producing bacteria in the gut, result in a decrease in the GABA content, which thus further alters neurotransmitter synthesis and affects the neurotransmission and behavior of fish [104]. Studies on how micro- and nanoplastics affect the function of the immune system through the gut microbiota are incredibly limited. However, Kang et al. found that micro- and nanoplastics affect gut mucus secretion; this was noted when there was a significant increase in the amount of mucus in the exposed group when compared to the control group. Because mucus contains a variety of microbiota, it plays a key role in a variety of physiological functions, including immune function [94].

Table 3. Gut microbiota changes in metabolites due to microplastic exposure.

Species	Properties of Microplastics Used	Changes in Gut Microbiota Metabolome	Reference
Simulation in vitro with human cell Caco-2 and gut microbiota	Polyethylene microplastics (30–140 μm)	• Significantly lower expression in the metabolic pathway of pentose phosphate metabolism • Mannose metabolism gene expression is down-regulated, and fructose and mannose metabolism is inhibited • Decreased secondary bile acid synthesis	[97]
Marine medaka (*Oryzias melastigma*)	Polystyrene microplastics (2.5 μm)	• Significant reduction in carbohydrate metabolic pathways • Significant up-regulation of fat metabolic pathways	[98]
Mealworms (*Tenebrio molitor*)	Polystyrene microplastics	• Interference with the starch and sucrose metabolism of intestinal bacteria • Interference with the histidine metabolism of gut microbiota (related to pentose phosphate metabolism, alanine metabolism, and glutamate metabolism pathways) • Disrupts glyoxylate and dicarboxylic acid metabolism (key energy metabolic pathways)	[99]
Mouse	Polystyrene microplastics (5 μm)	• Significant alterations in the metabolic pathways of tyrosine functional genes	[26]
Rare minnow (*Gobiocypris rarus*)	Polystyrene microplastics (1 μm)	• Altered amino acid metabolic pathways • Disorders of normal lipid metabolism	[100]
Mouse	Polystyrene microplastics (5 μm)	• Decreased TCA and triglyceride levels	[101]
Chicken	Polyethylene microplastics	• Significant alterations in amino acid metabolic pathways (accompanied by significant decreases in L-serine and ornithine levels)	[102]
Mouse	Polystyrene micro- and nanoplastics (5 μm and 70 nm)	• Significant damage to the intestine and decreased expression of tight-junction proteins • Decrease in short-chain-fatty-acid-producing bacteria and increase in Gram-negative bacteria	[103]
Discus fish (*Symphysodon aequifasciatus*)	Polystyrene nanoplastics (~80 nm)	• *Lactobacillus* and *Bifidobacterium* (i.e., the main GABA-producing bacteria) decreased • Altered neurotransmitter synthesis that resulted in behavioral changes in discus fish	[104]
Marine medaka (*Oryzias melastigma*)	Polystyrene micro- and nanoplastics (45 μm and 50 nm)	• The group exposed to 50 nm polystyrene nanoplastics exhibited stronger oxidative stress • The group exposed to 45 μm polystyrene microplastics exhibited intestinal damage, as well as an increase in gut mucus secretion and alterations to the gut microbiota	[94]

4.5. Micro- and Nanoplastics and Population

Although more and more scientists are studying micro- and nanoplastics, most of the studies on their toxicity involve ecological, non-mammalian, and laboratory mouse models, but relatively few studies have been conducted on human populations. Current research on micro- and nanoplastics relevant to humans is limited to the discovery of their accumulation in human tissues and in vitro experiments with human cells. After scientists discovered the presence of microplastics in placenta by Raman spectroscopy in 2021 [18], in 2023, Zhu et al. used infrared spectroscopy to characterize microplastics in placenta. A total of 11 types were found in the placenta, with particle sizes ranging from 20.34 μm to 307.29 μm, most of which were predominantly fragments (<100 μm) and fibers (200–307.29 μm) [105]. Furthermore, systemic toxicity of microplastics to the placenta was postulated by a machine learning approach [106]. One researcher found that polycarbonate, polyethylene terephthalate, and polystyrene exhibited the highest toxic effects on all enzymes. These microplastics can effectively identify the site of enzyme activity and pose a significant risk to the placenta through inhibition of key enzymes. Whether these micro- and nanoplastics in the placenta make it into the fetus is critical. Braun et al. collected maternal placenta and meconium to test for the presence of microplastics [107]. The results showed that polyethylene, polypropylene, polystyrene, and polyurethane were present in placenta and meconium, and their particle sizes were all > 50 μm. In addition to the placenta, the researchers found through a small prospective cohort study that microplastics in fetuses may also come from breastfeeding, feeding bottles, and plastic toy use [108]. These microplastics accumulate in the fetus, and liu et al. found that they can affect the composition of fetal gut microbiota and cause microbiota disturbance [109]. The gut microbiota was mainly composed of Proteus, Bacillus, and Villus, and the Chao index of gut microbiota and polystyrene was negatively correlated. The concentration of microplastics was positively correlated with *Pseudomonas aeruginosa*, *streptococcus*, and *Clostridium*. In addition, Huang et al. detected the presence of microplastics in almost all sputum samples collected from people with respiratory diseases [110]. In addition, polyurethane was the main microplastic detected, followed by polyester, polyvinyl chloride, etc. According to the data analysis, it was revealed that the exposure of microplastics may be related to smoking or not smoking and invasive air tube inspection. Furthermore, microplastics have been found in human testicles, semen, and intestines [111,112]. In addition, scientists used vitro experiments to study the possible effects of micro- and nanoplastics on humans. Annangi used human nasal epithelial cells to study the possible effects of polystyrene nanoplastics in the air and found that after exposure to nanoplastics, reactive oxygen species increase, mitochondrial membranes lose potential, and autophagy accumulation regulates the autophagy process [113]. Furthermore, human fibroblasts exposed to polystyrene nanoplastics also showed mitochondrial damage and disrupted the expression of caspase3, caspase9, cytochrome c, and other related proteins to induce cell apoptosis [114]. Polystyrene nanoplastics can be internalized by human alveolar basal epithelial cells and human colorectal adenocarcinoma cells [115] and damage the cell vitality of liver cancer cells [116]. All of this indicates the potential threat of micro- and nanoplastics to human health.

In conclusion, not only microplastics but also nanoplastics are found to be widespread. It is well known that the concentration and size of contaminants can significantly affect their toxicity. Scientists have realized the seriousness of the problem and begun conducting in-depth research. First, microplastics' different particle sizes affect their accumulation in different tissues. Deng et al. used 5 μm and 20 μm polystyrene microplastics to investigate whether different particle sizes would affect their accumulation in tissues and organs. They found that although microplastics of different sizes were accumulated in liver, kidney, and intestine, their tissue accumulation kinetics and distribution were strongly dependent on their particle sizes [24]. After 4 weeks of exposure, the concentrations of 5 μm microplastics in liver, kidney, and intestine reached $4.42 \times 10^6 \pm 4.23 \times 10^5$ items/g, $1.38 \times 10^7 \pm 1.36 \times 10^6$ items/g, and $2.03 \times 10^7 \pm 2.00 \times 10^6$ items/g, respectively. However, the concentrations of 20 μm microplastics were relatively low: $1.73 \times 10^5 \pm 1.68 \times 10^4$ items/g, $1.78 \times 10^5 \pm 1.91 \times 10^4$ items/g, and $1.77 \times 10^5 \pm 1.86 \times 10^4$ items/g. Yang et al. also found that the absorption rate constant of microplastics with smaller particles is larger, while that of larger particles is smaller [74]. Compared to microplastics, nanoplastics have smaller particle sizes and are more likely to enter organisms through various routes. For example, when zebrafish embryos were exposed to 100 nm, 500 nm, and 1 μm micro- and nanoplastics, large amounts of nanoplastics were found to be deposited on the chorionic surface and yolk sac of the embryos in the 100 nm and 500 nm exposed groups, and they even were observed in the brains of larvae [117]. However, in the 1 μm microplastic exposure group, microplastics were only deposited on the surface of the embryonic chorionic membrane and did not enter the embryo interior and larval brain tissue. More interestingly, Zhang et al. studied the accumulation of 70 nm, 200 nm, and 500 nm polystyrene nanoplastics using human alveolar basal epithelial cells (A549) and human colorectal adenocarcinoma cells (Caco-2) [115]. It was found that all 3 sizes of nanoplastics could be internalized, but the numbers of nanoparticles of 200 nm and 500 nm were lower than that of 70 nm nanoplastics. These studies show that smaller micro- and nanoplastics seem to more easily enter organisms and accumulate in tissues, so smaller nanoplastics should deserve more attention. In addition to the differences in bioaccumulation, the scientists also found that smaller plastic particles seem to cause more serious biotoxicity and biohazards. When human hepatoma cells (HepG2 cells) were exposed to 50 nm, 100 nm, 1 μm, and 5 μm micro- and nanoplastics, scientists found that smaller aminated particles (50 nm, 100 nm) were more harmful to cell viability than larger aminated particles (1 μm, 5 μm) [116]. In addition, Wang et al. also found that the smaller the particle size of nanoplastics was, the more easily they were bound to superoxide dismutase (SOD) in cells to form complexes, and the smaller nanoplastics (100 nm) induced more significant changes in SOD activity (20% increase in activity). However, nanoplastics with larger particle size (200 nm and 1 μm) had little effect [118].

Furthermore, exposure concentration is also an important factor affecting the toxicity of micro- and nanoplastics. For example, Banerjee et al. found that not only the particle size but also the concentration was an important factor in their toxicity [116]. The effects of aminated nanoplastics (50 nm and 100 nm) on cell activity increased with the increase in the concentration, and the toxicity was greatest when the maximum exposure concentration was reached (100 μg/mL). Interestingly, however, Hu et al. found that nanoplastics had a double effect on the effects of Pseudomonas aeruginosa (PAO1) [119]. Using 0.1 mg/L, 20 mg/L, and 50 mg/L polystyrene nanoplastics to culture PAO1, the researchers found that 20 mg/L and 50 mg/L polystyrene nanoplastics significantly inhibited the nitrate reduction process, and the expression of related denitrification genes was also significantly down-regulated. However, denitrification of PAO1 was promoted when it was exposed to 0.1 mg/L polystyrene nanoplastics. The double effect was also reflected when Macrobrachium nipponense was exposed to different concentrations of polystyrene nanoplastics [120]. The study found that with the increase in the concentration, the activity of SOD, CAT, and other antioxidant enzymes generally decreased, while the expression of antioxidant-related genes showed a trend of first increasing and then decreasing. The expressions of SOD and CAT genes in the 5 mg/L and 10 mg/L exposure groups were significantly higher than those in the control group, and the expressions of SOD and CAT genes were significantly lower than those in the control group when concentration increased to 20 mg/L and 40 mg/L. In addition, the trend of immunoenzyme activity in Macrobrachium nipponense was consistent with that of antioxidation-related genes. These results suggest that low concentrations of polystyrene micro- and nanoplastics appear to enhance the survival of organisms in the environment, but high exposure has toxic effects.

5. Preventive Strategies

Most of the current studies focus on the dangers of plastics to organisms and humans, but the ultimate goal is to understand how to prevent these hazards (Table 4).

Anthocyanin is a safe and non-toxic natural pigment extracted from bayberry, mulberry, strawberry, blueberry, and other berries. It functions as an antioxidant and anti-inflammatory, as well as having anti-apoptosis, anti-cancer, anti-diabetes, and neuroprotective properties [121,122]. Cyanidin-3-glucoside (C3G) is one of the functional factors of anthocyanins, and it has been reported to alleviate the toxic effects of polystyrene by promoting the excretion of polystyrene from feces and reducing the harm of locomotion behavior in terms of head and body bending time [121–126]. The mechanisms by which C3G alleviates the toxic reactions caused by polystyrene include the activation of autophagy, the remodeling of intestinal flora, and changing metabolic pathways. Autophagy is necessary for C3G to alleviate polystyrene-induced cytotoxicity. By activating the Sirt1-Foxo1 signaling pathway, C3G increases LC3 levels in cells (an increased expression level can activate autophagy) and decreases p62 levels (a decreased expression level promotes autophagosome formation), which causes autophagy and promotes polystyrene degradation [122]. Through 16S rRNA high-throughput sequencing, Chen found that the intestinal microbiota composition was changed after PS and C3G treatment in mice [123]. As a result, the numbers of *Dubosiella* used as intestinal probiotics decreased after 6 weeks of PS administration, while the decreasing trend was significantly alleviated after C3G intervention. Compared with the polystyrene-exposed group, the level of probiotic genes in the C3G-treated group was significantly increased. C3G was also involved in energy metabolism and improved mitochondrial dysfunction. Studies have found that C3G can correct polystyrene-induced mitochondrial dysfunction and increase the ATP content of cells and nematodes by activating the AMPK/SIRT1/PGC-1α signaling pathway [125]. Moreover, the accumulation of polystyrene in the body will cause oxidative stress, thus producing excessive ROS and superoxide anions (O^{2-}). By alleviating the effects of oxidative stress, C3G can delay age-related physiological decline and aging, as well as prolong the lifespan of nematodes [122].

Table 4. Approaches used to counteract the toxic effects of microplastic exposure.

Species	Chemicals and Strategies	Prevention Approaches and Effects	Reference
Mouse and Caco2 cells	Polystyrene and C3G	• Triggers autophagy by activating the Sirt1-Foxo1-1 signaling pathway to alleviate polystyrene-induced toxicity • The co-localization of polystyrene and lysosomes was observed, suggesting that PS is encapsulated and degraded • The co-localization of autophagy genes and PS was found, suggesting that autophagy is involved in the beneficial effects of C3G	[122]
Mouse	Polystyrene and C3G	• C3G remodels the gut microbiota in mice and affects the gene abundance of bacterial functional pathways • Differential metabolic pathways and metabolites were discovered • Significantly increased levels of probiotics	[123]
Mouse	Polystyrene and C3G	• Effectively reduces tissue accumulation and increases polystyrene excretion from feces • C3G regulates intestinal microbiome disturbance and regulates inflammatory function genes • Triggered alterations in functional pathways in response to xenobiotic polystyrene, thus reducing bacterial functional genes associated with disease and inflammation	[124]
C. elegans and Caco2 cells	Polystyrene and C3G	• Recovery of polystyrene-induced ATP reduction, achieved by activating the AMPK/SIRT1/PGC-1α signaling pathway and by improving mitochondrial dysfunction • Increased fecal polystyrene efflux	[125]
C. elegans	Polystyrene and C3G	• C3G can ameliorate polystyrene-induced oxidative stress and shorten its lifespan • C3G can significantly enhance the expression of DAF-16 pathway-related genes	[126]
C. elegans	Polystyrene and FMT	• Promotes intracellular GSH production by activating the PMK-1/SKN-1 pathway and also reduces the production of ROS and O^{2-} induced by polystyrene • FMT significantly alleviated the harm caused by the polystyrene-induced inhibition of nematode body lengths and motility behaviors	[127]

Fecal microbiota translocation (FMT) is another effective strategy to deal with polystyrene toxicity. In an experiment regarding microbiota transplantation, it was found that the GSH content in the polystyrene-exposed group was decreased and the consumption of GSH was increased, while the group treated with FMT exhibited accelerated generation of GSH, which thus alleviated oxidative stress by increasing the expression level of GSH synthase [127]. Overall, prevention and treatment strategies for addressing the public health problems caused by microplastics are still limited and there is an urgent need for further exploration.

6. Conclusions and Perspectives

Microplastic pollution has pervaded the environment. Human exposure and the cumulative uptake of these microplastics are expected to increase over time, and this phenomenon has aroused wide concern among scientists. A growing number of scientists, however, have studied the potential hazards of microplastics when exposed to ecosystems, invertebrates and vertebrates, and laboratory mouse models. Knowledge about the toxicity of microplastics is still limited.

(i) Most of the current studies on the toxicity of micro- and nanoplastics have focused on the ecological environment and non-mammalian and laboratory mouse models. So far, what we know about micro- and nanoplastics and human health includes the fact that micro- and nanoplastics have accumulated in human tissues and organs, and relatively little research has been done on the harm they cause. However, the accumulation of micro- and nanoplastics in human tissues having not been discovered until recent years, the ethical limitations of collecting human specimens, and our current limited understanding of the toxicity of micro- and nanoplastics and the biomarkers that reflect their toxicity have limited scientists to conducting epidemiological studies. Determining whether micro- and nanoplastics have direct or indirect relationships with the occurrence and development of human diseases still requires scientists to continue efforts and exploration.

(ii) There are limited data on the ecological, biological, and human toxicity of micro- and nanoplastics under environmentally relevant conditions. Exposure concentrations of the microplastics used in the laboratory study were significantly higher than those associated with the environment, so the scientists speculate that the laboratory results may overstate the harm caused by micro- and nanoplastics at the environmentally associated concentrations. In addition, extensive studies are still needed to elucidate the pathological mechanisms by which microplastics cause toxic hazards at the cellular and tissue levels and the health consequences of long-term exposure.

(iii) In addition, factors affecting the toxicological role of microplastics, such as sex differences, the dose–response relationship, exposure frequency, and the type and size of microplastics have not yet been thoroughly investigated. Therefore, it is urgent to conduct more in-depth research on the factors influencing microplastics' toxicity, microplastics-related knowledge, and potential risks, so as to provide a scientific basis for policy makers to cooperate with each other, solve this pressing environmental problem, and protect human health.

However, we believe that this review may have some limitations. In summary, there is a lack of population studies on micro- and nanoplastics, so although our aim was to study the toxic hazards of micro- and nanoplastics in the population, unfortunately, based on our literature search, there are few data available on toxic effects of microplastics at the population level. The relatively few data on population studies in this review still do not allow for a better assessment of the hazards of micro- and nanoplastics in humans.

Supplementary Materials: The following supporting information can be downloaded at https://www.mdpi.com/article/10.3390/metabo13060739/s1. Table S1: Toxic effects of microplastic exposure to environmental organisms; Table S2: Toxic effects of microplastic exposure to marine invertebrates and vertebrates.

Author Contributions: Conceptualization: H.N. and X.L. (Xueqing Li); literature search: H.N., Y.J., L.H., Y.L., Y.H. and L.W.; writing—original draft: H.N., S.L., M.X., Z.C., X.W. and X.L. (Xiaoming Lou); writing—review and editing: all authors. All authors have read and agreed to the published version of the manuscript.

Funding: This research received no external funding.

Conflicts of Interest: The authors declare no conflict of interest.

References

1. Geyer, R.; Jambeck, J.R.; Law, K.L. Production, use, and fate of all plastics ever made. *Sci. Adv.* **2017**, *3*, e1700782. [CrossRef]
2. Wu, M.; Tu, C.; Liu, G.; Zhong, H. Time to Safeguard the Future Generations from the Omnipresent Microplastics. *Bull. Environ. Contam. Toxicol.* **2021**, *107*, 793–799. [CrossRef]
3. Rillig, M.C. Microplastic in terrestrial ecosystems and the soil? *Environ. Sci. Technol.* **2012**, *46*, 6453–6454. [CrossRef]
4. Andrady, A.L. Microplastics in the marine environment. *Mar. Pollut. Bull.* **2011**, *62*, 1596–1605. [CrossRef] [PubMed]
5. Thompson, R.C.; Olsen, Y.; Mitchell, R.P.; Davis, A.; Rowland, S.J.; John, A.W.G.; McGonigle, D.; Russell, A.E. Lost at sea: Where is all the plastic? *Science* **2004**, *304*, 838. [CrossRef] [PubMed]
6. Wang, J.; Liu, X.; Li, Y.; Powell, T.; Wang, X.; Wang, G.; Zhang, P. Microplastics as contaminants in the soil environment: A mini-review. *Sci. Total Environ.* **2019**, *691*, 848–857. [CrossRef]
7. Yang, D.; Shi, H.; Li, L.; Li, J.; Jabeen, K.; Kolandhasamy, P. Microplastic Pollution in Table Salts from China. *Environ. Sci. Technol.* **2015**, *49*, 13622–13627. [CrossRef]
8. Free, C.M.; Jensen, O.P.; Mason, S.A.; Eriksen, M.; Williamson, N.J.; Boldgiv, B. High-levels of microplastic pollution in a large, remote, mountain lake. *Mar. Pollut. Bull.* **2014**, *85*, 156–163. [CrossRef]
9. Kosuth, M.; Mason, S.A.; Wattenberg, E.V. Anthropogenic contamination of tap water, beer, and sea salt. *PLoS ONE* **2018**, *13*, e0194970. [CrossRef] [PubMed]
10. Liu, H.; Yang, X.; Liu, G.; Liang, C.; Xue, S.; Chen, H.; Ritsema, C.J.; Geissen, V. Response of soil dissolved organic matter to microplastic addition in Chinese loess soil. *Chemosphere* **2017**, *185*, 907–917. [CrossRef] [PubMed]
11. Weithmann, N.; Möller, J.N.; Löder, M.G.J.; Piehl, S.; Laforsch, C.; Freitag, R. Organic fertilizer as a vehicle for the entry of microplastic into the environment. *Sci. Adv.* **2018**, *4*, eaap8060. [CrossRef]
12. Abbasi, S.; Keshavarzi, B.; Moore, F.; Turner, A.; Kelly, F.J.; Dominguez, A.O.; Jaafarzadeh, N. Distribution and potential health impacts of microplastics and microrubbers in air and street dusts from Asaluyeh County, Iran. *Environ. Pollut.* **2019**, *244*, 153–164. [CrossRef]
13. Obbard, R.W.; Sadri, S.; Wong, Y.Q.; Khitun, A.A.; Baker, I.; Thompson, R.C. Global warming releases microplastic legacy frozen in Arctic Sea ice. *Earth's Future* **2014**, *2*, 315–320. [CrossRef]
14. Mattsson, K.; Ekvall, M.T.; Hansson, L.-A.; Linse, S.; Malmendal, A.; Cedervall, T. Altered behavior, physiology, and metabolism in fish exposed to polystyrene nanoparticles. *Environ. Sci. Technol.* **2015**, *49*, 553–561. [CrossRef]
15. Karami, A.; Golieskardi, A.; Choo, C.K.; Larat, V.; Karbalaei, S.; Salamatinia, B. Microplastic and mesoplastic contamination in canned sardines and sprats. *Sci. Total Environ.* **2018**, *612*, 1380–1386. [CrossRef]
16. Mason, S.A.; Welch, V.G.; Neratko, J. Synthetic Polymer Contamination in Bottled Water. *Front. Chem.* **2018**, *6*, 407. [CrossRef] [PubMed]
17. Schwabl, P.; Köppel, S.; Königshofer, P.; Bucsics, T.; Trauner, M.; Reiberger, T.; Liebmann, B. Detection of Various Microplastics in Human Stool: A Prospective Case Series. *Ann. Intern. Med.* **2019**, *171*, 453–457. [CrossRef]
18. Ragusa, A.; Svelato, A.; Santacroce, C.; Catalano, P.; Notarstefano, V.; Carnevali, O.; Papa, F.; Rongioletti, M.C.A.; Baiocco, F.; Draghi, S.; et al. Plasticenta: First evidence of microplastics in human placenta. *Environ. Int.* **2021**, *146*, 106274. [CrossRef] [PubMed]
19. Amato-Lourenço, L.F.; Carvalho-Oliveira, R.; Júnior, G.R.; Dos Santos Galvão, L.; Ando, R.A.; Mauad, T. Presence of airborne microplastics in human lung tissue. *J. Hazard. Mater.* **2021**, *416*, 126124. [CrossRef]
20. Leslie, H.A.; van Velzen, M.J.M.; Brandsma, S.H.; Vethaak, A.D.; Garcia-Vallejo, J.J.; Lamoree, M.H. Discovery and quantification of plastic particle pollution in human blood. *Environ. Int.* **2022**, *163*, 107199. [CrossRef]
21. Pitt, J.A.; Kozal, J.S.; Jayasundara, N.; Massarsky, A.; Trevisan, R.; Geitner, N.; Wiesner, M.; Levin, E.D.; Di Giulio, R.T. Uptake, tissue distribution, and toxicity of polystyrene nanoparticles in developing zebrafish (*Danio rerio*). *Aquat. Toxicol.* **2018**, *194*, 185–194. [CrossRef]
22. Brun, N.R.; van Hage, P.; Hunting, E.R.; Haramis, A.-P.G.; Vink, S.C.; Vijver, M.G.; Schaaf, M.J.M.; Tudorache, C. Polystyrene nanoplastics disrupt glucose metabolism and cortisol levels with a possible link to behavioural changes in larval zebrafish. *Commun. Biol.* **2019**, *2*, 382. [CrossRef]

23. Lu, Y.; Zhang, Y.; Deng, Y.; Jiang, W.; Zhao, Y.; Geng, J.; Ding, L.; Ren, H. Uptake and Accumulation of Polystyrene Microplastics in Zebrafish (*Danio rerio*) and Toxic Effects in Liver. *Environ. Sci. Technol.* **2016**, *50*, 4054–4060. [CrossRef]
24. Deng, Y.; Zhang, Y.; Lemos, B.; Ren, H. Tissue accumulation of microplastics in mice and biomarker responses suggest widespread health risks of exposure. *Sci. Rep.* **2017**, *7*, 46687. [CrossRef] [PubMed]
25. Wan, Z.; Wang, C.; Zhou, J.; Shen, M.; Wang, X.; Fu, Z.; Jin, Y. Effects of polystyrene microplastics on the composition of the microbiome and metabolism in larval zebrafish. *Chemosphere* **2019**, *217*, 646–658. [CrossRef]
26. Jin, Y.; Lu, L.; Tu, W.; Luo, T.; Fu, Z. Impacts of polystyrene microplastic on the gut barrier, microbiota and metabolism of mice. *Sci. Total Environ.* **2019**, *649*, 308–317. [CrossRef]
27. Galloway, T.S. Micro-and nano-plastics and human health. In *Marine Anthropogenic Litter*; Springer: Cham, Switzerland, 2015; pp. 343–366.
28. Liebezeit, G.; Liebezeit, E. Non-pollen particulates in honey and sugar. *Food Addit. Contam. Part A* **2013**, *30*, 2136–2140. [CrossRef]
29. Mintenig, S.M.; Löder, M.G.J.; Primpke, S.; Gerdts, G. Low numbers of microplastics detected in drinking water from ground water sources. *Sci. Total Environ.* **2019**, *648*, 631–635. [CrossRef]
30. Schymanski, D.; Goldbeck, C.; Humpf, H.-U.; Fürst, P. Analysis of microplastics in water by micro-Raman spectroscopy: Release of plastic particles from different packaging into mineral water. *Water Res.* **2018**, *129*, 154–162. [CrossRef] [PubMed]
31. Van Cauwenberghe, L.; Janssen, C.R. Microplastics in bivalves cultured for human consumption. *Environ. Pollut.* **2014**, *193*, 65–70. [CrossRef] [PubMed]
32. Li, J.; Green, C.; Reynolds, A.; Shi, H.; Rotchell, J.M. Microplastics in mussels sampled from coastal waters and supermarkets in the United Kingdom. *Environ. Pollut.* **2018**, *241*, 35–44. [CrossRef] [PubMed]
33. Castle, L. Chemical migration into food: An overview. In *Chemical Migration and Food Contact Materials*; Woodhead Publishing Ltd.: Cambridge, UK, 2007; pp. 1–13.
34. Bumbudsanpharoke, N.; Choi, J.; Ko, S. Applications of Nanomaterials in Food Packaging. *J. Nanosci. Nanotechnol.* **2015**, *15*, 6357–6372. [CrossRef] [PubMed]
35. Addo Ntim, S.; Norris, S.; Goodwin, D.G.; Breffke, J.; Scott, K.; Sung, L.; Thomas, T.A.; Noonan, G.O. Effects of consumer use practices on nanosilver release from commercially available food contact materials. *Food Addit. Contam. Part A* **2018**, *35*, 2279–2290. [CrossRef]
36. Prata, J.C.; da Costa, J.P.; Lopes, I.; Duarte, A.C.; Rocha-Santos, T. Environmental exposure to microplastics: An overview on possible human health effects. *Sci. Total Environ.* **2020**, *702*, 134455. [CrossRef]
37. Gasperi, J.; Wright, S.L.; Dris, R.; Collard, F.; Mandin, C.; Guerrouache, M.; Langlois, V.; Kelly, F.J.; Tassin, B. Microplastics in air: Are we breathing it in? *Curr. Opin. Environ. Sci. Health* **2018**, *1*, 1–5. [CrossRef]
38. Browne, M.A.; Crump, P.; Niven, S.J.; Teuten, E.; Tonkin, A.; Galloway, T.; Thompson, R. Accumulation of microplastic on shorelines woldwide: Sources and sinks. *Environ. Sci. Technol.* **2011**, *45*, 9175–9179. [CrossRef]
39. Kole, P.J.; Löhr, A.J.; Van Belleghem, F.G.A.J.; Ragas, A.M.J. Wear and Tear of Tyres: A Stealthy Source of Microplastics in the Environment. *Int. J. Environ. Res. Public Health* **2017**, *14*, 1265. [CrossRef]
40. Verschoor, A.; De Poorter, L.; Roex, E.; Bellert, B. Quick scan and prioritization of microplastic sources and emissions. *RIVM Lett. Rep.* **2014**, *156*, 1–41.
41. Liao, Z.; Ji, X.; Ma, Y.; Lv, B.; Huang, W.; Zhu, X.; Fang, M.; Wang, Q.; Wang, X.; Dahlgren, R.; et al. Airborne microplastics in indoor and outdoor environments of a coastal city in Eastern China. *J. Hazard. Mater.* **2021**, *417*, 126007. [CrossRef]
42. Guideline, A. *Interactions Affecting the Achievement of Acceptable Indoor Environments*; ASHRAE: Atlanta, GA, USA, 2011.
43. Vianello, A.; Jensen, R.L.; Liu, L.; Vollertsen, J. Simulating human exposure to indoor airborne microplastics using a Breathing Thermal Manikin. *Sci. Rep.* **2019**, *9*, 8670. [CrossRef]
44. Jenner, L.C.; Rotchell, J.M.; Bennett, R.T.; Cowen, M.; Tentzeris, V.; Sadofsky, L.R. Detection of microplastics in human lung tissue using μFTIR spectroscopy. *Sci. Total Environ.* **2022**, *831*, 154907. [CrossRef]
45. Marine, P.I.T.C.T.; Litter, M. *Are We Polluting the Environment through Our Personal Care?* UNEP: Nairobi, Kenya, 2015.
46. Sun, Q.; Ren, S.-Y.; Ni, H.-G. Incidence of microplastics in personal care products: An appreciable part of plastic pollution. *Sci. Total Environ.* **2020**, *742*, 140218. [CrossRef]
47. Praveena, S.M.; Shaifuddin, S.N.M.; Akizuki, S. Exploration of microplastics from personal care and cosmetic products and its estimated emissions to marine environment: An evidence from Malaysia. *Mar. Pollut. Bull.* **2018**, *136*, 135–140. [CrossRef]
48. Hernandez, L.M.; Yousefi, N.; Tufenkji, N. Are there nanoplastics in your personal care products? *Environ. Sci. Technol. Lett.* **2017**, *4*, 280–285. [CrossRef]
49. Sykes, E.A.; Dai, Q.; Tsoi, K.M.; Hwang, D.M.; Chan, W.C.W. Nanoparticle exposure in animals can be visualized in the skin and analysed via skin biopsy. *Nat. Commun.* **2014**, *5*, 3796. [CrossRef]
50. de Souza Machado, A.A.; Lau, C.W.; Till, J.; Kloas, W.; Lehmann, A.; Becker, R.; Rillig, M.C. Impacts of Microplastics on the Soil Biophysical Environment. *Environ. Sci. Technol.* **2018**, *52*, 9656–9665. [CrossRef]
51. López-Rojo, N.; Pérez, J.; Alonso, A.; Correa-Araneda, F.; Boyero, L. Microplastics have lethal and sublethal effects on stream invertebrates and affect stream ecosystem functioning. *Environ. Pollut.* **2020**, *259*, 113898. [CrossRef]
52. Green, D.S.; Jefferson, M.; Boots, B.; Stone, L. All that glitters is litter? Ecological impacts of conventional versus biodegradable glitter in a freshwater habitat. *J. Hazard. Mater.* **2021**, *402*, 124070. [CrossRef] [PubMed]

53. Wang, J.; Lv, S.; Zhang, M.; Chen, G.; Zhu, T.; Zhang, S.; Teng, Y.; Christie, P.; Luo, Y. Effects of plastic film residues on occurrence of phthalates and microbial activity in soils. *Chemosphere* **2016**, *151*, 171–177. [CrossRef] [PubMed]
54. Bosker, T.; Bouwman, L.J.; Brun, N.R.; Behrens, P.; Vijver, M.G. Microplastics accumulate on pores in seed capsule and delay germin Response of soil dissolved organic matter to microplastic addition in Chinese loess soil ation and root growth of the terrestrial vascular plant Lepidium sativum. *Chemosphere* **2019**, *226*, 774–781. [CrossRef] [PubMed]
55. Nolte, T.M.; Hartmann, N.B.; Kleijn, J.M.; Garnæs, J.; van de Meent, D.; Jan Hendriks, A.; Baun, A. The toxicity of plastic nanoparticles to green algae as influenced by surface modification, medium hardness and cellular adsorption. *Aquat. Toxicol.* **2017**, *183*, 11–20. [CrossRef]
56. Zhang, C.; Chen, X.; Wang, J.; Tan, L. Toxic effects of microplastic on marine microalgae Skeletonema costatum: Interactions between microplastic and algae. *Environ. Pollut.* **2017**, *220*, 1282–1288. [CrossRef] [PubMed]
57. Zhou, Y.; Liu, X.; Wang, J. Ecotoxicological effects of microplastics and cadmium on the earthworm *Eisenia foetida*. *J. Hazard. Mater.* **2020**, *392*, 122273. [CrossRef] [PubMed]
58. Jiang, X.; Chang, Y.; Zhang, T.; Qiao, Y.; Klobučar, G.; Li, M. Toxicological effects of polystyrene microplastics on earthworm (*Eisenia fetida*). *Environ. Pollut.* **2020**, *259*, 113896. [CrossRef]
59. Tlili, S.; Jemai, D.; Brinis, S.; Regaya, I. Microplastics mixture exposure at environmentally relevant conditions induce oxidative stress and neurotoxicity in the wedge clam *Donax trunculus*. *Chemosphere* **2020**, *258*, 127344. [CrossRef]
60. Sussarellu, R.; Suquet, M.; Thomas, Y.; Lambert, C.; Fabioux, C.; Pernet, M.E.J.; Le Goïc, N.; Quillien, V.; Mingant, C.; Epelboin, Y.; et al. Oyster reproduction is affected by exposure to polystyrene microplastics. *Proc. Natl. Acad. Sci. USA* **2016**, *113*, 2430–2435. [CrossRef] [PubMed]
61. Pitt, J.A.; Trevisan, R.; Massarsky, A.; Kozal, J.S.; Levin, E.D.; Di Giulio, R.T. Maternal transfer of nanoplastics to offspring in zebrafish (*Danio rerio*): A case study with nanopolystyrene. *Sci. Total Environ.* **2018**, *643*, 324–334. [CrossRef]
62. Bourdages, M.P.T.; Provencher, J.F.; Baak, J.E.; Mallory, M.L.; Vermaire, J.C. Breeding seabirds as vectors of microplastics from sea to land: Evidence from colonies in Arctic Canada. *Sci. Total Environ.* **2021**, *764*, 142808. [CrossRef]
63. Provencher, J.F.; Gaston, A.J.; Mallory, M.L.; O'Hara, P.D.; Gilchrist, H.G. Ingested plastic in a diving seabird, the thick-billed murre (*Uria lomvia*), in the eastern Canadian Arctic. *Mar. Pollut. Bull.* **2010**, *60*, 1406–1411. [CrossRef]
64. Xia, X.; Sun, M.; Zhou, M.; Chang, Z.; Li, L. Polyvinyl chloride microplastics induce growth inhibition and oxidative stress in Cyprinus carpio var. larvae. *Sci. Total Environ.* **2020**, *716*, 136479. [CrossRef] [PubMed]
65. Wang, J.; Li, Y.; Lu, L.; Zheng, M.; Zhang, X.; Tian, H.; Wang, W.; Ru, S. Polystyrene microplastics cause tissue damages, sex-specific reproductive disruption and transgenerational effects in marine medaka (*Oryzias melastigma*). *Environ. Pollut.* **2019**, *254*, 113024. [CrossRef] [PubMed]
66. Qiao, R.; Sheng, C.; Lu, Y.; Zhang, Y.; Ren, H.; Lemos, B. Microplastics induce intestinal inflammation, oxidative stress, and disorders of metabolome and microbiome in zebrafish. *Sci. Total Environ.* **2019**, *662*, 246–253. [CrossRef] [PubMed]
67. Lei, L.; Wu, S.; Lu, S.; Liu, M.; Song, Y.; Fu, Z.; Shi, H.; Raley-Susman, K.M.; He, D. Microplastic particles cause intestinal damage and other adverse effects in zebrafish *Danio rerio* and nematode *Caenorhabditis elegans*. *Sci. Total Environ.* **2018**, *619–620*, 1–8. [CrossRef]
68. Qiang, L.; Cheng, J. Exposure to microplastics decreases swimming competence in larval zebrafish (*Danio rerio*). *Ecotoxicol. Environ. Saf.* **2019**, *176*, 226–233. [CrossRef]
69. Chen, L.; Hu, C.; Lok-Shun Lai, N.; Zhang, W.; Hua, J.; Lam, P.K.S.; Lam, J.C.W.; Zhou, B. Acute exposure to PBDEs at an environmentally realistic concentration causes abrupt changes in the gut microbiota and host health of zebrafish. *Environ. Pollut.* **2018**, *240*, 17–26. [CrossRef] [PubMed]
70. Xu, X.Y.; Lee, W.T.; Chan, A.K.Y.; Lo, H.S.; Shin, P.K.S.; Cheung, S.G. Microplastic ingestion reduces energy intake in the clam *Atactodea striata*. *Mar. Pollut. Bull.* **2017**, *124*, 798–802. [CrossRef]
71. Wen, B.; Zhang, N.; Jin, S.-R.; Chen, Z.-Z.; Gao, J.-Z.; Liu, Y.; Liu, H.-P.; Xu, Z. Microplastics have a more profound impact than elevated temperatures on the predatory performance, digestion and energy metabolism of an Amazonian cichlid. *Aquat. Toxicol.* **2018**, *195*, 67–76. [CrossRef]
72. Greven, A.-C.; Merk, T.; Karagöz, F.; Mohr, K.; Klapper, M.; Jovanović, B.; Palić, D. Polycarbonate and polystyrene nanoplastic particles act as stressors to the innate immune system of fathead minnow (*Pimephales promelas*). *Environ. Toxicol. Chem.* **2016**, *35*, 3093–3100. [CrossRef]
73. Karami, A.; Romano, N.; Galloway, T.; Hamzah, H. Virgin microplastics cause toxicity and modulate the impacts of phenanthrene on biomarker responses in African catfish (*Clarias gariepinus*). *Environ. Res.* **2016**, *151*, 58–70. [CrossRef]
74. Yang, Y.-F.; Chen, C.-Y.; Lu, T.-H.; Liao, C.-M. Toxicity-based toxicokinetic/toxicodynamic assessment for bioaccumulation of polystyrene microplastics in mice. *J. Hazard. Mater.* **2019**, *366*, 703–713. [CrossRef]
75. Zhao, L.; Shi, W.; Hu, F.; Song, X.; Cheng, Z.; Zhou, J. Prolonged oral ingestion of microplastics induced inflammation in the liver tissues of C57BL/6J mice through polarization of macrophages and increased infiltration of natural killer cells. *Ecotoxicol. Environ. Saf.* **2021**, *227*, 112882. [CrossRef] [PubMed]
76. Li, S.; Shi, M.; Wang, Y.; Xiao, Y.; Cai, D.; Xiao, F. Keap1-Nrf2 pathway up-regulation via hydrogen sulfide mitigates polystyrene microplastics induced-hepatotoxic effects. *J. Hazard. Mater.* **2021**, *402*, 123933. [CrossRef] [PubMed]
77. Lu, L.; Wan, Z.; Luo, T.; Fu, Z.; Jin, Y. Polystyrene microplastics induce gut microbiota dysbiosis and hepatic lipid metabolism disorder in mice. *Sci. Total Environ.* **2018**, *631–632*, 449–458. [CrossRef]

78. Li, B.; Ding, Y.; Cheng, X.; Sheng, D.; Xu, Z.; Rong, Q.; Wu, Y.; Zhao, H.; Ji, X.; Zhang, Y. Polyethylene microplastics affect the distribution of gut microbiota and inflammation development in mice. *Chemosphere* **2020**, *244*, 125492. [CrossRef] [PubMed]
79. Schirinzi, G.F.; Pérez-Pomeda, I.; Sanchís, J.; Rossini, C.; Farré, M.; Barceló, D. Cytotoxic effects of commonly used nanomaterials and microplastics on cerebral and epithelial human cells. *Environ. Res.* **2017**, *159*, 579–587. [CrossRef]
80. Hwang, J.; Choi, D.; Han, S.; Choi, J.; Hong, J. An assessment of the toxicity of polypropylene microplastics in human derived cells. *Sci. Total Environ.* **2019**, *684*, 657–669. [CrossRef]
81. Stock, V.; Böhmert, L.; Lisicki, E.; Block, R.; Cara-Carmona, J.; Pack, L.K.; Selb, R.; Lichtenstein, D.; Voss, L.; Henderson, C.J.; et al. Uptake and effects of orally ingested polystyrene microplastic particles in vitro and in vivo. *Arch. Toxicol.* **2019**, *93*, 1817–1833. [CrossRef]
82. Guimarães, A.T.B.; Freitas, Í.N.; Mubarak, N.M.; Rahman, M.M.; Rodrigues, F.P.; Rodrigues, A.S.d.L.; Barceló, D.; Islam, A.R.M.T.; Malafaia, G. Exposure to polystyrene nanoplastics induces an anxiolytic-like effect, changes in antipredator defensive response, and DNA damage in Swiss mice. *J. Hazard. Mater.* **2023**, *442*, 130004. [CrossRef]
83. Luo, T.; Wang, C.; Pan, Z.; Jin, C.; Fu, Z.; Jin, Y. Maternal Polystyrene Microplastic Exposure during Gestation and Lactation Altered Metabolic Homeostasis in the Dams and Their F1 and F2 Offspring. *Environ. Sci. Technol.* **2019**, *53*, 10978–10992. [CrossRef]
84. Vethaak, A.D.; Legler, J. Microplastics and human health. *Science* **2021**, *371*, 672–674. [CrossRef]
85. Wang, Y.; Mao, Z.; Zhang, M.; Ding, G.; Sun, J.; Du, M.; Liu, Q.; Cong, Y.; Jin, F.; Zhang, W.; et al. The uptake and elimination of polystyrene microplastics by the brine shrimp, *Artemia parthenogenetica*, and its impact on its feeding behavior and intestinal histology. *Chemosphere* **2019**, *234*, 123–131. [CrossRef] [PubMed]
86. Cui, Y.; Wang, Q.; Chang, R.; Zhou, X.; Xu, C. Intestinal Barrier Function-Non-alcoholic Fatty Liver Disease Interactions and Possible Role of Gut Microbiota. *J. Agric. Food Chem.* **2019**, *67*, 2754–2762. [CrossRef] [PubMed]
87. Sekirov, I.; Russell, S.L.; Antunes, L.C.M.; Finlay, B.B. Gut microbiota in health and disease. *Physiol. Rev.* **2010**, *90*, 859–904. [CrossRef] [PubMed]
88. Ley, R.E.; Peterson, D.A.; Gordon, J.I. Ecological and evolutionary forces shaping microbial diversity in the human intestine. *Cell* **2006**, *124*, 837–848. [CrossRef]
89. Tang, W.H.W.; Kitai, T.; Hazen, S.L. Gut Microbiota in Cardiovascular Health and Disease. *Circ. Res.* **2017**, *120*, 1183–1196. [CrossRef]
90. Round, J.L.; O'Connell, R.M.; Mazmanian, S.K. Coordination of tolerogenic immune responses by the commensal microbiota. *J. Autoimmun.* **2010**, *34*, J220–J225. [CrossRef]
91. Buffie, C.G.; Bucci, V.; Stein, R.R.; McKenney, P.T.; Ling, L.; Gobourne, A.; No, D.; Liu, H.; Kinnebrew, M.; Viale, A.; et al. Precision microbiome reconstitution restores bile acid mediated resistance to Clostridium difficile. *Nature* **2015**, *517*, 205–208. [CrossRef]
92. Agus, A.; Planchais, J.; Sokol, H. Gut Microbiota Regulation of Tryptophan Metabolism in Health and Disease. *Cell Host Microbe* **2018**, *23*, 716–724. [CrossRef]
93. Liu, Z.; Yu, P.; Cai, M.; Wu, D.; Zhang, M.; Chen, M.; Zhao, Y. Effects of microplastics on the innate immunity and intestinal microflora of juvenile Eriocheir sinensis. *Sci. Total Environ.* **2019**, *685*, 836–846. [CrossRef] [PubMed]
94. Kang, H.-M.; Byeon, E.; Jeong, H.; Kim, M.-S.; Chen, Q.; Lee, J.-S. Different effects of nano- and microplastics on oxidative status and gut microbiota in the marine medaka *Oryzias melastigma*. *J. Hazard. Mater.* **2021**, *405*, 124207. [CrossRef]
95. Zhao, Y.; Qin, Z.; Huang, Z.; Bao, Z.; Luo, T.; Jin, Y. Effects of polyethylene microplastics on the microbiome and metabolism in larval zebrafish. *Environ. Pollut.* **2021**, *282*, 117039. [CrossRef]
96. Zhu, D.; Chen, Q.-L.; An, X.-L.; Yang, X.-R.; Christie, P.; Ke, X.; Wu, L.-H.; Zhu, Y.-G. Exposure of soil collembolans to microplastics perturbs their gut microbiota and alters their isotopic composition. *Soil Biol. Biochem.* **2018**, *116*, 302–310. [CrossRef]
97. Huang, W.; Yin, H.; Yang, Y.; Jin, L.; Lu, G.; Dang, Z. Influence of the co-exposure of microplastics and tetrabromobisphenol A on human gut: Simulation in vitro with human cell Caco-2 and gut microbiota. *Sci. Total Environ.* **2021**, *778*, 146264. [CrossRef]
98. Yan, W.; Hamid, N.; Deng, S.; Jia, P.-P.; Pei, D.-S. Individual and combined toxicogenetic effects of microplastics and heavy metals (Cd, Pb, and Zn) perturb gut microbiota homeostasis and gonadal development in marine medaka (*Oryzias melastigma*). *J. Hazard. Mater.* **2020**, *397*, 122795. [CrossRef] [PubMed]
99. Peng, B.-Y.; Sun, Y.; Xiao, S.; Chen, J.; Zhou, X.; Wu, W.-M.; Zhang, Y. Influence of Polymer Size on Polystyrene Biodegradation in Mealworms (): Responses of Depolymerization Pattern, Gut Microbiome, and Metabolome to Polymers with Low to Ultrahigh Molecular Weight. *Environ. Sci. Technol.* **2022**, *56*, 17310–17320. [CrossRef] [PubMed]
100. Hou, M.; Xu, C.; Zou, X.; Xia, Z.; Su, L.; Qiu, N.; Cai, L.; Yu, F.; Wang, Q.; Zhao, X.; et al. Long-term exposure to microplastics induces intestinal function dysbiosis in rare minnow (*Gobiocypris rarus*). *Ecotoxicol. Environ. Saf.* **2022**, *246*, 114157. [CrossRef]
101. Jiang, P.; Yuan, G.-H.; Jiang, B.-R.; Zhang, J.-Y.; Wang, Y.-Q.; Lv, H.-J.; Zhang, Z.; Wu, J.-L.; Wu, Q.; Li, L. Effects of microplastics (MPs) and tributyltin (TBT) alone and in combination on bile acids and gut microbiota crosstalk in mice. *Ecotoxicol. Environ. Saf.* **2021**, *220*, 112345. [CrossRef] [PubMed]
102. Li, A.; Wang, Y.; Kulyar, M.F.-E.A.; Iqbal, M.; Lai, R.; Zhu, H.; Li, K. Environmental microplastics exposure decreases antioxidant ability, perturbs gut microbial homeostasis and metabolism in chicken. *Sci. Total Environ.* **2023**, *856*, 159089. [CrossRef]
103. Qiao, J.; Chen, R.; Wang, M.; Bai, R.; Cui, X.; Liu, Y.; Wu, C.; Chen, C. Perturbation of gut microbiota plays an important role in micro/nanoplastics-induced gut barrier dysfunction. *Nanoscale* **2021**, *13*, 8806–8816. [CrossRef]

104. Huang, J.-N.; Wen, B.; Xu, L.; Ma, H.-C.; Li, X.-X.; Gao, J.-Z.; Chen, Z.-Z. Micro/nano-plastics cause neurobehavioral toxicity in discus fish (*Symphysodon aequifasciatus*): Insight from brain-gut-microbiota axis. *J. Hazard. Mater.* **2022**, *421*, 126830. [CrossRef]
105. Zhu, L.; Zhu, J.; Zuo, R.; Xu, Q.; Qian, Y.; An, L. Identification of microplastics in human placenta using laser direct infrared spectroscopy. *Sci. Total Environ.* **2023**, *856*, 159060. [CrossRef]
106. Enyoh, C.E.; Duru, C.E.; Ovuoraye, P.E.; Wang, Q. Evaluation of nanoplastics toxicity to the human placenta in systems. *J. Hazard. Mater.* **2023**, *446*, 130600. [CrossRef]
107. Braun, T.; Ehrlich, L.; Henrich, W.; Koeppel, S.; Lomako, I.; Schwabl, P.; Liebmann, B. Detection of Microplastic in Human Placenta and Meconium in a Clinical Setting. *Pharmaceutics* **2021**, *13*, 921. [CrossRef] [PubMed]
108. Liu, S.; Lin, G.; Liu, X.; Yang, R.; Wang, H.; Sun, Y.; Chen, B.; Dong, R. Detection of various microplastics in placentas, meconium, infant feces, breastmilk and infant formula: A pilot prospective study. *Sci. Total Environ.* **2022**, *854*, 158699. [CrossRef] [PubMed]
109. Liu, S.; Liu, X.; Guo, J.; Yang, R.; Wang, H.; Sun, Y.; Chen, B.; Dong, R. The Association Between Microplastics and Microbiota in Placentas and Meconium: The First Evidence in Humans. *Environ. Sci. Technol.* **2022**. [CrossRef]
110. Huang, S.; Huang, X.; Bi, R.; Guo, Q.; Yu, X.; Zeng, Q.; Huang, Z.; Liu, T.; Wu, H.; Chen, Y.; et al. Detection and Analysis of Microplastics in Human Sputum. *Environ. Sci. Technol.* **2022**, *56*, 2476–2486. [CrossRef]
111. Zhao, Q.; Zhu, L.; Weng, J.; Jin, Z.; Cao, Y.; Jiang, H.; Zhang, Z. Detection and characterization of microplastics in the human testis and semen. *Sci. Total Environ.* **2023**, *877*, 162713. [CrossRef]
112. Hou, Z.; Meng, R.; Chen, G.; Lai, T.; Qing, R.; Hao, S.; Deng, J.; Wang, B. Distinct accumulation of nanoplastics in human intestinal organoids. *Sci. Total Environ.* **2022**, *838*, 155811. [CrossRef]
113. Annangi, B.; Villacorta, A.; López-Mesas, M.; Fuentes-Cebrian, V.; Marcos, R.; Hernández, A. Hazard Assessment of Polystyrene Nanoplastics in Primary Human Nasal Epithelial Cells, Focusing on the Autophagic Effects. *Biomolecules* **2023**, *13*, 220. [CrossRef] [PubMed]
114. Li, Y.; Guo, M.; Niu, S.; Shang, M.; Chang, X.; Sun, Z.; Zhang, R.; Shen, X.; Xue, Y. ROS and DRP1 interactions accelerate the mitochondrial injury induced by polystyrene nanoplastics in human liver HepG2 cells. *Chem. Biol. Interact.* **2023**, *379*, 110502. [CrossRef]
115. Zhang, Y.-X.; Wang, M.; Yang, L.; Pan, K.; Miao, A.-J. Bioaccumulation of differently-sized polystyrene nanoplastics by human lung and intestine cells. *J. Hazard. Mater.* **2022**, *439*, 129585. [CrossRef]
116. Banerjee, A.; Billey, L.O.; McGarvey, A.M.; Shelver, W.L. Effects of polystyrene micro/nanoplastics on liver cells based on particle size, surface functionalization, concentration and exposure period. *Sci. Total Environ.* **2022**, *836*, 155621. [CrossRef]
117. Zhou, R.; Zhou, D.; Yang, S.; Shi, Z.; Pan, H.; Jin, Q.; Ding, Z. Neurotoxicity of polystyrene nanoplastics with different particle sizes at environment-related concentrations on early zebrafish embryos. *Sci. Total Environ.* **2023**, *872*, 162096. [CrossRef]
118. Wang, Y.; Shi, H.; Li, T.; Yu, L.; Qi, Y.; Tian, G.; He, F.; Li, X.; Sun, N.; Liu, R. Size-dependent effects of nanoplastics on structure and function of superoxide dismutase. *Chemosphere* **2022**, *309*, 136768. [CrossRef]
119. Hu, Y.; Kang, Y.; Huang, F.; Su, Y.; Zhou, X.; Wang, A.-J.; Gao, S.-H. Distinct responses of Pseudomonas aeruginosa PAO1 exposed to different levels of polystyrene nanoplastics. *Sci. Total Environ.* **2022**, *852*, 158214. [CrossRef]
120. Li, Y.; Liu, Z.; Li, M.; Jiang, Q.; Wu, D.; Huang, Y.; Jiao, Y.; Zhang, M.; Zhao, Y. Effects of nanoplastics on antioxidant and immune enzyme activities and related gene expression in juvenile Macrobrachium nipponense. *J. Hazard. Mater.* **2020**, *398*, 122990. [CrossRef] [PubMed]
121. Yan, F.; Chen, X.A.; Zheng, X. Protective effect of mulberry fruit anthocyanin on human hepatocyte cells (LO2) and *Caenorhabditis elegans* under hyperglycemic conditions. *Food Res. Int.* **2017**, *102*, 213–224. [CrossRef] [PubMed]
122. Chen, W.; Chu, Q.; Ye, X.; Sun, Y.; Liu, Y.; Jia, R.; Li, Y.; Tu, P.; Tang, Q.; Yu, T.; et al. Canidin-3-glucoside prevents nano-plastics induced toxicity via activating autophagy and promoting discharge. *Environ. Pollut.* **2021**, *274*, 116524. [CrossRef] [PubMed]
123. Chen, W.; Tu, P.; Ye, X.; Tang, Q.; Yu, T.; Zheng, X. Cyanidin-3-O-glucoside impacts fecal discharge of polystyrene microplastics in mice: Potential role of microbiota-derived metabolites. *Toxicol. Appl. Pharmacol.* **2022**, *453*, 116212. [CrossRef] [PubMed]
124. Chen, W.; Zhu, R.; Ye, X.; Sun, Y.; Tang, Q.; Liu, Y.; Yan, F.; Yu, T.; Zheng, X.; Tu, P. Food-derived cyanidin-3-O-glucoside reverses microplastic toxicity promoting discharge and modulating the gut microbiota in mice. *Food Funct.* **2022**, *13*, 1447–1458. [CrossRef] [PubMed]
125. Chen, W.; Ye, X.; Tang, Q.; Yu, T.; Tu, P.; Zheng, X. Cyanidin-3-O-glucoside reduces nanopolystyrene-induced toxicity and accumulation: Roles of mitochondrial energy metabolism and cellular efflux. *Environ. Sci. Nano* **2022**, *9*, 2572–2586. [CrossRef]
126. Chen, W.; Chen, Z.; Shan, S.; Wu, A.; Zhao, C.; Ye, X.; Zheng, X.; Zhu, R. Cyanidin-3-O-glucoside promotes stress tolerance and lifespan extension of *Caenorhabditis elegans* exposed to polystyrene via DAF-16 pathway. *Mech. Ageing Dev.* **2022**, *207*, 111723. [CrossRef] [PubMed]
127. Chu, Q.; Zhang, S.; Yu, X.; Wang, Y.; Zhang, M.; Zheng, X. Fecal microbiota transplantation attenuates nano-plastics induced toxicity in *Caenorhabditis elegans*. *Sci. Total Environ.* **2021**, *779*, 146454. [CrossRef] [PubMed]

MDPI AG
Grosspeteranlage 5
4052 Basel
Switzerland
Tel.: +41 61 683 77 34

Metabolites Editorial Office
E-mail: metabolites@mdpi.com
www.mdpi.com/journal/metabolites

www.ingramcontent.com/pod-product-compliance
Lightning Source LLC
Chambersburg PA
CBHW041018270926
51850CB00024B/1391
* 9 7 8 3 7 2 5 8 6 1 7 3 6 *